TESLA

TESLA

INVENTOR OF THE ELECTRICAL AGE

W. BERNARD CARLSON

PRINCETON UNIVERSITY PRESS
PRINCETON AND OXFORD

Published by Princeton University Press, 41 William Street, Princeton, New Jersey
    08540

In the United Kingdom: Princeton University Press, 6 Oxford Street, Woodstock,
    Oxfordshire OX20 1TW

press.princeton.edu

Jacket and frontispiece photograph: Nikola Tesla, c.1894. Bain News Service. USA
    Reproduction Number: LC-DIG-ggbain-04851 (digital file from original neg.):
    LC-B2- 1026-9 [P&P]. Courtesy of Library of Congress Prints and
    Photographs Division Washington, D.C. 20540

Library of Congress Cataloging-in-Publication Data

Carlson, W. Bernard.
Tesla : inventor of the electrical age / W. Bernard Carlson.
    pages cm
Summary: "Nikola Tesla was a major contributor to the electrical revolution that
    transformed daily life at the turn of the twentieth century. His inventions, patents,
    and theoretical work formed the basis of modern AC electricity, and contributed
    to the development of radio and television. Like his competitor Thomas Edison,
    Tesla was one of America's first celebrity scientists, enjoying the company of
    New York high society and dazzling the likes of Mark Twain with his electrical
    demonstrations. An astute self-promoter and gifted showman, he cultivated a
    public image of the eccentric genius. Even at the end of his life when he was living
    in poverty, Tesla still attracted reporters to his annual birthday interview, regaling
    them with claims that he had invented a particle-beam weapon capable of bringing
    down enemy aircraft. Plenty of biographies glamorize Tesla and his eccentricities,
    but until now none has carefully examined what, how, and why he invented. In
    this groundbreaking book, W. Bernard Carlson demystifies the legendary inventor,
    placing him within the cultural and technological context of his time, and focusing
    on his inventions themselves as well as the creation and maintenance of his celebrity.
    Drawing on original documents from Tesla's private and public life, Carlson shows
    how he was an "idealist" inventor who sought the perfect experimental realization
    of a great idea or principle, and who skillfully sold his inventions to the public
    through mythmaking and illusion. This major biography sheds new light on
    Tesla's visionary approach to invention and the business strategies behind his most
    important technological breakthroughs"—Provided by publisher.
Summary: "This is a biography of one of the major 20th-century scientists, Nikola
    Tesla. It is interdisciplinary, containing accounts of U.S. manufacturing in the early
    1900s and other contemporary cultural materials"— Provided by publisher.
Includes bibliographical references and index.
ISBN 978-0-691-05776-7 (hardback : acid-free paper)
1. Tesla, Nikola, 1856–1943. 2. Electrical engineers—United States--Biography. 3.
    Inventors—United States—Biography. I. Title.
TK140.T4C37 2013   621.3092—dc23
[B]   2012049608

British Library Cataloging-in-Publication Data is available

Publication of this book has been aided by a grant from the Alfred P. Sloan Foundation

This book has been composed in Baskerville 10 Pro and Outage Cut

To Jane, who has believed from the very beginning
For Tom Hughes, to whom the debt can never be repaid

CONTENTS

List of Illustrations   ix

INTRODUCTION
Dinner at Delmonico's   1

CHAPTER ONE
An Ideal Childhood (1856–1878)   12

CHAPTER TWO
Dreaming of Motors (1878–1882)   34

CHAPTER THREE
Learning by Doing (1882–1886)   60

CHAPTER FOUR
Mastering Alternating Current (1886–1888)   76

CHAPTER FIVE
Selling the Motor (1888–1889)   100

CHAPTER SIX
Searching for a New Ideal (1889–1891)   117

CHAPTER SEVEN
A Veritable Magician (1891)   129

CHAPTER EIGHT
Taking the Show to Europe (1891–1892)   143

CHAPTER NINE
Pushing Alternating Current in America (1892–1893)   158

CHAPTER TEN
Wireless Lighting and the Oscillator (1893–1894)   176

CHAPTER ELEVEN
Efforts at Promotion (1894–1895)   193

CHAPTER TWELVE
Looking for Alternatives (1895–1898)   214

CHAPTER THIRTEEN
Stationary Waves (1899–1900)   262

CHAPTER FOURTEEN
Wardenclyffe (1900–1901)   302

CHAPTER FIFTEEN
The Dark Tower (1901–1905)   331

CHAPTER SIXTEEN
Visionary to the End (1905–1943)   368

EPILOGUE   396

Note on Sources   415
Abbreviations and Sources   421
Notes   423
Acknowledgments   473
Index   477

ILLUSTRATIONS

FIGURE 0.1. "Showing the Inventor [Tesla] in the Effulgent
Glory of Myriad Tongues of Electric Flame After He Has
Saturated Himself with Electricity."   2

FIGURE 1.1. Tesla's father, Milutin.   15

FIGURE 1.2. Tesla's birthplace in Smiljan in Lika.   16

FIGURE 2.1. Faraday's principle of electromagnetic induction. 36

FIGURE 2.2. Diagram illustrating the right-hand rule.   36

FIGURE 2.3. Hippolyte Pixii's magneto with the first
commutator from 1832.   38

FIGURE 2.4. Simplified view of an electric generator.   39

FIGURE 2.5. Simplified view of a commutator in an electric
generator.   40

FIGURE 2.6. Gramme generator for classroom
demonstrations.   42

FIGURE 2.7. Arago's spinning disk and Babbage and
Hershel's modification.   53

FIGURE 2.8. Eddy currents in a disk spinning in a magnetic
field.   54

FIGURE 2.9. Baily's electric motor from 1879.   56

FIGURE 3.1. First transformers developed by Zipernowsky,
Bláthy, and Déri in 1884–85.   62

FIGURE 3.2. Tesla when he was in Paris in 1883.   63

FIGURE 3.3. Tesla's system in which the generator produced three separate alternating currents that were delivered to the motor over six different wires.   65

FIGURE 3.4. Tesla's AC motor in Strasbourg in 1882.   67

FIGURE 3.5. Edison Machine Works on Goerck Street in New York.   71

FIGURE 3.6. Tesla in 1885.   74

FIGURE 4.1. Tesla's thermoelectric motor, 1886.   77

FIGURE 4.2. Tesla's pyromagnetic generator, 1886–87.   83

FIGURE 4.3. Tesla's AC motor, 1887.   86

FIGURE 4.4. Tesla's egg of Columbus apparatus, circa 1887.   91

FIGURE 4.5. Tesla's experimental arrangement for experimenting with AC motors, fall 1887.   93

FIGURE 4.6. Tesla motor built in 1887–88.   94

FIGURE 5.1. Ferraris' AC motor, circa 1885.   109

FIGURE 6.1. Apparatus used by Hertz to study electromagnetic waves.   121

FIGURE 6.2. Diagram of a Tesla coil.   122

FIGURE 7.1. Circuit used by Tesla for his 1891 lecture at Columbia College.   135

FIGURE 7.2. Tesla demonstrating his wireless lamps before the American Institute of Electrical Engineers in May 1891.   136

FIGURE 7.3. Diagram showing how Tesla grounded his transmitter and receiver.   139

FIGURE 8.1. Apparatus used by Tesla in his 1892 London lecture to illuminate Sir William Thomson's name.   149

FIGURE 8.2. Single-wire motor demonstrated by Tesla in his 1892 London lecture.   150

FIGURE 8.3. Tesla lecturing before the French Physical Society and the International Society of Electricians in 1892.   153

FIGURE 10.1. Receiver used by Tesla to detect electromagnetic waves in the mid-1890s.   180

FIGURE 10.2. Tesla's oscillator, or combination steam engine and generator.   182

FIGURE 10.3. Circuit used by Tesla to deliver wireless power to his lamps in his South Fifth Avenue laboratory, circa 1894.   189

FIGURE 10.4. Receiving coil for Tesla's resonant transformer, as used in the South Fifth Avenue laboratory, circa 1894.   190

FIGURE 10.5. Three phosphorescent bulbs as used in the South Fifth Avenue laboratory, circa 1894.   192

FIGURE 11.1. Tesla, circa 1894–95.   194

FIGURE 11.2. Robert Underwood Johnson and Tesla in the South Fifth Avenue laboratory.   198

FIGURE 11.3. Katharine Johnson.   199

FIGURE 11.4. Photograph of Tesla taken by phosphorescent light.   200

FIGURE 11.5. Mark Twain in Tesla's laboratory in 1894.   202

FIGURE 11.6. Tesla's vision of transmitting power and messages contrasted with that of other inventors in the 1890s.   210

FIGURE 11.7. Tesla coil for ascertaining and discharging the electricity of the Earth.   212

FIGURE 12.1. Tesla's laboratory at East Houston Street.   219

FIGURE 12.2. Tesla seated in his laboratory in 1896.   220

FIGURE 12.3. Shadowgraph made by Tesla of a human foot in a shoe in 1896.   223

FIGURE 12.4. Diagram showing interior of Tesla's first radio-controlled boat in 1898.   226

FIGURE 12.5. Diagram showing Tesla's radio-controlled boat and transmitter in 1898.   228

FIGURE 12.6. Newspaper sketch from 1898 showing how Tesla planned to demonstrate his radio-controlled boat at the Paris Exposition.   236

FIGURE 12.7. Pencil sketch of Tesla at a café.   238

FIGURE 12.8. Richmond P. Hobson.   241

FIGURE 12.9. NT, "Electrical Transformer."   250

FIGURE 12.10. Demonstration done in Tesla's Houston Street laboratory to show the feasibility of conducting high-frequency currents through a low-pressure gas, 1898.   251

FIGURE 12.11. "System of Transmission of Electrical Energy."   253

FIGURE 12.12. "Tesla's proposed arrangement of balloon stations for transmitting electricity without wires."   254

FIGURE 12.13. Tesla's system of electrical power transmission issuing streamers in the East Houston Street laboratory in 1898.   255

FIGURE 13.1. Colorado Springs in the early twentieth century.   263

FIGURE 13.2 View of Tesla's experimental station in Colorado Springs.   266

FIGURE 13.3. Interior of the experimental station showing the components that provided power to the primary coil of the magnifying transmitter.   268

FIGURE 13.4. Diagram showing how Jupiter's moon Io passes through a torus of charged particles.   277

FIGURE 13.5. Notebook sketch of circuit typically used by Tesla at Colorado Springs.   280

FIGURE 13.6. Diagram of tuned circuit used by Tesla in Colorado Springs.   287

FIGURE 13.7. The magnifying transmitter at Colorado Springs with several secondary coils energized by the primary coils on the circular wall.   288

FIGURE 13.8. Three incandescent lamps located outside the experimental station and powered by the magnifying transmitter.   290

FIGURE 13.9. "Experiment to illustrate the transmission of electrical energy without wire."   291

FIGURE 13.10. Unidentified assistant at main power switch in the experimental station.   293

FIGURE 13.11. "Discharge of 'extra coil' issuing from many wires fastened to the brass ring."   296

FIGURE 13.12. Tesla seated in magnifying transmitter, with discharge passing from the secondary coil to another coil.   298

FIGURE 14.1. Tesla's laboratory at Wardenclyffe.   319

FIGURE 14.2. The Machine Shop at Wardenclyffe.   320

FIGURE 14.3. The Engine and Dynamo Room at Wardenclyffe.   321

FIGURE 14.4. The Electrical Room at Wardenclyffe.   322

FIGURE 14.5. The laboratory and tower at Wardenclyffe.   323

FIGURE 14.6. Patent diagram for the Wardenclyffe Tower showing one version of the elevated terminal as well as the circuitry Tesla planned to use.   324

FIGURE 14.7. The tower at Wardenclyffe showing
hemispherical terminal on top.   325

FIGURE 15.1. "Tesla's Wireless Transmitting Tower, 185 feet
high, at Wardenclyffe, N. Y., from which the city of New
York will be fed with electricity, . . ."   340

FIGURE 15.2. DeForest Wireless Automobile operating in
the New York financial district in 1903.   351

FIGURE 15.3. Tesla in 1904.   355

FIGURE 15.4. Tesla's prospectus from February 1904.   356

FIGURE 15.5. Tesla's vision of the Earth as filled with an incompressible fluid.   363

FIGURE 16.1. Tesla turbine.   370

FIGURE 16.2. Tesla's turbine test apparatus at the Edison Waterside Station, New York, in 1912.   372

FIGURE 16.3. Tesla in his office in the Woolworth Building,
circa 1916.   374

FIGURE 16.4. Tesla at one of his birthday interviews, 1935.   381

FIGURE 16.5. Tesla's plan for a high potential generator
to be used as a particle beam weapon.   383

FIGURE 16.6. Diagram showing the projector in Tesla's
particle beam weapon.   385

TESLA

There is something within me that might be illusion as it is often [the] case with young delighted people, but if I would be fortunate to achieve some of my ideals, it would be on the behalf of the whole of humanity.

NIKOLA TESLA, 1892

INTRODUCTION

DINNER AT DELMONICO'S

It was a hot summer night in New York in 1894, and the reporter had decided that it was time to meet the Wizard.

The reporter, Arthur Brisbane, was an up-and-coming newspaperman from Joseph Pulitzer's *New York World*. He had covered the mystery of Jack the Ripper in London, the Homestead Strike in Pittsburgh, and the first execution by electrocution at Sing Sing. Brisbane had an eye for detail and could tell a story that would intrigue a hundred thousand readers. He would go on to edit the *New York Journal* for William Randolph Hearst, help start the Spanish-American War, and define tabloid journalism.[1]

Brisbane specialized in writing features for the *World*'s new Sunday edition, and he had profiled prime ministers and popes, prizefighters and actresses. Now people were telling him to do a story about an inventor, Nikola Tesla. His name was on everyone's lips: "Every scientist knows his work and every foolish person included in . . . New York society knows his face." Not only would his inventions be used to generate electricity at the new plant under construction at Niagara Falls, but Tesla had taken 250,000-volt shocks through his body to demonstrate the safety of alternating current (AC). During such demonstrations, Tesla became "a most radiant creature, with light flaming at every pore of his skin, from the tips of his fingers and from the end of every hair on

1

his head" (Figure 0.1). A dozen reliable sources had told Brisbane that "there was not the slightest doubt about his being a very great man." "Our foremost electrician," people said. "Greater than Edison."[2] Brisbane was curious. Who was this man? What made him tick? Could Tesla be made into a good story for thousands of readers?

The reporter had heard that the Wizard frequently dined at the most fashionable restaurant in Manhattan, Delmonico's on Madison Square. Delmonico's chefs had invented signature dishes such as Lobster Newberg, Chicken à la King, and Baked Alaska. But even more than the food, Delmonico's was the hub of New York society, the place to see and be seen. This is where the old social aristocracy, Ward McAllister's Four Hundred, dined alongside the *nouveau riche* of Wall Street and the rising middle class. It was where balls and cotillions, poker games and stag parties, ladies' luncheons and post-theater suppers were held. Without Delmonico's, observed the *New York Herald*, "the whole social machinery of entertaining would . . . come to a standstill."[3] Clearly, thought Brisbane, this Wizard had both ambition and style. What made him tick?

FIGURE 0.1. "Showing the Inventor in the Effulgent Glory of Myriad Tongues of Electric Flame After He Has Saturated Himself with Electricity."

From Arthur Brisbane, "Our Foremost Electrician," *New York World,* 22 July 1894, in TC 9:44–48, on 46.

Brisbane found Tesla at Delmonico's late that summer evening, talking with Charles Delmonico, whose Swiss great-uncles had founded the restaurant in 1831. Having lived previously in Prague, Budapest, and Paris, Tesla found it easy to chat with the urbane Charley Delmonico. Most likely Tesla had put in a long day at his downtown laboratory and had stopped by for his supper before going home to his hotel, the Gerlach, around the corner.

The reporter carefully took in the physical appearance of the Wizard:

Nikola Tesla is almost the tallest, almost the thinnest and certainly the most serious man who goes to Delmonico's regularly.

He has eyes set very far back in his head. They are rather light. I asked him how he could have such light eyes and be a Slav. He told me that his eyes were once much darker, but that using his mind a great deal had made them many shades lighter. . . .

He is very thin, is more than six feet tall and weighs less than a hundred and forty pounds. He has very big hands. Many able men do—Lincoln is one instance. His thumbs are remarkably big, even for such big hands. They are extraordinarily big. This is a good sign. The thumb is the intellectual part of the hand. . . .

Nikola Tesla has a head that spreads out at the top like a fan. His head is shaped like a wedge. His chin is as pointed as an ice-pick. His mouth is too small. His chin, though not weak, is not strong enough.

As he studied Tesla's outward appearance, Brisbane began to assess his psychological makeup:

His face cannot be studied and judged like the faces of other men, for he is not a worker in practical fields. He lives his life up in the top of his head, where ideas are born, and up there he has plenty of room. His hair is jet black and curly. He stoops—most men do when they have no peacock blood in them. He lives inside himself. He takes a profound interest in his own work. He has that supply of self-love and self-confidence which usually goes with success. And he differs from most men who are written about and talked about in the fact that he has something to tell.

Like other reporters, Brisbane collected the usual background facts—that Tesla had been born in 1856 to a Serbian family in Smiljan, a small mountain village on the military frontier of the Austro-Hungarian Empire (in what is now Croatia), that he had started inventing as a boy, and that he had studied engineering at a school in Graz, Austria. Anxious to get ahead, Tesla had immigrated to America and arrived penniless in New York in 1884.

It was Tesla's meteoric rise since 1884 that made for great newspaper copy. After working briefly for Edison, Tesla had struck out on his own, set up a laboratory, and invented a new AC motor that used a rotating magnetic field. Even though Tesla tried to explain to Brisbane the principle behind the rotating magnetic field, the reporter had to conclude that it was "a thing which may be described but not understood." Instead, Brisbane highlighted how the entrepreneurs behind the massive hydroelectric project at Niagara had rejected Edison's direct current (DC) system and instead chosen Tesla's ideas for generating and transmitting electric power by employing multiphase AC. Tesla's work in power engineering was widely respected, but Brisbane might well have added that Tesla had lectured before distinguished scientific organizations and been awarded honorary degrees by Columbia and Yale. In ten short years, the inventor sitting in front of Brisbane had gone from being penniless and unknown to being America's foremost inventor. Here was one of the great rags-to-riches stories.

But what about the future, asked Brisbane, as the Wizard was only thirty-eight years old. Ah, "the electricity of the future"—here was a topic Tesla loved to discuss:

> When Mr. Tesla talks about the electrical problems upon which he is really working he become[s] a most fascinating person. Not a single word that he says can be understood. He divides time up into billionths of seconds, and supplies power enough from nothing to do all the work in the United States. He believes that electricity will solve the labor problem. . . . It is certain, according to Mr. Tesla's theories that the hard work of the future will be the pressing of electric buttons. A few centuries from now the criminal . . . will be sentenced to press fifteen electric buttons every day. His fellows, long since disused to work, will look upon his toil with pity and horror.

Brisbane listened with rapt attention as Tesla described how he was perfecting new electric lights using high-frequency AC to replace Edison's incandescent lamps. "The present incandescent system, compared to the Tesla idea," thought Brisbane, "is as primitive as an ox cart with two solid wooden wheels compared to modern railroading." The Wizard, though, was even more excited about his ideas for the wireless transmission of power and messages: "You may think me a dreamer and very far gone," he said, "if I should tell you what I really hope for. But I can tell you that I look forward with absolute confidence to sending messages through the earth without any wires.

I have also great hopes of transmitting electric force in the same wave without waste. Concerning the transmission of messages through the earth I have no hesitation in predicting success."

For hours the reporter talked with the Wizard, as "all that he said was interesting, both the electrical things and the others." Tesla spoke of his Serbian background and his love of poetry. He told Brisbane that he valued hard work but that marriage and love interfered with success. He didn't believe in mental telepathy, or "psychical electricity," but was fascinated by how the human mind works. "I talked with this Mr. Tesla of Smiljan," wrote Brisbane, "until the feeble daylight found Mr. Delmonico's scrub-ladies scrubbing his marble floor." They parted friends. Brisbane wrote a front-page story that made Tesla a household name and went on to become one of the most powerful newspaper editors in America.

So what happened to the Wizard? Although he could not know it at the time, Tesla was at his zenith that summer in 1894. Over the previous ten years, he had enjoyed a meteoric rise and was greatly admired by his fellow engineers and scientists. As the *Electrical Engineer* (London) proclaimed, "No man in our age has achieved such a universal scientific reputation in a single stride as this gifted young electrical engineer."[4] Such brilliance, such promise; what happened?

Over the following decade, 1894 to 1904, Tesla continued to invent, developing a high-frequency, high-voltage transformer (now known as a Tesla coil), new electric lamps, a combination steam engine and electric generator, and a host of other devices. Learning that Heinrich Hertz had detected invisible electromagnetic waves in 1885–86, Tesla was among the first to experiment with how to use these waves to create new technology, including an amazing radio-controlled boat. Tesla's grand dream, of course, was to transmit power and messages through the earth, thus rendering obsolete the existing electrical, telephone, and telegraph networks. In pursuit of this dream, he built experimental stations in Colorado Springs and Wardenclyffe, Long Island, ever confident that his system was feasible and that millions of dollars would roll in. Although Tesla boldly predicted as early as 1899 that he would transmit messages across the Atlantic, Guglielmo Marconi did it first in 1901, and so Marconi went into the history books as the inventor of radio. Between 1903 and 1905, Tesla could no longer find backers for his inventions, he encountered problems with his equipment, and he ultimately suffered a nervous breakdown. Though he lived until 1943, by 1904 Tesla's best days were behind him. As Laurence A. Hawkins wrote in 1903, "Ten years ago, if public opinion in this country had been required to name the electrician of greatest promise, the answer would without doubt have been 'Nikola Tesla.' To-day his name provokes at best a regret that so great a promise should have been unfulfilled."[5]

In writing about Tesla, one must navigate between unfair criticism and excessive enthusiasm. On the one hand, we can follow Hawkins's lead and denigrate Tesla for not completing his inventions after 1894, especially his plan for wireless power. Surely someone so determined to pursue wireless power and challenge the status quo of big business and technological systems must have been either wrong or crazy. Yes, Tesla got it right with AC, but he sure got it wrong with radio, and that's why Marconi beat him out. For me, this approach sets up a misleading dichotomy: when inventors get it right, they are heralded as geniuses, and when they get it wrong, they must be insane.

On the other hand, it's easy to celebrate Tesla as a figure second only to Leonardo da Vinci in terms of technological virtuosity. Tesla has dedicated fans who believe that he single-handedly invented electricity and electronics.[6] As one fan stated on his webpage, "Tesla invented just about everything. As you work on your computer, remember Tesla. His Tesla Coil supplies the high voltage for the picture tube you use. The electricity for your computer comes from a Tesla-designed AC generator, is sent through a Tesla transformer, and gets to your house through 3-phase Tesla power."[7] I agree wholeheartedly that we need to understand how Tesla invented these key devices and that we should assess his role in the electrical revolution that reshaped society between 1880 and 1920.[8] But in doing so, we should be careful that we do not convert Tesla into a "superman" with fantastic intellectual powers.[9]

Previous biographies of Tesla have tended to be celebratory.[10] In this book, I want to strike a balance between celebrating and criticizing Tesla; as suggested, he had a spectacular ascent (1884–94) followed by an equally dramatic descent (1895–1905). The task for a Tesla biographer is to piece together his life so that both the ascent and descent make sense. Indeed, the factors that made for an individual's success should also explain that person's failures. One measure of a good historical explanation is symmetry—that the framework used sheds light on both success and failure.

Moreover, while previous biographies have focused largely on Tesla's personality, this book seeks to take measure of both the man *and* his creative work. Throughout the book, I will seek to answer three basic questions: *How did Tesla invent? How did his inventions work? And what happened as he introduced his inventions?* To answer these questions, I will draw on Tesla's correspondence, business records, legal testimony, publications, and surviving artifacts. Some readers may be disappointed that their favorite Tesla story is not here and that there may be more technical discussion than they would like. However, as a historian, I have to tell Tesla's story based on the documents, not on the wishes and dreams we might like to project onto heroes like Tesla. In many ways, Brisbane had it right when he said that the purpose of his story was "to discover this great new electrician thoroughly; to interest Americans in [Tesla's] personality so that they may study his future achievements with proper care."

CONCEPTS AND THEMES

To tell the story of Tesla's dramatic rise and fall, then, we need a framework that allows us to piece the story together. In particular, since Tesla was an inventor, we need a way to think about invention. From my perspective, it's all too easy to associate invention with imponderables such as genius, mystery, and luck; in contrast, I view invention as a process that we can analyze and understand.[11]

Invention refers to the activities by which individuals create new devices or processes that serve human needs and wishes. To do so, an inventor must often investigate phenomena in nature. In some cases, an inventor need only observe nature closely to discover what will work, but in other cases, he or she must tease out new insights by experiment or ingenious manipulation. Because nature does not readily yield up her secrets, one could say that an inventor "negotiates" with nature.[12]

At the same time, invention is not simply discovering how to make something; an inventor must also connect his or her invention with society. In some situations, needs are well-known and society readily takes up a new invention. Since railroads in the mid-nineteenth century needed stronger rails and armies wanted stronger cannon barrels, there was a ready demand for Henry Bessemer's new steel-making process in 1856. In other situations, though, there is no preexisting need and an inventor must convince society of an invention's value. For example, when Alexander Graham Bell invented the telephone in 1876, he found few people willing to buy it; indeed, it took the Bell Telephone Company decades to convince Americans that every home should have a telephone. Bell and his successor companies had to invent not only the telephone but also a marketing strategy that reflected the interests of users. In this sense, inventors "negotiate" with society.[13]

What makes invention interesting is that inventors stand astride the natural and social worlds. On the one hand, they must be willing to engage nature, to find out what will work; on the other hand, inventors must also interact with society, exchanging their inventions for money, fame, or resources. To succeed, inventors must be creative on both sides—in how they negotiate with both nature and society.

In moving between nature and society, inventors develop their own worldview and creative method, reflecting their personality, education, experience, and context. Inventors find their own ways to probe nature, fashion their discoveries into working devices, and ultimately convince other people that their creation is useful or valuable. As Tesla's story unfolds, you will see that his approach was influenced by his religious background, his friends and backers, and his problems with emotional depression. As Thomas Hughes has suggested, inventors—like artists—evolve a unique style.[14]

Tesla's style as an inventor can be described as a tension between ideal and illusion. I have borrowed this tension from the allegory of the cave found in *The Republic* by Plato.[15] Plato developed this allegory to illustrate the difference between ignorance and enlightenment, between how ordinary people and philosophers perceived the world and truth. To explain how ordinary people had a limited understanding of the truth, Plato imagined a group of individuals trapped in the cave who were shackled to chairs and their heads locked in braces so they could not turn around and see how light (or truth) came into the cave. Trapped in this way, they spent their lives debating the flickering shadows projected on the wall by people and things passing in front of a fire behind them. For Plato, then, ordinary people could only deal with illusions. In contrast, the philosopher for Plato was like a prisoner who, freed from the shackles, came to understand that the shadows on the wall were not reality at all, as he could now perceive the true form of reality in the way that the fire and the moving objects created the shadows. Plato's philosophers could look directly at the fire and even the sun outside the cave to know the truth. Only philosophers, concluded Plato, could fathom universal truths, *ideals*.

As we shall see, Tesla was like Plato's philosopher, someone who chose to seek out and understand ideals. As Tesla told one biographer, he was inspired by a saying from Sir Isaac Newton: "I simply hold the thought steadily in my mind's eye until a clear light dawns upon me."[16] In harnessing nature for his inventions, Tesla spent a great deal of time and energy trying to discern the fundamental principle on which to base an invention and then worked to manifest that ideal as a working device. With his AC motor, the ideal was the rotating magnetic field; similarly, the ideal of electromagnetic resonance lay behind his devices related to broadcasting electric power without wires.

On several occasions, Tesla elaborated on his idealist approach to invention; here is how he described it to his fellow electrical engineers when he was awarded the Edison Medal in 1917:

> I have unconsciously evolved what I consider a new method of materializing inventive concepts and ideas, which is exactly opposite to the purely experimental of which undoubtedly Edison is the greatest and most successful exponent. The moment you construct a device to carry into practice a crude idea you will find yourself inevitably engrossed with the details and defects of the apparatus. As you go on improving and reconstructing, your force of concentration diminishes and you lose sight of *the great underlying principle*. You obtain results, but at the sacrifice of quality.
>
> My method is different. I do not rush into constructive work. When I get an idea, I start right away *to build it up in my mind*. I change the structure, I make improvements, I experiment, I run the device in my mind. It is absolutely the same to me whether I operate my turbine in thought or test it actually in my shop. It makes no difference, the results are the same. In this way, you see, I can rapidly develop and perfect an invention, without touching anything. When I have gone so far that I have put into the device every possible improvement I can think

of, that I can see no fault anywhere, I then construct this final product of my brain. Every time my device works as I conceive it should and my experiment comes out exactly as I plan it [emphasis added].[17]

I suspect that Tesla came to this idealist approach partly through his religious background. As Chapter 1 will reveal, Tesla's father and uncles were all priests in the Serbian Orthodox Church and Tesla absorbed something of that faith's beliefs that through the Son of God, the Word or Logos, everything in Creation is endowed with an underlying principle.[18] In this sense, Tesla was much like the great British scientist Michael Faraday, whose research in electricity and chemistry was strongly influenced by his religious beliefs; Faraday was a member of the Sandemanian Church, a Christian sect founded in 1730 that gave Faraday a strong sense of the unity of God and nature.[19]

In taking an idealist approach to invention, Tesla was exhibiting what the economist Joseph Schumpeter called subjective, as opposed to objective, rationality (see Chapter 2). For Schumpeter, engineers and managers come up with incremental innovations by going out and assessing existing needs whereas entrepreneurs and inventors introduce radical and disruptive innovations by responding to ideas that come from within.[20] With objective rationality, the individual shapes ideas in response to the outside world (the market) whereas with subjective rationality, the individual reshapes the outside world to conform to his or her internal ideas. With both the rotating magnetic field and electromagnetic resonance, we will see that the ideals came from within and Tesla struggled to reorder the social world in order to make his inventions a reality.

Tesla's style as an idealist inventor was both similar to and different than that of other inventors. Tesla was very much like Alexander Graham Bell, who called himself a "theoretical inventor" since he preferred to edit and shape inventions in his mind. In contrast, Thomas Edison was almost opposite in style, preferring to develop his ideas by physical means, either by sketching or manipulating devices on the workbench.[21]

Having identified the ideal behind an invention, Tesla was willing to write it up as an article or patent, and he took great delight in demonstrating it to the public. However, Tesla was not especially interested in the nitty-gritty work of converting his inventions into profitable products. Moreover, he was often frustrated that ordinary people did not grasp the ideals underlying his inventions, and so he resorted to illusions to convince them of the value of his creations. Tesla came to believe that along with identifying the ideal for an invention, he also had to create the right illusion—about the exciting and revolutionary changes that his invention would bring about for society. Through demonstrations, technical papers, and newspaper interviews, Tesla sought to capture the imagination of the public as well as the entrepreneurs who would purchase and develop his inventions. Illusions were the means by which Tesla negotiated with society and secured the resources he needed to convert his ideals into real machines.

In using the term "illusion" here, I must emphasize that Tesla was not attempting to deceive potential backers by lying or giving them inaccurate information. Rather, the interaction between an inventor and his backers is analogous to what takes place between an actor and the audience: the actor may say certain things and make certain gestures, but it is the audience who interprets the statements and gestures and shapes them into an impression. In doing so, members of the audience merge what the performer offers with what they know from the larger culture.[22] In his public lectures, Tesla provided his audiences with just the right sort of information—a blend of wizardry, scientific facts, and social commentary—such that they drew the conclusion that his invention would change the world. What Tesla did was encourage people to see in his inventions whole new worlds of possibility. In fact, I would argue that all inventors and entrepreneurs have to generate illusions about their creations—that we can never know in advance what impact an invention will have and so the discussion about a new technology often turns on illusion. As the science-fiction writer Arthur C. Clarke aptly noted, "Any sufficiently advanced technology will appear as magic."[23]

Inventors, then, succeed by harnessing nature in a new device and connecting that device to people's hopes and wishes. Many inventors and entrepreneurs strive to create the right illusion for untested technologies and novel business plans, but Tesla was extraordinary in linking inventions and cultural wishes.[24] What is unfortunate is that during the second decade of his career (1894–1904)—when he was at the height of his creative powers—Tesla concentrated more on creating illusions than converting his ideals into working machines. Tesla's story, as we shall see, was a struggle between ideal and illusion.

CHAPTER ONE
AN IDEAL CHILDHOOD
[1856–1878]

> Our first endeavors are purely instinctive, promptings
> of an imagination vivid and undisciplined. As we grow
> older reason asserts itself and we become more and more
> systematic and designing. But those early impulses,
> tho[ugh] not immediately productive, are of the greatest
> moment and may shape our very destinies.
>
> **NIKOLA TESLA,** *My Inventions* (1919)

Inventors must live with an exquisite tension. On the one hand, they must be in touch with their inner feelings, insights, and impulses—what Tesla calls the "promptings of an imagination vivid and undisciplined"—since these are often the source of new ideas and inventions. On the other hand, inventors can convert an insight into a practical invention only by connecting it to the larger world of markets and needs, and they do this by systematic thinking and design. Inventors must merge the subjective (what they know from inside themselves) with the objective (what they learn about the outside world).[1] How did Tesla learn in his childhood to cultivate his imagination and not let reason overwhelm it?

We are able to investigate this overarching question about creative tension because Tesla described his emotional and intellectual development in an autobiography he published in 1919.[2] But before we can examine his inner life, we must begin by exploring where Tesla was born and who his parents were.

STRANGERS IN A STRANGE LAND

Nikola Tesla was born in 1856 in Smiljan in the province of Lika in what is today Croatia. At that time, Croatia was the military frontier district of the Austro-Hungarian Empire and the area was sometimes referred to as the Krajina. Yet Tesla's father, Milutin, and mother, Djuka, were both Serbs, and Serbia is located farther south in the Balkans, in what was then the Ottoman Empire. How was it that the Tesla family was living in Croatia in the mid-nineteenth century? How did they cope with being strangers in a strange land?

As the journalist Tim Judah has observed, "The Serbs [have] always been a people on the move."[3] Descendants of the Slavs who migrated south from what is modern-day Germany and Poland, the Serbs have moved periodically across the Balkan Peninsula, sometimes in search of better farmland and sometimes in response to violence and invasion. During the height of their power in the fifteenth and sixteenth centuries, the Ottoman Turks swept north through much of the Balkan Peninsula, displacing several Christian populations. The Turks pushed the Serbs from their homeland (now modern Serbia and part of Kosovo), with the result that some Serbs migrated to Croatia.[4] Anxious to defend their Balkan frontier from the Ottoman Turks, the Austrian authorities encouraged the Serbs to settle in Croatia and join the army since the Serbs were sworn enemies of the Turks. Unlike other parts of the Austrian Empire, Croatia was firmly controlled by military officers and every twelfth male subject in the region was required to serve in the army. As a result, the Austrians came to regard Croatia as a source of troops that it used not only to protect their Balkan border but also to fight in other wars.[5]

Tesla's ancestors migrated from Western Serbia to Lika in the 1690s. The Serbs struggled to farm this hard land, which was mountainous and sparsely populated. According to Tesla, the soil was so rocky that the Likan Serbs were fond of saying "that when God distributed the rocks over the earth He carried them in a sack, and that when he was above our land the sack broke."[6]

The name Tesla, in Serbo-Croatian, has two meanings. Typically it refers to an adze or a small ax with a blade at right angles to the handle. However, it can also be used to describe a person with protruding teeth, a facial characteristic common in the Tesla family.

Tesla's grandfather, also named Nikola, was born in 1789 in Lika. During his childhood, Croatia was ceded by the Austrians to Napoleon and became part of the French empire as the Illyrian provinces.[7] Like other Likan Serbs, grandfather Nikola pursued a military career; during the Napoleonic Wars, he joined the French army, rose to the rank of sergeant, and married Ana Kalinic, the daughter of a colonel.

After Napoleon's defeat in 1815, the Illyrian provinces reverted to the Austrian Empire. To keep the Turks out and maintain tight control over the local population of Croats and Serbs, the Austrians continued to operate the province as a military frontier. Although the official religion of the Austrian Empire was Roman Catholic, the Austrians allowed the Serbs to have their own Orthodox churches in Croatia.

In the years after the Napoleonic Wars, grandfather Nikola returned to Lika, where he made the transition from the French army to serving the Austrian Empire. Nikola and Ana had two sons, Milutin (1819–79) and Josif, and three daughters, Stanka, Janja, and one whose name has been lost. Both sons were first sent to a German-language public school and then to the Austrian Military Officers' Training School (probably the Theresian Military Academy in Wiener Neustadt). Josif thrived in this environment and became a professor at a military academy in Austria. A skilled mathematician, Josif wrote several standard works on mathematics.[8]

In contrast to his father and brother, Milutin did not find military life to his liking. Following a reprimand at school for not keeping his brass buttons polished, he quit and instead chose to become a priest in the Serbian Orthodox Church. Milutin enrolled in the Orthodox Seminary in Plaski and graduated in 1845 as the top student in his class.

In 1847, Milutin married Djuka (Georgina) Mandic (1822–92), the twenty-five-year-old daughter of a priest, Nikola Mandic from Gracac. Just as the Tesla family pursued military careers, so most of the men in the Mandic clan joined the clergy; not only was Djuka's father a priest but so were her grandfather and brothers. Several of Djuka's brothers were very successful; while brother Nikolai became the Archbishop of Sarajevo and Metropolitan of the Serbian Orthodox Church in Bosnia, Pajo rose to the rank of general-staff colonel in the Austrian army, and Trifun became a well-known hotelier and landowner.[9]

Shortly after marrying Djuka, Milutin was assigned to a parish of forty households in Senj on the Adriatic coast of Croatia. There, in a stony church perched on a steep cliff, they made a home, and three children were born to Djuka: Dane (1848–63), Angelina (b. 1850), and Milka (b. 1852).

At Senj, Milutin was expected to build up the congregation as well as represent Serbs before "foreign and Catholic persons." Tall and pale, Milutin had high cheekbones and a sparse beard, which gave him a serious visage (Figure 1.1). His congregation found him to be an energetic preacher, and for his sermon "On Labor," he was awarded the Red Sash by his bishop. As an idealistic young priest, Milutin was willing to challenge the Austrian authorities. In 1848 he asked the local military commander to allow Serbian soldiers to attend Orthodox services on Sundays, but the Austrians refused and insisted that the Serbs continue to attend Catholic mass.[10]

Perhaps reflecting his father's experience in the Army of Napoleon, Milutin's worldview combined progressive thought and nationalism. Throughout the territories conquered by Napoleon, the French swept out old ideas about feudalism and absolute monarchy, introduced science and rationality, promoted education by setting up high schools (gymnasia), and stimulated ethnic groups to dream of autonomy.[11] None of these ideas, of course, would have sat well with either the Austrians or the Ottoman Turks. Like other educated Serbs in the mid-nineteenth century, Milutin believed that the condition of Serbs would improve only if they were able to preserve their traditions and create their own nation separate from both the Austrians

FIGURE 1.1. Tesla's father, Milutin. From NTM.

and the Turks. As Milutin wrote in an 1852 letter, "By God! Nothing is as sacred to me as my church and my forefathers' law and custom, and nothing so precious as liberty, well-being and advancement of my people and my brothers, and for these two, the church and the people, wherever I am, I'll be ready to lay down my life."[12]

But despite his zeal, Milutin found Senj to be a difficult assignment. His salary was barely enough to make ends meet and the damp seaside air affected his health. Consequently, Milutin requested a transfer, and in 1852 he was sent to the church of St. Apostles Peter and Paul in Smiljan in Lika.

Translated, Smiljan means "the place of sweet basil," and the Tesla family found this village much more congenial. The parish of Saints Peter and Paul served seventy to eighty households (about a thousand people) and consisted of a white church located at the foot of the Bog-danic mountain, beside a running brook called Vaganac. Although picturesque, the church was isolated, with the nearest neighbors two miles away. Besides the church, there was a fine house for the family and an allotment of fertile farmland (Figure 1.2).[13] To permit Milutin to travel to families throughout the parish, a Turkish pasha from Bosnia presented him with a magnificent Arabian stallion as a reward for helping some local Muslims.[14]

FIGURE 1.2. Tesla's birthplace in Smiljan Lika as it appeared in the 1930s. From KSP, Smithsonian Institution.

At Smiljan, Djuka had the resources to create a comfortable home for her family. "My mother was indefatigable," remembered Tesla.

> She worked regularly from four o'clock in the morning till eleven in the evening. From four to breakfast time—six a.m.—while others slumbered, I never closed my eyes but watched my mother with intense pleasure as she attended quickly—sometimes running—to her many self-imposed duties. She directed the servants to take care of all our domestic animals, she milked the cows, she performed all sorts of labor unassisted, set the table, prepared breakfast for the whole household. Only when it was ready to be served did the rest of the family get up. After breakfast everybody followed my mother's inspiring example. All did their work diligently, liked it, and so achieved a measure of contentment.[15]

With the household in Djuka's capable hands, Milutin's health improved, and he resumed preaching with vigor. Milutin began assembling a library with volumes on religion, mathematics, science, and literature in several languages. He recited poetry with ease and boasted that if a particular classic were lost, he could recover it from his memory. Milutin's most prized possession was an edition of the *Sluzhebnik* or Serbian book of liturgy printed in Venice in 1519. Tesla inherited this book from his father and carried it with him to America.[16]

Milutin also started writing articles for several Serbian newspapers and magazines, including the Serbian *Diary* of Novi Sad, the *Srbobran* published in Zagreb, and a Serbo-Dalmatian magazine from Zadar. Concerned that illiteracy would prevent the Serbs from making social and political progress, Milutin called for a school where Serbs could be taught in their own language.[17] Hence, Milutin was something of a reformer who sought ways to improve the daily life of the Serbian people.

CHILD OF LIGHT

It was in these felicitous circumstances at Smiljan that Tesla was born at midnight between the ninth and tenth of July 1856 (Old

Style).[18] Family legend has it that a violent thunderstorm was raging at that moment, which frightened the village midwife. Fearful, the midwife said, "He'll be a child of the storm." In response, his mother replied, "No, of light." Tesla was baptized at home on the day he was born, suggesting that the family was concerned that he was weak as a newborn. As required by Austrian law, the infant was enlisted in the First Lika Regiment, the Ninth Medak Company, headquartered in Raduč, with the expectation that he would serve from age fifteen onward.[19] As a very young child, Tesla enjoyed playing with his older siblings as well as his younger sister Marica (b. 1859). Together they would run around the churchyard or farmyard with the pigeons, chickens, geese, and sheep kept by the family.[20] But Tesla's favorite companion was the family's black cat, Macak. Macak followed young Nikola everywhere, and they spent many happy hours rolling on the grass.

It was Macak the cat who introduced Tesla to electricity on a dry winter evening. "As I stroked Macak's back," he recalled, "I saw a miracle that made me speechless with amazement. Macak's back was a sheet of light and my hand produced a shower of sparks loud enough to be heard all over the house." Curious, he asked his father what caused the sparks. Puzzled at first, Milutin finally answered, "Well, this is nothing but electricity, the same thing you see through the trees in a storm." His father's answer, equating the sparks with lightning, fascinated the young boy. As Tesla continued to stroke Macak, he began to wonder, "Is nature a gigantic cat? If so, who strokes its back?" "It can only be God," he concluded.

This first observation was followed by yet another remarkable event. As the room grew darker and the candles were lit, Macak got up and took a few steps. "He shook his paws as though he were treading on wet ground," remembered Tesla in 1939,

> I looked at him attentively. Did I see something or was it an illusion? I strained my eyes and perceived distinctly that his body was surrounded by a halo like the aureola [*sic*] of a saint!
>
> I cannot exaggerate the effect of this marvelous night on my childish imagination. Day after day I have asked myself "what is electricity?" and found no answer. Eighty years have gone by since that time and I still ask the same question, unable to answer it.[21]

Just as legend has it that young James Watt was intrigued by how steam could raise the lid of a kettle, so Macak the cat provided the initial inspiration for Tesla to spend a lifetime studying electricity.

AN ACTIVE IMAGINATION

At an early age, Tesla began to tinker, inspired by his mother, Djuka. While the peasants around her in Lika used crude tools that had remained unchanged for centuries, Djuka fashioned better devices in order to run her household efficiently. As her son fondly remembered,

> My mother was an inventor of the first order and would, I believe, have achieved great things had she not been so remote from modern life and its multifold opportunities. She invented and constructed all kinds of tools and devices and wove the finest designs from thread which [*sic*] was spun by her. . . . She worked indefatigably, from break of day till late at night, and most of the wearing apparel and furnishings of the home was [*sic*] the product of her hands. When she was past sixty, her fingers were still nimble enough to *tie three knots in an eyelash*.[22]

Following his mother's example, Tesla made things as a youngster. One early invention involved an effort, as Tesla put it, "to harness the energies of nature to the service of man." Hoping to create a flying machine, Tesla fashioned a spindle with four rotors on one end and a disk on the other. Intuitively he thought the spinning rotors might create enough lift to carry the whole device into the air, much like a modern helicopter. To power the device, Tesla planned to fasten june bugs to the rotors until a strange boy, who was the son of a retired officer in the Austrian army, came along. Much to Tesla's disgust, the boy gobbled up the june bugs. Tesla abandoned the project and resolved never to touch another insect again in his life.[23] This aborted flying machine was followed by other creative endeavors. Like many a curious lad, Tesla took mechanical clocks apart only to discover how much more difficult it was to put them back together. He made his own wooden sword and imagined himself a great Serbian warrior. "At that time I was under the sway of the Serbian national poetry and full of admiration for the feats of the heroes," recalled Tesla. "I used to spend hours in mowing down my enemies in the form of corn-stalks which ruined the crops and netted me several spankings from my mother."[24]

While on the outside Tesla appeared to be a typical happy boy, on the inside his powerful imagination could at times be out of control. As he described in his autobiography: "Up to the age of eight years, . . . [m]y feelings came in waves and surges and vibrated unceasingly between extremes. My wishes were of consuming force and like the heads of the hydra, they multiplied. I was opprest [*sic*] by thoughts of pain in life and death and religious fear. I was swayed by superstitious belief and lived in constant dread of the spirit of evil, of ghosts and ogres and other unholy monsters of the dark."

Even more disturbing, Tesla found it difficult to distinguish images from reality:

> In my boyhood I suffered from a peculiar affliction due to the appearance of images, often accompanied by strong flashes of light, which marred the sight of real objects and interfered with my thought and action. They were pictures of things and scenes which I had really seen, never of those I imagined. When a word was spoken to me the image of the object it designated would present itself vividly to my vision and sometimes I was quite unable to distinguish whether what I saw was tangible or not. This caused me great discomfort and anxiety. . . . They seem to have been unique altho[ugh] I was probably predisposed as I know that my brother experienced a similar trouble. . . . They certainly were not hallucinations such as are produced in diseased and anguished minds, for in other respects I was normal and composed. To give an idea of my distress, suppose that I had witnest [*sic*] a funeral or some such nerve-racking spectacle. Then, inevitably, in the stillness of night, a vivid picture of the scene would thrust itself before my eyes and persist despite all my efforts to banish it. Sometimes it would even remain fixt [*sic*] in space tho[ugh] I pushed my hand thr[ough] it.[25]

Unable to control these images, Tesla felt weak and powerless.

A DEATH IN THE FAMILY

Adding to his emotional difficulties, Tesla lived in the shadow of his older brother, Dane, who was regarded by his parents as extraordinarily gifted. As the eldest son, Dane was expected to follow his father and uncles into the clergy. But in 1863, Dane was killed by his father's high-spirited Arabian horse, and Nikola, at the age of seven, was an eyewitness to the tragedy.[26]

Distraught by the loss of his favorite son, Milutin uprooted the family from Smiljan and moved to the larger nearby town of Gospić, the county seat for the Lika-Senj as well as an administrative center for the Austrian Military Frontier.[27] There Milutin preached for the next sixteen years under the onion-shaped dome of the Church of the Great Martyr George. Although he continued his pastoral duties and taught religion in the local schools, Milutin wrote fewer articles and embraced fewer causes. He developed "the odd habit of talking to himself and would often carry on an animated conversation and indulge in heated argument," changing his voice so that it sounded as if several different people were talking. Milutin never got over the death of Dane, and well before his time he came to be called "Old Man Milovan."[28]

For Tesla, both the death of his brother and the sudden move to Gospić were deeply disturbing. He loved his home in the country and missed seeing the animals in the farmyard. He had just finished his first year of school in Smiljan and was overwhelmed by the hurly-burly of the larger town. "In our new house I was but a prisoner," he wrote, "watching the strange people I saw thru the window blinds. My bashfulness was such that I would rather have faced a roaring lion than one of the city dudes who strolled about."[29] Tesla was so fond of his home village that when he filed his first patents in America, he listed himself as from Smiljan in Lika, not Gospić.

The sudden death of his brother irrevocably altered Tesla's relationship with his parents, particularly with his father. Grieving for Dane, on whom they had pinned all of their hopes, Milutin and Djuka were unable to appreciate the promise of their other son. "Anything I did that was creditable merely caused my parents to feel their loss more keenly," recalled Tesla. "So I grew up with little confidence in myself." (Alexander Graham Bell's family was deeply affected by the sudden death in 1870 of Bell's older and younger brothers, Melville James and Ted; in this case, the Bell family came to together with high expectations for the remaining son.)[30] Like many children, Tesla sought to win back the love of his parents by striving to be perfect. Hoping now that his second son would become a priest, Milutin drilled him with "all sorts of exercises—as, guessing one another's thoughts, discovering the defects of some form or expression, repeating long sentences or performing mental calculations. These daily lessons were intended to strengthen memory and reason and especially to develop the critical sense, and were undoubtedly very beneficial."[31] Yet as Tesla described them in his recollections, one senses that he performed as a duty to his father.

About this time, Tesla discovered the pleasures of reading in his father's library. But rather than being pleased that his second son had a passion for reading, Milutin was angered by it. "He did not permit it and would fly into a rage when he caught me in the act," explained Tesla. "He hid the candles when he found that I was reading in secret. He did not want me to spoil my eyes." But this did not stop Tesla, who surreptitiously obtained tallow and cast his own candles. With these homemade candles he would read all night, often until dawn.[32]

The worst moment with his father, however, came one Sunday when Tesla was helping in church by ringing the bells. As he recalled in his autobiography, "There was a wealthy lady in town, a good but pompous woman, who used to come to the church gorgeously painted up and attired with an enormous train and attendants. One Sunday I had just finished ringing the bell in the belfry and rushed downstairs when this grand dame was sweeping out and I jumped on her train. It tore off with a ripping noise which sounded like a salvo of musketry fired by raw recruits. My father was livid with rage. He gave me a gentle slap on the cheek, the only corporal punishment he ever administered to me but I almost feel it now. The embarrassment and confusion that followed are indescribable."[33]

Unable to please his father, Tesla "contracted many strange likes, dislikes and habits," or what now might be called obsessions. He developed a violent aversion to earrings on women and pearls, although other jewelry was tolerable. He refused to touch the hair of other people and was disturbed by smells such as camphor. "When I drop little squares of paper in a dish filled with liquid, I always sense a peculiar and awful taste in my mouth," he noted, and "I counted the steps in my walks and calculated the cubical contents of soup plates, coffee cups and pieces of food—otherwise my meal was unenjoyable. All repeated acts or operations I performed had to be divisible by three and if I mist [sic] I felt impelled to do it all over again, even if it took hours."[34] These obsessions plagued Tesla throughout his life, and although he struggled to understand their cause, they undoubtedly interfered with his relationships with other people.

AN ACT OF WILL

With his parents distracted by grief over Dane and disappointed in him, Tesla as a boy "was compelled to concentrate attention upon myself" and became introspective. While he suffered at first, he soon discovered that being able to look inside himself was a great blessing and a means to achievement.

Looking inside himself, Tesla underwent a profound change when he was twelve years old. In the course of his reading, he came across a Serbian translation of a novel titled *Abafi* (1836) by the well-known Hungarian writer Miklós Jósika. Set in Jósika's native Transylvania in the sixteenth century, this historical novel recounted the struggles of Prince Sigismund Báthory (1572–1613) as he defended his principality against the Hungarians, Turks, and Austrians. Into this setting—replete with "ruined castles, ancient customs, shining armour, Turkish pashas, and bold intrigues at court"—Jósika introduced a fictitious young nobleman, Olivér Abafi, who emerges as the hero of the story. Abafi starts out as frivolous and unruly, but as the novel progresses, he grows in moral stature, eventually sacrificing himself for his prince and country. As one contemporary reviewer observed, Jósika used Abafi to demonstrate how "a young man absorbed by debauchery and love of pleasure, who, by firmness of will and energy of resolution, exalts himself into one of the most respected and exemplary heroes of his country, that inflexibility of purpose can overcome every thing."[35]

Inspired by Abafi's transformation, the novel awakened Tesla's willpower and he realized that he could exercise control over his feelings. "At first my resolutions faded like snow in April," he remembered, "but in a little while I conquered my weakness and felt a pleasure I never knew before—that of doing as I willed. In the course of time this vigorous mental exercise became second nature. At the outset my wishes had to be subdued but gradually desire and will grew to be identical."[36]

As he developed his willpower, Tesla sought to control the visions that had been troubling him. These visions, noted Tesla, "usually occurred when I found myself in a dangerous or distressing situation, or when I was greatly exhilarated. In some instances I have seen all the air around me filled with tongues of living flame." To banish these images when they tormented him, Tesla had tried to concentrate on something else, but since he had seen little of the world, he soon ran out of things to substitute. Now, however, he discovered that it was better to work with the images, to let his imagination roam freely and thus to channel them:

> Then I instinctively commenced to make excursions beyond the limits of the small world of which I had knowledge, and I saw new scenes. These were at first very blurred and indistinct, and would flit away when I tried to concentrate my attention upon them, but by and by I succeeded in fixing them; they gained in strength and distinctness and finally assumed the concreteness of real things. I soon discovered that my best comfort was attained if I simply went on in my vision farther and farther, getting new impressions all the time, and so I began to travel—of course, in my mind. Every night (and sometimes during the day), when alone, I would start on my journeys—see new places, cities and countries—live there, meet people and make friendships and acquaintances and, however unbelievable, it is a fact that they were just as dear to me as those in actual life and not a bit less intense in their manifestations.[37]

Although he did not realize it at the time, by developing his self-control and learning to channel his powerful imagination, Tesla had begun to acquire the mental skills that would serve him well as an inventor. Not only would he be able to freely explore new ideas in his mind, but he would also have the discipline and concentration he would need to shape and mentally edit these ideas into actual devices (see Chapter 12).[38]

Along with learning to channel the images, Tesla evolved his own rational explanation for them. He had noticed that frequently the troubling images seemed to come not from within himself but as a result of something he had seen out in the world. At first he thought this might have been coincidental,

> but soon I convinced myself that it was not so. A visual impression, consciously or unconsciously received, invariably preceded the appearance of the image. Gradually the desire arose in me to find out, every time, what caused the images to appear, and the satisfaction of this desire soon became a necessity. The next observation I made was that, just as these images followed as a result of something I had seen, so also the thoughts which I conceived were suggested in like manner. Again, I experienced the same desire to locate the image which caused the thought, and this search for the original visual impression soon grew to be a second nature. My mind became automatic, as it were, and in the course of years of continued, almost unconscious performance, I acquired the ability of locating . . . instantly the visual impression which started the thought.

Following these observations, Tesla decided that every thought or action that he took could be attributed to some sort of external stimulation, be it something he saw, heard, tasted, or touched. If this was true, he then concluded he was "an automaton endowed with power of movement, which merely responds to external stimuli beating upon my sense organs, and thinks and acts and moves accordingly." Although flesh and blood, he was simply no more than a machine whose output was determined by the inputs—a "meat machine," as he once put it.[39] Since this mechanistic view did away with the need for free will or a soul, one wonders if Tesla ever discussed this theory with his father; such views would have certainly placed greater distance between Milutin and his son.

As Tesla gained control of his inner life, he also began to look to the wider world for approval and less to his father. This is illustrated by what happened when the citizens of Gospić got a new fire engine. Under the leadership of a young merchant, the citizens had organized a fire department with uniforms and a red-and-black pumping engine. To demonstrate the engine, the fire department paraded proudly through the streets and down to the river. There, sixteen firemen began to furiously pump the engine's handles up and down, but no water came out of the hose. As he watched the scene unfold, Tesla admitted that "My knowledge of the mechanism was nil and I knew next to nothing of air pressure, but instinctively I felt for the suction hose in the water [i.e., the river] and found that it had collapsed." Recognizing that this blockage was causing the problem, Tesla waded into the water and eliminated the kink in the input hose. Immediately the fire engine began to work and water gushed from the hose on the other end. Grateful that he had saved the day, the firemen hoisted Nikola on their shoulders and celebrated him as a hero. Here Tesla learned that solving technical problems could lead to recognition and approval.[40]

EDUCATION AT THE GYMNASIA

Upon arriving in Gospić, Tesla attended the local elementary or normal school for three years. In one of the classrooms, he found demonstration models of waterwheels and turbines. Fascinated by these devices, Tesla duplicated several and tested them in a local stream. Tesla proudly showed these wheels to one of his uncles, but this uncle did not appreciate the boy's mechanical ingenuity and scolded him for wasting his time with such activities. Tesla nevertheless continued to think about turbines, and when he read a description of Niagara Falls, he dreamed of using a giant wheel to capture the power of the falls. "I told my uncle that I would go to America and carry out this scheme," Tesla recalled, and "Thirty years later I saw my ideas carried out at Niagara" (see Chapter 9).[41]

At age ten, Tesla entered the Real Gymnasium in Gospić, the nineteenth-century equivalent of junior high school. Like his father and Uncle Josif, Tesla excelled at mathematics. Taking advantage of his ability to visualize things in his mind's eye, he rapidly performed calculations that drew praise from his mathematics professor. But while he did well in mathematics, Tesla found the required drawing class difficult. This was surprising because other members of his family could draw easily, and Tesla attributed his difficulty to his preference for undisturbed thought. In addition, Tesla was also left-handed as a youngster, which may have prevented him from being able to carry out the assignments since they were often designed for right-handed students. His grades in drawing were so low that his father had to intercede with the school authorities in order that Tesla might continue at the school. Hence it is not surprising that Tesla avoided making drawings throughout his career as an inventor, even when they might have helped convey his ideas to other people.[42]

During his second year at the *gymnasium* in Gospić, Tesla became obsessed with creating a flying machine. Often in his imagination he would journey to distant places by flying, but he did not know how it happened. Impressed with how the vacuum created inside the fire engine had been able to lift water from the river and pump it under pressure into a hose, Tesla wrestled in his mind with a way to combine a vacuum with the fact that the air on the atmosphere is under a pressure of fourteen pounds per square inch.[43] After weeks of mental engineering, Tesla came up with a design which biographer John O'Neill described in the following way:

> He figured that a pressure of fourteen pounds should turn a cylinder at high speed and he could arrange to get advantage of such pressure by surrounding one half of a cylinder with a vacuum and having the remaining half of its surface exposed to air pressure. He carefully built a box of wood. At one end was an opening into which a cylinder was fitted with a very high order of accuracy, so that the box would be airtight; and on one side of the cylinder the edge of the box made a right-angle contact. On the cylinder's other side the box made a tangent, or flat, contact. This arrangement was made because he wanted the air pressure to be exerted at a tangent to the surface of the cylinder—a situation that he knew would be required in order to produce rotation. If he could get that cylinder to rotate, all he would have to do in order to fly would be to attach a propeller to a shaft from the cylinder, strap the box to his body and obtain continuous power from his vacuum box that would lift him through the air.[44]

To test this idea, Tesla carefully constructed a wooden model. As he pumped the air out of the inside cylinder, the shaft turned slightly, making him delirious with joy. "Now I had something concrete," he later wrote, "a flying machine with nothing more than a rotating shaft, flapping wings, and—a vacuum of unlimited power! From that time on I made my daily [imaginary] aerial excursions in a vehicle of comfort and luxury as might have befitted King Solomon." Of course, such a device would have been a perpetual motion machine, and years later Tesla realized that the atmospheric pressure acted at right angles to the surface of the cylinder and that the slight rotary effect he observed was due to a leak in his apparatus. "Tho[ugh] this knowledge came gradually it gave me a painful shock," Tesla recalled, indicating that he had really hoped that he could build an actual machine that would connect his dreams with reality.[45]

Tesla completed his studies at the *gymnasium* in Gospić in 1870, but just as he did so, he "was prostrated with a dangerous illness or rather, a score of them, and my condition became so desperate that I was given up by physicians."[46] One wonders if these vague problems were related to overly intense images since it is around this time (age twelve) that Tesla overcame them through a combination of willpower and learning to channel the images.

In recuperating from this illness, Tesla read constantly. Because of his voracious appetite for books, the local public library sent Tesla all the volumes that had not been catalogued and allowed him to read and classify them. Among the new books that he encountered were several novels by Mark Twain. Tesla found them unlike anything he had read previously, "so captivating as to make me utterly forget my hopeless state."[47] Years later Tesla became friends with Twain, and when Tesla told him this story, Twain burst into tears.

Once he regained his strength, Tesla resumed his studies at the higher Real Gymnasium in Karlovac (or Carlstadt), Croatia. There Tesla stayed with his father's sister, Stanka, and her husband, Colonel Bankovic, "an old war-horse having participated in many battles." Located at the confluence of four rivers, Karlovac was low and marshy, and Tesla contracted malaria, which he treated with copious amounts of quinine.

Milutin had not wavered in his determination to have a son follow him into the priesthood and sent his son to study in Karlovac so that he could prepare for the seminary. While this prospect filled him with dread, Tesla found that he was increasingly attracted to physics, particularly to the study of electricity. At Karlovac, his favorite teacher was the professor of physics, who illustrated his lectures with demonstration models, some of which were his own design. Among these, Tesla was captivated by the radiometer invented by the British scientist William Crookes. Consisting of four tinfoil vanes on a pivot inside a vacuum bulb, Tesla was thrilled to see the vanes spin rapidly in bright light. Recalling his professor demonstrating this remarkable device, Tesla said, "It is impossible for me to convey an adequate idea of the intensity of feeling I experienced in witnessing his exhibitions of these mysterious phenomena. Every impression produced a thousand echoes in my mind. I wanted to know more of this wonderful force." In response, he read everything he could find about electricity and began experimenting with batteries, induction coils, and electrostatic generators. Though he loved these investigations, Tesla knew that his parents wanted him to go into the priesthood and so "resigned [myself] to the inevitable with [an] aching heart."[48]

A FATHER'S PROMISE

Upon completing his studies at Karlovac, Tesla intended to return home to Gospić, but before he could do so, he received a message from his father instructing him to go on a shooting expedition in the mountains. Since his father did not approve of hunting, these instructions puzzled Tesla and he decided to ignore them and return home. There he discovered that the town was in the grip of a cholera epidemic, which was why Milutin had suggested the hunting trip. Upon arriving home, Tesla fell ill, and he struggled for nine months, bedridden and weak, to recover. His condition deteriorated, developing "into dropsy, pulmonary trouble, and all sorts of diseases until finally my coffin was ordered."[49]

During one particularly bad spell when it looked like Tesla was near death, his father rushed to his side and encouraged him to rally his strength. Looking up at his father's pallid and anxious face, Tesla said, "Perhaps I may get well if you will let me study engineering." Although it went against his wishes, Milutin did not want to lose another son. "You will go to the best technical institution in the world," his father solemnly promised, and Tesla "knew that he meant it. A heavy weight was lifted from my mind." On the strength of this promise, along with a little help from an herbal cure—"a bitter decoction of a peculiar bean"—Tesla came back to life "like another Lazarus to the utter amazement of everybody."[50]

Although Tesla was anxious to begin his engineering studies, he and his family now faced another hurdle: Tesla had reached the age where, as a Serb living in the Krajina, he was expected to serve in the Austrian army for three years. Although they might have been able to get him posted to one of his brother-in-law's regiments, Milutin was concerned that his son was still not strong enough to survive army life. Consequently, although avoiding conscription was a serious offense, Milutin decided that Tesla should disappear from Gospić and hide in the mountains while he and his brothers came up with a plan for his son's future. For nine months, from the early fall of 1874 until the following summer, Tesla roamed in the mountains of Croatia, "loaded with a hunter's outfit and a bundle of books."[51]

Tramping in the forest, Tesla grew stronger physically and mentally. As he hiked, he worked on several visionary inventions. For instance, he developed a scheme whereby he would ship letters and packages between continents via a pipe under the ocean. The mail would be put into spherical containers and then shot through the pipe by means of hydraulic pressure. Although he carefully planned how his pumping plant could impart a high velocity to the water in the pipe, he failed to realize that the higher the velocity of the fluid, the greater the resistance of the pipe walls to the fluid flow; as a result, he was forced to abandon this splendid idea.

Another scheme involved building a ring around the Earth's equator in order to improve passenger travel. By applying the appropriate reactionary forces, Tesla thought, the ring could be made stationary while Earth continued to rotate. People would then travel up to the ring, wait for their destination to appear below, and then drop back down to Earth. Tesla thought such a plan would enable people to travel about a thousand miles an hour, but he acknowledged that it would be impossible to build the ring. As impractical as these schemes were, they reveal that Tesla from the start envisioned systems that embraced the whole Earth, a theme that figures prominently in his work on wireless power.

In conjuring up these schemes, Tesla realized the power of his ability to generate mental images. Not only could he use his imagination to undertake fantastic journeys, but he could also direct this talent toward creating new machines. "I observed to my delight that I could visualize with the greatest facility," he later claimed. "I needed no models, drawings or experiments. I could picture them all as real in my mind." Moreover, for Tesla, working with mental images meant that he could concentrate on identifying and exploring the ideal behind an invention.[52]

But how did Tesla know that it was important to seek out the ideal underlying an invention? I suspect that this willingness to seek the ideal grew out of the religious beliefs he acquired from his father and uncles in the Serbian Orthodox Church.

Like all Christians, the Orthodox believe in the Trinity, that God is three persons in one: the Father, the Son, and the Spirit. As in Western Christianity, they further believe that through the Son, "the Word became flesh and dwelt among us" (John 1:14) and that through the Incarnation, Jesus lived on Earth and died for our sins. However, in Orthodox Christianity, the fact that the Son of God is the Word takes on a deeper meaning; as Bishop Kallistos Ware explains,

> The second person of the Trinity is the Son of God, his "Word" or Logos. . . . He it is who was born on earth as man, from the Virgin Mary in the city of Bethlehem. But as Word or Logos he is also at work before the Incarnation. He is the principle of order and purpose that permeates all things, drawing them to unity in God, and so making the universe into a "cosmos," a harmonious and integrated whole. The Creator-Logos has imparted to each thing its own indwelling *logos* or inner principle, which makes that thing to be distinctly itself, and which at the same time draws and directs that thing towards God. Our human task as craftsmen or manufacturers is to discern this *logos* dwelling in each thing and to render it manifest; we seek not to dominate but to co-operate.[53]

For Orthodox Christians, then, the material universe is not only orderly but everything in it—natural and manmade—has an underlying divine principle, a logos that can be discovered by humans. Indeed, one of the ways that humans can praise God—whether as craftsmen, manufacturers, or inventors—is to seek out the logos in all things. Hence, Orthodox beliefs about the Son of God as the Word or Logos would have prompted Tesla to seek out the ideal in his inventions.

To be sure, though Tesla later in life called himself a Christian, he does not appear to have gone to Orthodox Church or practiced his faith. However, that does not mean his religious background had nothing to do with his approach to invention. Indeed, growing up surrounded by Orthodox priests (his father and uncles), Tesla could not have helped but absorb some aspects of their worldview; his interest in finding an ideal underlying each invention is rooted in their faith.

LESSONS FROM HOME

When Tesla returned to Gospić after this sojourn in the mountains, he learned that his father had kept his promise and had secured for him a scholarship from the Military Frontier Administration Authority (Grenzlandsverwaltungsbehoerde). The scholarship would pay 420 gulden a year for three years and would permit Tesla to attend the Joanneum Polytechnic School in Graz, Austria. Upon completion of his studies, Tesla would owe the Military Authority eight years of service.[54]

As Tesla prepared to leave Gospić to begin his studies in Graz, his mother presented him with a shoulder bag that she had made. Colorful and beautifully embroidered, the bag was typical of the textiles produced in Tesla's home province of Lika. Tesla treasured this bag and carried it with him throughout his life.[55]

Just as Tesla took this bag as a tangible remembrance of his family and homeland, we can ask what intangible things he carried away as he left home for Graz. As Serbs living in the Austrian Military Frontier, his grandparents on both sides had learned how to survive as strangers in a strange land; they had learned how to make their peace with the Austrian authorities by moving into the professions—the clergy and the military—that were open to them. We can see that coming from this background, Tesla would be well prepared to adapt to living in America, that he would have the emotional and intellectual wherewithal needed to move up quickly as an immigrant in New York in the 1880s. At the same time, one wonders if the experience of growing up in the "outside" group in Croatia also meant that Tesla learned to be careful and suspicious around strangers and, for that reason, often chose to keep to himself as an adult.

From his mother and father, Tesla carried away traits that would serve him well as an inventor. From his mother, he inherited not only mechanical ingenuity but also an awareness of the satisfaction that came from creating useful things. Although his relationship with his father was strained, Tesla absorbed some of his father's values as a social reformer. In particular, as he grew older, Tesla became less interested in making money from his inventions and more concerned with how they would help humanity. Much like his father, who hoped education and political autonomy would improve the life of the Serbs, Tesla came to believe that his inventions—such as his radio-controlled boat and wireless power—would end war and usher in a new and prosperous age.

But most of all from his childhood, Tesla came away with the intellectual abilities essential to invention. He had been born with an unusually powerful visual imagination—so powerful that at times he could not differentiate between images and reality. As an adolescent, however, Tesla learned how to control this imagination, to channel and direct it. At first he simply went on elaborate journeys in his mind, but he gradually discovered that he could control his imagination to envision new machines. To do so, Tesla learned that he had to strike a balance between letting his imagination run wild and disciplining it so that he could work out the details of a new machine. Drawing on his Orthodox religious background, too, he knew that there had to be an underlying principle, the ideal, behind an invention. Thrilled with how he could use his imagination to find those principles and envision new technology, Tesla knew in his heart that he wanted to be an inventor. Hence, as he slipped his peasant bag over his shoulder and set out for Graz, Tesla left his home in Lika with the heritage, traits, and skills that would allow him to pursue his dream of being an inventor.

CHAPTER TWO

DREAMING OF MOTORS

[1878–1882]

Tesla arrived in Graz in the fall of 1875 to begin his studies at the Joanneum Polytechnic School. The Joanneum was founded in 1811 as a gift
from Archduke John to the counts of Styria (an Austrian province),
and in 1864 it became a Technische Hochschule. Along with institutes
at Vienna, Prague, and Brno, the Joanneum was one of four schools in
the Austrian Empire that offered engineering degrees.[1]

Though the school offered a course of study in civil engineering,
Tesla initially enrolled in mathematics and physics, with the intention
of becoming a professor.[2] In so doing, he would have been following
in the footsteps of his Uncle Josif, and so Tesla may have chosen mathematics and physics in order to please his father. Much as he wanted
to support his remaining son, Milutin probably found it hard to picture what Tesla would do as an engineer whereas being a professor or
teacher of mathematics may have seemed like a more plausible career.[3]

AN INTRODUCTION TO ELECTRICITY

At the Joanneum, Tesla excelled at mathematics, but his favorite lectures were given by Professor Jacob Pöschl in physics. "Professor

Pöschl," recalled Tesla, "was peculiar; it was said of him that he wore the same coat for twenty years. But what he lacked in personal magnetism he made up in the perfection of his exposition. I never saw him miss a word or gesture, and his demonstrations and experiments came off with clocklike precision."[4]

Sitting in Pöschl's lectures, Tesla was introduced in a systematic way to electricity. If Pöschl was typical of other nineteenth-century lecturers in electricity, he might well have provided a historical overview, beginning with the ancient Greeks and progressing up to the latest developments with dynamos and electric lighting. In order to understand Tesla's subsequent electrical inventions, let's review the major topics just as Pöschl would have for Tesla circa 1876.

Although the ancient Greeks were aware that static electricity could be produced by rubbing amber with silk, our modern understanding of electricity dates from the late seventeenth and eighteenth centuries. Several investigators—such as Henry Cavendish and Benjamin Franklin—systematically studied static electricity. These natural philosophers concentrated on how different bodies could be electrically charged and sparks given off. In the early nineteenth century, electrical science expanded dramatically from the study of static charge to investigating what was then called dynamic electricity, or how charge could flow through a conductor. Building on the work of Luigi Galvani, Alessandro Volta demonstrated in 1800 how one could generate a flow of charge by alternating layers of two kinds of metal with papers soaked in acid. Known as a pile, Volta's layers of metal and acid-soaked paper were the first electric battery. While chemists and philosophers energetically debated what caused electricity to be produced in Volta's pile, other scientists used it to conduct new experiments.[5]

Among these scientists was Hans Christian Oersted, who in 1820 discovered a relationship between electricity and magnetism. Oersted connected a wire to a Voltaic pile and then placed a magnetic compass under the wire. To Oersted's amazement, the compass needle was deflected only when he connected or disconnected the wire from the pile. Oersted's experiments were repeated by André-Marie Ampere, who established that it was a flow of charge—a current—that was interacting with the magnetism of the needle and causing motion. But what was the exact relationship between the current, magnetism, and motion?

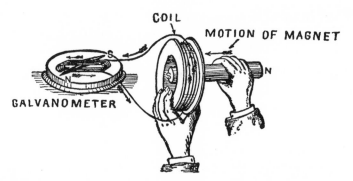

FIGURE 2.1. Faraday's principle of electromagnetic induction.

By moving the bar magnet in and out of the coil, Faraday was able to induce an electrical current which would have caused the needles in the galvanometer to swing back and forth. From *Hawkins Electrical Guide* (New York: Theo. Audel, 1917), 1:131, fig. 130.

In 1831, Michael Faraday answered this question. Using a donut-shaped coil of wire and a bar magnet, Faraday demonstrated the laws of electromagnetic induction. Faraday showed that if one moved the magnet in and out of the donut, one could induce or generate a current in the donut coil. Conversely, if one sent a current through the coil, the magnet would move (Figure 2.1). However, to get either effect—to generate current or produce motion—the configuration of the coil and the bar magnet had to be at right angles with each other. In fact, the current induced would be at a third right angle, perpendicular to both the coil and the magnet. Engineers today refer to this as the right-hand rule (Figure 2.2).

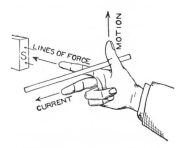

FIGURE 2.2. Diagram illustrating the right-hand rule.

Electrical engineers use this rule to remember how an electric current is induced when a conductor is moved through a magnetic field. If a conductor (such as the rod shown) moves in the direction of the thumb, it cuts the lines of force of the magnetic field that are in the direction of the forefinger. The current produced will move in the conductor in the direction of the middle finger.

From *Cyclopedia of Applied Electricity* (1905), Part II, fig. 5, p. 9.

Faraday further realized the significance of Oersted's observation that the compass needle was deflected only when the current was turned on or off; when the current was passing steadily through the wire, there was no deflection. Faraday hypothesized that both the magnet and the electric coil were each surrounded by an electromagnetic field (often depicted as a series of force lines) and that current or motion was produced when one of these fields was changing. When one turned the current on or off in Oersted's wire, one energized or de-energized the field surrounding the wire, and this change interacted with the magnetic field surrounding the compass needle, causing the needle to swing. As we shall see, this realization that a changing field can induce a current or produce motion was essential for Tesla's work on motors.

During the middle decades of the nineteenth century, it proved difficult for scientists to fully appreciate the nuances of Faraday's theory. However, by looking at the small models that Faraday used to demonstrate his ideas, experimenters and instrument-makers quickly grasped the essence of his ideas and fashioned a variety of generators and motors. For these hands-on investigators, Faraday's laws of electromagnetic induction boiled down to this: if one wanted to build an electric generator, then one moved a conductor through a magnetic field and a current was induced in the conductor. Likewise, if one wanted to make an electric motor, then one used an electric current to produce an electromagnetic field that would cause a magnet or conductor to move.[6]

In utilizing Faraday's discoveries about induction, experimenters soon added several new features to generators and motors. First, to generate electricity, they wanted to use rotary motion—from a hand crank or a steam engine. Conversely, they sought an electric motor that would use electric current to produce rotary motion. Second, investigators came to desire electrical machines that either produced or consumed a current similar to that which came from a battery; they wanted to work with a current that possessed a steady voltage, or what is called direct current (DC). This fascination with DC may have been fostered by the rapid development in the 1840s and 1850s of telegraph systems that sent signals by interrupting a direct current.

To secure both of these features—rotary motion and direct current—electrical experimenters utilized a commutator. In both generators and motors, there are generally two sets of electromagnetic coils: a fixed set known as the field coils or the stator and a rotating set known as the rotor. A commutator is simply the device by which electric current moves into or out of the rotor. Introduced by Hippolyte Pixii in Paris in 1832, the commutator came to be an essential feature of DC motors and generators (Figure 2.3).

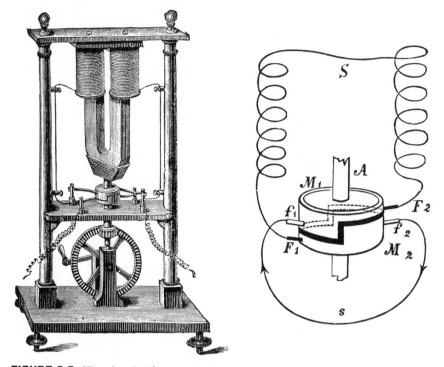

FIGURE 2.3. Hippolyte Pixii's magneto with the first commutator from 1832.

The left image shows the magneto and the right image depicts the commutator in detail. To operate this magneto, one turned the crank in the bottom portion of the machine. This caused the horseshoe magnet to rotate underneath the electromagnets at the top of the machine. As the horseshoe magnet moved, its magnetic field induced a current in the electromagnets. This current flowed through the wires on the vertical supports of the machine to the commutator located on the shaft between the crank and the horseshoe magnet. The electric current left the magneto via the two curly wires.

As shown in the right image, the commutator was located on shaft A which connected the hand crank and gears on the bottom of the magneto with the rotating magnet. The commutator consisted of two hollow cylindrical metal pieces (M_1, M_2) and four metal springs or brushes (F_1, F_2, f_1, and f_2). The pole pieces M_1 and M_2 were electrically insulated from each, as shown by the dark black line between them. S represents the current path of the two electromagnets at the top of the magneto while s represents the circuit outside the magneto.

As shaft A rotated, the four brushes slid along the surface of the pole pieces. As the horseshoe magnet rotated, it induced a current in circuit S that was delivered to the commuatator via F_1 and F_2. The current leaving the magneto was picked up by brushes f_1 and f_2. If the pole pieces M_1 and M_2 were properly positioned on the shaft, then brushes f_1 and f_2 passed over the insulation between the pole pieces at exact moment when the direction of the current was reversed in circuit S. In this way, the commutator converted the alternating current induced by horseshoe magnet in the electromagnets into a direct current.

From Alfred Ritter von Urbanitzky, *Electricity in the Service of Man* (London, 1886), figures 213 and 214 on pp. 228–229.

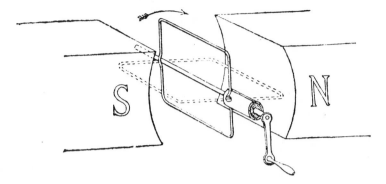

FIGURE 2.4. Simplified view of an electric generator.
 N and S are magnetic poles of the stator. The rotor is shown as the square loop of wire attached to the shaft and crank. The commutator is the two half-cylinders located between the loop and the crank. If one turned the crank, the rotor would rotate through the magnetic field of the stator and an alternating current would be induced in the rotor. This current would flow to the commutator where it would be converted into a direct current.
 From S. P. Thompson, *Dynamo-Electric Machinery*, 3 ed. (1888), fig. 10 on p. 36.

To understand how a commutator works, we need to look at the inner workings of first a DC generator and then a motor (Figure 2.4). Following Faraday's laws of electromagnetic induction, a generator produces current as the rotor spins and cuts across the magnetic field created by field coils. If we trace the path made by just one loop in the rotor coil, we can see that when that loop swings down through the magnetic field it will induce a current that flows in one direction (as specified by the right-hand rule in Figure 2.2). Similarly, as the loop continues its rotation, it will then swing up through the magnetic field and induce a current that will flow in the opposite direction. If one wishes to utilize this alternating current (AC), then one simply connects an individual slip ring to each end of the rotor loop and conducts the currents out of the generator. However, if like many nineteenth-century experimenters one wants a direct current, then one needs to collect all of the current flowing in one direction at one terminal of the generator and all of the current flowing in the opposite direction at the other terminal. This is accomplished by placing on the rotor shaft a commutator consisting of a metallic cylinder that is divided into segments insulated from each other (Figure 2.5). Stationary contacts or brushes rest on opposite sides of this cylinder and are

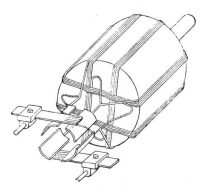

FIGURE 2.5. Simplified view of a commutator in an electric generator.

The commutator consists of the four semicircular pole pieces and the two horizontal brushes. Normally, there would be insulation between the pole pieces, but it is not included in this diagram. In this view, the pole pieces are connected to coils wound around a drum rotor. As the rotor rotates in a magnetic field (not shown), a current is induced in the coils and flows to the pole pieces. The brushes, touching in the pole pieces, collect the current and carry it out of the generator.

From S. P. Thompson, *Dynamo-Electric Machinery*, 3d ed. (1888), fig. 25 on p. 42.

positioned such that when the current generated in the rotor reverses its direction, the connections with the brushes are also reversed and the current delivered by the generator is always in the same direction.

In a DC motor, a commutator works in much the same fashion but its job is to deliver current to the rotor. Through the commutator, we can imagine an electric current flowing through a single loop in the rotor coil, and that current creates an electromagnetic field around that loop. At the same time, we can also send current through the field or stator coils of the motor and set up another electromagnetic field. Now, if one can get the electromagnetic field surrounding the rotor loop to be in the same direction as the field created by the stator coils, then the two fields will push each other apart and cause the rotor to turn. (Recall that in magnets, opposite poles attract and similar poles repel.) However, as the loop swings around to the other side, it will need a current flowing in an opposite direction in order to create a field that will be repulsed by the stator field. Hence, to get the rotor to rotate continuously, we need to regularly reverse the current so that different portions of the rotor coils will consistently have the proper field and be repulsed by the field created by stator coils. This current reversal is again provided by the commutator, which functions as a rotary switch and sends current in the appropriate direction to each portion of the rotor coil.

We have gone into some detail here about how commutators work in DC motors and generators because they are an essential element for rotating electrical machines. Nonetheless, commutators were (and continue to be) the Achilles' heel of DC machines; they were complicated to make and tended to wear out quickly. Commutators also often sparked if there was insufficient electrical insulation between the segments or if brushes were improperly adjusted and touched too many segments at one time. As we shall see in a moment, Tesla decided early on that commutators were the central problem in electrical machinery and he set out to eliminate them.

THE CHALLENGE OF SPARKING COMMUTATORS

It was during one of Pöschl's lectures in 1876–77 that Tesla first confronted the challenge of developing an AC motor.[7] The school had recently acquired a Gramme generator or dynamo from Paris (Figure 2.6). Developed by Belgian instrument-maker Zenobe T. Gramme, this machine was exciting for electrical experimenters because it produced a stronger and more steady direct current. By the end of the 1870s, Gramme dynamos were being used by several European inventors to power the first commercial arc-lighting systems.[8]

Professor Pöschl used his new Gramme dynamo to teach his students about electrical current. One popular use of the dynamo was demonstrating how electricity could transmit power over a distance. This feature was first revealed at the International Exhibition in Vienna in 1873 by Hippolyte Fontaine of the Gramme Company. Fontaine used one Gramme dynamo to generate an electric current that was then sent by wires to another dynamo, which served as a motor.[9] Electricians were excited by this demonstration because it revealed the potential for using electric motors in factories and transportation. Up to this point, electric motors had been limited because it was believed they could only be powered by expensive batteries, but now Fontaine had shown that they could be run by dynamos. Moreover, Fontaine had demonstrated for the first time that power could be transmitted from one place to another without having to use inefficient shafts, belts, or ropes to connect a steam engine with the machines. One could now have a system of power transmission in which one could generate electricity wherever convenient and then consume the power wherever it was needed.

To demonstrate the electrical transmission of power, Pöschl connected his Gramme dynamo to a battery in order to run it as a motor.[10]

FIGURE 2.b. Gramme generator for classroom demonstrations.
  From Alfred Ritter von Urbanitzky, *Electricity in the Service of Man* (London: Cassell, 1886), Fig. 232 on p. 251.

Although one can run a DC generator as a motor, it does require careful adjustment of the commutator brushes to prevent sparking. Pöschl had trouble adjusting the brushes on the Gramme dynamo, Tesla recalled: "While Prof. Pöschl was making demonstrations, running the machine as a motor, the brushes gave trouble, sparking badly, and I observed that it might be possible to operate a motor without these appliances. But he declared that it could not be done and did me the honor of delivering a lecture on the subject, the conclusion of which he remarked: 'Mr. Tesla may accomplish great things, but he certainly never will do this. It would be equivalent to converting a steady pulling force, like that of gravity, into a rotary effort. It is a perpetual motion scheme, an impossible idea.'"[11]

Although it may have been the good professor's intent to prevent Tesla's comments from distracting the other students from understanding how the motor worked, Pöschl used Tesla's interruption to make a more general point. Nineteenth-century scientists and engineers were well aware that the rotary motion needed to turn the machines of the Industrial Revolution was not readily available in Nature. Many forces—such as gravity, magnetism, or electric currents—commonly manifest themselves as linear forces, in the sense that they provide pushes or pulls in a single direction. To secure the desired rotary motion from these linear forces, one needed some sort of conversion device. For examples of these conversion devices, one need only look at how a waterwheel transforms the linear flow of a river or how the crank and flywheel on a steam engine converts the back-and-forth motion of the piston into rotary motion. For Pöschl, the commutator served as a conversion device, converting the linear electric current into a series of alternating pulses that caused the rotor to rotate. Because these conversion devices always absorbed some of the energy being converted from linear into rotary motion, Tesla's idea of a commutator-less motor must have seemed to Pöschl that Tesla would be cheating Nature, and hence Pöschl derisively called it a perpetual motion scheme.

Pöschl intended that his remarks should curb Tesla's flights of fancy but they instead stoked the fires of ambition. In watching the brushes spark and listening to Pöschl's rebuke, Tesla sensed a challenge. "Instinct," Tesla later observed, "is something which transcends knowledge. We have, undoubtedly, certain finer fibers that enable us to perceive truths when logical deduction, or any other willful effort of the brain is futile. For a time I wavered, imprest [sic] by the professor's authority, but soon became convinced I was right and undertook the task with all the fire and boundless confidence of youth."[12]

MENTAL ENGINEERING AN AC MOTOR

In order to take up the challenge of building a spark-free motor, Tesla abandoned his plans to become a teacher and switched in his second year at the Joanneum to the engineering curriculum. As would have been typical for engineering schools in Europe and America in the late 1870s, this curriculum focused not on electrical engineering but on civil engineering. When Tesla first described his education to reporters in the late 1880s, he stated that he had been trained at the Joanneum as a civil engineer.[13]

Though his engineering studies could have prompted Tesla to build a test model of a motor and conduct experiments, he instead chose to investigate the problem in his imagination: "I started by first picturing in my mind a direct-current machine, running it and following the changing flow of the currents in the rotor. Then I would imagine an alternator [an AC generator] and investigate the processes taking place in a similar manner. Next I would visualize systems comprising motors and generators and operate them in various ways. The images I saw were to me perfectly real and tangible."[14]

We see here that Tesla took two steps in conceptualizing his motor. First, although he started by thinking about a DC machine similar to the Gramme dynamo, he decided the solution would involve alternating current. One wonders why he made this shift from DC to AC since most of the work being done in electricity in the late 1870s utilized direct current. In Paris, two electricians, Paul Jablochkoff and Dieudonné François Lontin, were using AC to power several arc lights on a single circuit, but it is not clear that Tesla would have known about their work as a student in Graz.[15]

Instead of being inspired by existing machines, Tesla chose to use AC after carefully scrutinizing how a DC motor operated. As we discussed earlier, the rotor in a DC motor turns because, at any given time, the current flowing through the rotor's coils produces an electromagnetic field that opposes the electromagnetic field set up by the stator coil. To keep the rotor rotating, the commutator periodically reverses the current flowing through the rotor's windings; just as a portion of the rotor rotates from one side of the stator's magnetic field to the other, the commutator automatically reverses the current so that the electromagnetic field in that portion of the rotor is repelled by the stator field. Since the rotor's field was regularly alternating in a motor, thought Tesla, why not use an alternating current, supplied from a generator, to produce this field? By using AC, Tesla may have thought the commutator sparking would be reduced since the commutator would no longer need to reverse the current it delivered to the rotor.

Second, along with choosing to use AC in his motor, Tesla decided not to focus on just a motor but to "visualize systems comprising motors and generators." Again, it is unclear how a second-year engineering student would know enough to do this in 1878. At that time, electrical inventors were building systems that combined dynamos and arc lights, but they did not describe what they were doing as designing systems. However, one can surmise that Tesla abstracted the idea of treating the motor and generator as a system from what Pöschl may have told Tesla about the demonstration by Fontaine in Vienna. Fontaine had transmitted power by linking a dynamo and motor together, and it was perhaps the challenge of linking these two devices that fascinated Tesla. Tesla was not interested in building a motor that could be run by a battery; he wanted to create a motor that could work efficiently with a generator. As we shall see, Tesla's decision to think in terms of systems meant that he was not locked into thinking about motors in any particular way since he could manipulate not just parts of the motor but components of the system in which it resided. Thinking about the motor as part of a system proved to be central to his eventual success.

But for all his mental manipulations, Tesla could not come up with a workable system. "I began to think about and to work on a machine made according to the idea that had occurred to me," recalled Tesla. "Day and night, year after year, I worked incessantly."[16]

GROWING PAINS

During his first year at the Joanneum, Tesla had been a diligent student. "I had made up my mind to give my parents a surprise," wrote Tesla, "and during the whole first year I regularly started my work at three o'clock in the morning and continued until eleven at night, no Sundays or holidays excepted. As most of my fellow-students took things easily, naturally enough I eclipsed all records. In the course of that year I past thru nine exams and the professors thought I deserved more than the highest qualifications."

Armed with these flattering exam certificates, Tesla went home, excited to show his father what he had accomplished. Milutin, however, criticized these accomplishments. "That almost killed my ambition," said Tesla, but later, after his father had died, "I was pained to find a package of letters which the professors had written him to the effect that unless he took me away from the Institution I would be killed thru overwork." Frightened that he would lose his second son through overexertion, the father sought to dampen the young man's enthusiasm for study.[17]

Milutin's chastisement raised questions in Tesla's mind as to whether there was any emotional reward in studying so hard and that perhaps there was more to life than schoolwork. According to a former roommate, Kosta Kulišić, Tesla underwent a dramatic change in attitude toward the end of his second year at Graz. One day, Tesla encountered a member of one of the German cultural clubs who was clearly jealous that a Serb was doing so well in his studies. Lightly tapping Tesla on the shoulder with his cane, the German-speaking student said, "[W]hy waste time here; better go home and 'warm the chair,' so that profs can praise you even more." In response to this challenge, Tesla did not go back to his room to study but decided that he would show his fellow students that he could carouse just as well as they did. Tesla began hanging out with the other students at the Botanical Gardens, where he stayed late smoking and drinking coffee to excess. He learned to play dominoes and chess and became an accomplished billiards player. Most of all, though, he developed a passion for playing cards and gambling. "To sit down to a game of cards," Tesla later said, "was for me the quintessence of pleasure."[18]

Far more interested in carousing and gambling, Tesla returned to Graz in the fall of 1877 for a third year but stopped attending lectures, and university records indicate that he was not registered for the spring of 1878. This undoubtedly contributed to the cancellation of his military scholarship. In September 1878, Tesla wrote to a pro-Serbian newspaper in Novi Sad, the *Queen Bee*, to ask for help in securing another scholarship in order that he might continue his engineering studies in Vienna or Brno. Tesla told the newspaper that he had to relinquish his military scholarship because of illness but that he was now free "from that heavy obligation." In terms of qualifications, Tesla stated that he could now speak Italian, French, and English and he signed the letter "Nikola Tesla, technician."[19]

But the pro-Serbian group that published the *Queen Bee* turned down Tesla's request for a scholarship. Without telling his family, Tesla left Graz sometime in late 1878 and moved to Maribor in the Austrian province of Styria (today Slovenia). Maribor was 45 miles (72 kilometers) from Graz and 185 miles (298 kilometers) from his family in Gospić. In Maribor, Tesla found work as a draftsman in a tool and die shop run by a Master Drusko. In the evenings, Tesla spent his time in a pub called the Happy Peasant located near the train station. By chance, his old roommate Kulišić happened to pass through Maribor in January 1879 and was surprised to find Tesla sitting in the Happy Peasant playing cards for money. Kulišić was relieved to find his friend alive as Tesla had been quite depressed in Graz before he disappeared. When Kulišić asked him if he wanted to go back to Graz to complete his studies, Tesla responded coolly, "I like it here; I work for an engineer, receive sixty forints a month, and can earn a little more for every completed project."[20]

Kulišić left Tesla to his cards and engineering work but sent word to Tesla's family in Gospić that Tesla was living in Maribor. In March 1879, Milutin went to Maribor to plead with his son to come back and proposed that he might resume his studies in Prague. Milutin was especially angry that his son had taken up gambling, an activity he regarded as a senseless waste of time and money. When confronted by his father about his gambling, Tesla responded, "I can stop whenever I please but is it worth while to give up that which I would purchase with the joys of Paradise?" Tesla defied his father and refused to come home. Dejected, Milutin returned home and fell seriously ill.[21]

A few weeks after his father's visit, Tesla was arrested by the police in Maribor as a "vagrant" and deported to Gospić.[22] Heartbroken to see his son brought back by the police, Milutin passed away on 17 April 1879 (Old Style) at the age of sixty. The next day, priests came from all over the region and gave Milutin a "funeral liturgy fit for a saint."[23]

Not sure what to do, Tesla remained in Gospić after his father's death and continued to gamble. Like his father, his mother, Djuka, was worried but since she "knew that one's salvation could only be brought about thru his own efforts," she took a different approach. One afternoon, when Tesla had lost all his money but was still craving a game, she gave him a roll of bills, saying, "Go and enjoy yourself.

The sooner you lose all we possess the better it will be. I know that you will get over it." In response to his mother, Tesla faced his gambling addiction: "I conquered my passion then and there. . . . I not only vanquished but tore it from my heart so as not to leave even a trace of desire."[24]

Tesla eventually decided that he would honor his father's wishes and go back to school in Prague. To do so, he approached his maternal uncles, Petar and Pavle Mandic, who agreed to support him. Going to Prague made sense since Tesla had now decided to settle in Austria, and at the university in Prague he might further study the languages he would need to make his way in the Austrian Empire. In January 1880, Tesla moved to Prague to enroll in the Karl-Ferdinand University. Although he arrived too late to register for the spring term, he signed up in the summer and attended lectures in mathematics, experimental physics, and philosophy.[25]

Tesla also took a special course with Carl Stumpf titled "David Hume and the Investigation of the Human Intellect." Stumpf introduced Tesla to the concept of the mind as tabula rasa: that humans are born with a blank mind that is then shaped through life by sensory perceptions. This corresponded to notions he had already started formulating about how his own imagination worked, and he would later draw on Stumpf's ideas to develop his automaton or radio-controlled boat in the 1890s (see Chapter 12).[26]

In Prague, Tesla continued to puzzle over the problem of making an electric motor. "The atmosphere of that old and interesting city was favorable to invention," he remembered. "Hungry artists were plentiful and intelligent company could be found everywhere."[27] Stimulated by this environment, Tesla recalled, "I made a decided advance, which consisted in detaching the commutator from the machine and studying the phenomena in this new aspect, but still without result."[28] His idea here was to place the commutator on separate supports or arbors away from the frame of the motor, perhaps thinking that he could eliminate the sparking by increasing the distance between the rotor and the commutator. Although this line of thinking yielded no breakthrough, the process of envisioning these machines helped Tesla understand how motors worked. "Everyday I imagined arrangements on this plan without result," he noted, "but feeling that I was nearing a solution."[29]

INSIGHT IN BUDAPEST

While in Prague Tesla was supported by his maternal uncles, but they could not support him forever as a student. As one early biographical article put it, in Prague Tesla "began to feel the pinch, and to grow unfamiliar with the image of Francis Joseph the First," the ruling Austrian emperor whose portrait appeared on the currency. Eventually, when the money from his uncles stopped arriving, he "became a very fair example of high thinking and plain living; but he made up his mind to the struggle and determined to go through depending solely on his own resources."[30] In January 1881, Tesla left Prague and moved to Budapest.

Tesla chose Budapest because he had read in a Prague newspaper that Tivadar Puskás had received permission from Thomas Edison to build a telephone exchange there and that the project would be supervised by his brother, Ferenc. Since Ferenc had been a lieutenant in the Hussars, the light cavalry unit in which his uncle Pavle served, Tesla asked his uncle to recommend him to Ferenc so that he could get a job helping to build the new exchange.[31]

The Puskás family was part of the Transylvanian nobility, and Tivadar had studied law and technical subjects as a young man. A promoter and entrepreneur, Tivadar had traveled to America looking for opportunities. After trying his hand at gold mining in Colorado, he became interested in the telegraph and telephone.[32] In 1877, Puskás visited Edison at Menlo Park where he made quite an impression, arriving in a fancy carriage and flashing a roll of thousand-dollar bills. Edison took a liking to Puskás and showed him all of his current inventions, including the phonograph. Thrilled with everything he saw, Puskás offered to take out patents in Europe for Edison's telephone and phonograph at his own expense in return for a one-twentieth interest.[33] With such a deal, one wonders whether Puskás was hustling Edison or Edison was hustling Puskás. Puskás served as one of Edison's agents in Europe for many years and was actively involved in promoting the telephone, phonograph, and electric light.

Puskás proposed to Edison that he could establish telephone exchanges in major European cities. Up to this time, Edison and Alexander Graham Bell had been mainly thinking about installing telephones on private lines to link two locations, and Edison was intrigued by Puskás's plan for an exchange in which hundreds of subscribers could be connected with one another by means of a switchboard.[34] With Edison's blessing, Puskás set up a telephone exchange in Paris in 1879. Brother Ferenc helped with the Paris exchange and then traveled home to Budapest to establish an exchange there.

But Ferenc Puskás was not able to hire Tesla at once. In all likelihood, it took the Puskás brothers some time to arrange the financing for the Budapest exchange. Instead, through the help of the Puskás brothers or other friends, Tesla was hired as a draftsman in the Central Telegraph Office of the Hungarian government. Although the weekly salary of five dollars was meager, the position did introduce Tesla to practical electrical work. "I soon won the interest of the Inspector-in-Chief," Tesla later recalled, "and was thereafter employed on calculations, designs, and estimates in connection with new installations." Yet Tesla found much of the work boring, involving more routine drafting and calculation than he liked. "By the time he had extracted several hundred thousand square and cube roots for the public benefit," according to one account, "the limitations, financial and otherwise, of the position had become painfully apparent."[35]

Dissatisfied with his position in the telegraph office, Tesla quit and decided to concentrate on invention. Like many novice inventors, he was confident that he could quickly develop a great invention that would support him. "He proceeded at once to make inventions," noted a later story, "but their value was visible only to the eye of faith and they brought no grist to the mill."[36] Frustrated, Tesla suffered a "complete breakdown of the nerves" and fell into a deep depression.[37]

Convinced he was going to die, Tesla eventually recovered with the help of a new friend, Anthony Szigeti. In Budapest, he came "in contact with a number of young men in whom I became interested. One of these was Mr. Sigety [*sic*], who was a remarkable specimen of humanity. A big head with an awful lump on one side and a sallow complexion made him distinctly ugly, but from the neck down his body might have served for a statue of Apollo." Szigeti "was an athlete of extraordinary physical powers—one of the strongest men in Hungary. He dragged me out of the room and compelled me to make physical exercises . . . he saved my life."[38] Like Tesla, Szigeti enjoyed billiards, but he was also interested in electrical matters and encouraged Tesla to continue conceptualizing his motor. With Szigeti's help, Tesla gained

> [a] powerful desire to live and to continue the work. . . . My health returned and with it the vigor of the mind. . . . When I undertook the task it was not with a resolve such as men often make. With me it was a sacred vow, a question of life and death. I knew that I would perish if I failed. . . . Back in the deep recesses of the brain was the solution, but I could not yet give it outward expression.[39]

To help Tesla recover his strength, Szigeti convinced Tesla to walk with him each evening in the Városliget (City Park), and during these walks they discussed Tesla's ideas for an improved motor. In his 1919 autobiography, Tesla stated that the solution for his motor problem came to him during one of these walks as a Eureka moment:

> One afternoon, which is ever present in my recollection, I was enjoying a walk with my friend in the City Park and reciting poetry. At that age I knew entire books by heart, word for word. One of these was Goethe's "Faust." The sun was just setting and reminded me of the glorious passage:
>
> > The glow retreats, done is the day of toil;
> > It yonder hastes, new fields of life exploring;
> > Ah, that no wing can lift me from the soil
> > Upon its track to follow, follow soaring!
> >
> > A glorious dream! though now the glories fade.
> > Alas the wings that lift the mind no aid
> > Of wings to lift the body can bequeath me.
>
> As I uttered these inspiring words the idea came like a flash of lightning and in an instant the truth was revealed. I drew with a stick on the sand the diagrams shown six years later in my address before the American Institute of Electrical Engineers [AIEE], and my companion understood them perfectly.[40] The images were wonderfully sharp and clear and had the solidity of metal and stone, so much so that I told

him: "See my motor here; watch me reverse it." I cannot begin to describe my emotions.[41]

Through Goethe's imagery of the sun retreating and rushing forward and of the invisible wings lifting the mind but not the body, Tesla envisioned the idea of using a rotating magnetic field in his motor.

As dramatic as this story of sunsets and Goethe is, we must interpret this moment carefully. Yes, it is the way that Tesla recounted the invention of his AC motor in his 1919 autobiography, but in sworn patent testimony given in 1903, Tesla said *nothing* about having a Eureka moment with Szigeti in the park. From a legal standpoint, it would have been helpful to establish the moment of invention as taking place in 1882 since it would have buttressed Tesla's claims to have been the first to invent an AC motor.[42] Instead, Tesla's patent testimony suggests that it took him time to work out all of his ideas. Moreover, given what Tesla seems to have known in 1882, it is not likely that he understood in Budapest everything that he included in his 1888 AIEE lecture.

Nevertheless, it is clear that he had a major breakthrough in Budapest. Based on what he knew before Budapest and the experiments he did subsequently in 1883 and 1887 (see Chapters 3 and 4), his breakthrough consisted of three related insights. First, Tesla realized that he could get the rotor in his motor to rotate not by delivering any current to it but by taking advantage of induced eddy currents. Second, he realized that he could induce the eddy currents in the rotor by creating a rotating magnetic field in the stator windings. And third, Tesla had a hunch that somehow the rotating magnetic field could be produced using AC.

To discuss these insights that Tesla had while walking in the park, it is helpful to talk about a device frequently discussed in nineteenth-century electricity texts, Arago's wheel. Let me emphasize that there is no evidence that Tesla knew about Arago's wheel and used it in his thinking about motors, but this device can assist us in visualizing what Tesla accomplished in Budapest.[43]

In 1824, the French scientist François Arago became fascinated by the curious behavior of a compass needle when one spun a copper disk underneath it. If the copper disk was spun quickly enough, the compass needle not only stopped pointing north but began to rotate as well (Figure 2.7). Shortly after Arago reported his discovery, Charles

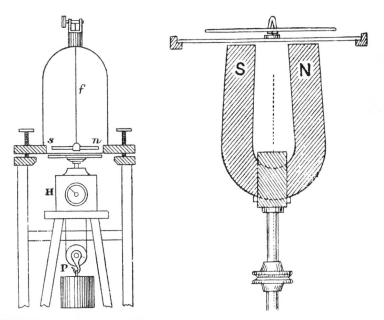

FIGURE 2.7. Arago's spinning disk (left) and Babbage and Hershel's modification (right). From S. P. Thompson, *Polyphase Electric Currents*, 2 ed., (1900), figs. 315 and 316 on p. 423.

Babbage and Charles Herschel in England demonstrated the reverse phenomenon: that if one rotated a horseshoe magnet underneath a pivoted copper disk, the disk would spin. Natural philosophers were puzzled by Arago's wheel and wondered about the relationship between magnetism and motion.

As with the puzzle of Oersted's experiment, it was once again Faraday who explained the riddle of Arago's wheel—its motion was caused by electromagnetic induction. By experiment, Faraday demonstrated that as the magnet rotated underneath the copper disk, the movement of the magnetic field induced swirls of current in the disk (Figure 2.8). Calling them eddy currents, Faraday pointed out that these currents produced an electric field opposite to the magnetic field, and as a result of this repulsion, the copper disk moved.[44]

Returning to Tesla, the first insight that he probably had in the park in Budapest was to realize that he did not have to deliver any current to the rotor in his motor. Just as eddy currents caused Arago's copper disk to rotate, so through his own mental engineering Tesla realized that the magnetic field of the stator in his motor could induce

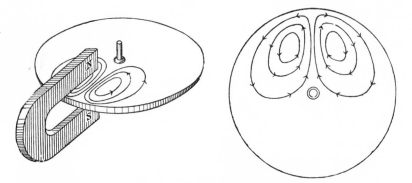

FIGURE 2.8. Eddy currents in a disk spinning in a magnetic field.
From S. P. Thompson, *Polyphase Electric Currents*, 2d ed. (1900), fig. 319 on p. 425.

eddy currents in the rotor and cause it to rotate. Borrowing Goethe's imagery, the induced currents were the invisible wings that would lift the rotor and set it spinning.

Since the currents would be induced in the rotor, there was no need to use a commutator to deliver currents to the rotor. Hence, he really could eliminate the commutator and its sparking. Tesla's decision not to feed currents into the rotor was an important departure from prevailing practice since most electricians in the early 1880s assumed that one needed to have electromagnets in both the rotor and stator in order to get a motor to produce any significant mechanical force or torque.[45]

Once he knew that induced currents would cause the rotor to turn, Tesla came quickly to his second and most important insight: to produce currents in the rotor, one needed a rotating magnetic field. Just as Babbage and Herschel had rotated a horseshoe magnet underneath their copper disk, so Tesla now realized that the key to his motor would be to create a rotating magnetic field in the stator windings. As the magnetic field rotated around a copper disk rotor, it should cause the disk to rotate.

In arriving at this second insight, it is important to note that Tesla did so by reversing standard practice. Up to this time, most electrical experts had designed DC motors in which the magnetic field of the stator was kept constant and the magnetic poles in the rotor were changed by means of a commutator. Instead, Tesla chose to do the opposite: rather than changing the magnetic poles in the rotor, why not change the magnetic poles in the stator? Tesla saw that if the magnetic field in the stator rotated, it would induce an opposing magnetic field in the rotor and thus cause the rotor to turn. As we will see, this willingness to reverse standard practice—to be a maverick—was one of the hallmarks of Tesla's style as an inventor.

Unlike Babbage and Herschel, however, Tesla did not want to create a rotating magnetic field by mechanically turning a magnet underneath the rotor; an effective motor converted electricity into motion, not motion into motion. How then could Tesla use an electric current to create a rotating magnetic field?

This brings us to Tesla's third insight in the park. Based on his extensive mental engineering, Tesla had a hunch that somehow one or more alternating currents could be used to create a rotating magnetic field. If so, his thinking would have paralleled that of an English physicist, Walter Baily, who reported in 1879 how he had used two electric currents to cause Arago's wheel to rotate. Instead of a horseshoe magnet, Baily placed four electromagnets underneath his copper disk (see Figure 2.9). Baily linked the coils in series, joining one with the other diagonally across from it. He then connected each pair of electromagnets to a rotating switch that controlled the current delivered from two separate batteries to the pairs of electromagnets. As Baily rotated his switch, the electromagnets were sequentially energized to become either north or south magnetic poles, with the effect that the magnetic field underneath the copper disk rotated. As a scientist, Baily seems to have been satisfied to know that electric currents could be used to turn Arago's wheel, and he regarded his motor as a scientific toy.[46]

Again, there is no evidence that Tesla knew about Baily's motor when he was in Budapest in 1882. Rather, Baily's motor helps us visualize the significant insight that Tesla came to through his own mental engineering—that there must be a way to use one or more alternating currents to create a rotating magnetic field. Perhaps he came to this hunch about using alternating currents while reflecting on Goethe's imagery of the sun retreating and then rushing forward. Indeed, it is remarkable that Tesla came to this insight as a young man of twenty-six using the power of his imagination and without reference to devices like Arago's wheel and Baily's motor.

Thirty years later, when the patent litigation was all over and Tesla wrote about the invention of his motor in Budapest, he insisted that the idea came to him fully developed: "When an idea presents itself it

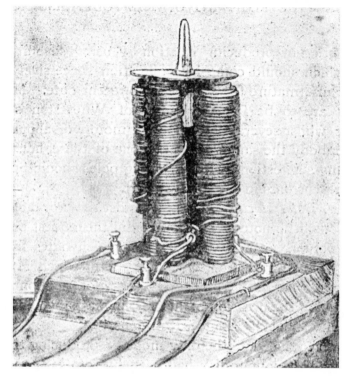

FIGURE 2.9. Baily's electric motor, 1879.
From S. P. Thompson, *Polyphase Electric Currents,* 2d ed., (1900), fig. 330 on p. 438.

is, as a rule, crude and imperfect. Birth, growth and development are phases normal and natural. It was different with my invention. In the very moment I became conscious of it, I saw it fully developed and perfected. . . . My imaginings were equivalent to realities."[47]

However, despite these claims, it is not likely that Tesla understood everything about his AC motor at the time. In particular, he probably did not understand how to actually use two or more alternating currents. Given that Tesla did not have any firsthand experience building electrical machines before his walk in the park, it is not likely that he would have known how to fashion a rotating switch such as Baily's in order to control the current from two batteries. Equally important, I find it doubtful that Tesla—or any other electrical inventor, for that matter—understood in 1882 how several alternating currents with different phases could create a rotating magnetic field.[48] In many ways, the limits of Tesla's breakthrough in Budapest only become clear when we examine closely the first motor that he built in 1883 in Strasbourg (see Chapter 3).

Nonetheless, that walk in the park in Budapest was an intellectual turning point for Tesla. There, with Szigeti by his side, gazing into the sunset, Tesla did understand something about how a rotating magnetic field could be used in a motor. More than likely the vision was incomplete, but Tesla discerned enough to know that he was on to something big. He had discovered the first grand ideal of his career and he fully intended to exploit it.

The walk was an emotional turning point for Tesla, for he now knew his path. In Budapest, he had solved the problem posed by Pöschl's sparking motor, and in so doing, Tesla became convinced of his own creative powers. "I had carried out what I had undertaken and pictured myself achieving wealth and fame," he later wrote. "But more than all this was to me the revelation that I was an inventor. This was the one thing I wanted to be. Archimedes was my ideal. I [had] admired the works of artists, but to my mind, they were only shadows and semblances. The inventor, I thought, gives to the world creations which are palpable, which live and work."[49]

CREATIVE DESTRUCTION AND
SUBJECTIVE RATIONALITY

Before we leave Tesla and Szigeti in the park, we should take a moment to reflect on the nature of Tesla's insight that afternoon, not just from a technical viewpoint but from a cognitive perspective as well. To do so we need to connect Tesla with the economist Joseph Schumpeter's ideas about innovation and the creative destruction of capitalism.

Schumpeter was fascinated by the role that innovation played in the modern economy, and he emphasized in his writings that there were two kinds of innovative activity. On the one hand, there are the creative responses of entrepreneurs and inventors who introduce radically new products, processes, and services and in so doing wreak the creative destruction that Schumpeter regarded as a central characteristic of capitalism. More recently, Clayton Christensen has

characterized Schumpeter's creative responses as "disruptive innovations" in the sense that selected firms sometimes pursue technologies
that disrupt the pattern of established industries and alter the everyday life of consumers.[50]

On the other hand, there are the adaptive responses of managers and engineers who undertake the steady and incremental work
of establishing the corporate structures, manufacturing procedures,
and marketing plans that allow products and services to be produced
and consumed.[51] Clearly the success of any economy—especially the
United States in Tesla's time, from 1870 to 1920—has depended on
getting the right mix of creative and adaptive innovations. Yet getting
the right mix is neither automatic nor obvious, and so one of the great
questions confronting historians of business and technology is understanding how creative and adaptive responses come together.

Tesla's creative insight in the park gives us an opportunity to
develop a second idea that Schumpeter had about innovation. He
suggested that there were two kinds of thinking underlying the creative and adaptive responses of entrepreneurs and managers, what
he called two kinds of rationality. For the businessman or manager,
there was *objective* rationality in the sense that the manager went out,
looked at the market, measured demand, and acted accordingly; it
was objective in the sense that the logic of what to do came from
the world "out there." In contrast, Schumpeter thought that entrepreneurs employed *subjective* rationality; for them, the guiding logic
came from within—from their own thoughts, feelings, and wishes—
and their actions were based on efforts to impose this internal logic
on the outside world.

To explain subjective rationality, Schumpeter described a hypothetical encounter between a businessman and an efficiency engineer.
Because the businessman paid attention to delivering what customers
wanted, he had little interest in the engineer's suggestions for improving the efficiency of his operation that were based on theory and calculation. With his eye on the external signals from the marketplace, the
businessman could not appreciate the internal logic of the engineer
that drew on science and mathematics; at the same time, the engineer
failed to grasp the importance of consumer demand. "I mention this
class of cases," Schumpeter concluded,

not only because they are in themselves important and the source of much inadequate interpretation, but also because at least the engineer's rationality is so excellent an example of subjective rationality and for the importance of attending to it. [An] Engineer's rationality turns on ends perceived with ideal clearness. It goes about devising means by ideally rational and conscious efforts. It reacts promptly to a purely rational new impulse—e.g., a new calculation published in a professional periodical. It is comparatively free from extraneous considerations. That is to say, it functions in a particular way because of the "conscious" quality of its intentional struggle for rationality.[52]

In my view, one can easily substitute "inventor" for "engineer" in the above quotation. Many inventors work from an internal logic that makes sense to them and they strive to manifest those internal ideas in terms of a new device.

As Schumpeter rightly observes, we have not adequately interpreted the role that subjective rationality plays in economic life. Rather than trace how inventors or entrepreneurs develop disruptive technologies, both scholars and laypeople assume that the sources of new ideas are unknowable and we attribute them to intuition, genius, or "gut feelings."

However, Tesla's career poses an opportunity to better understand what we mean by subjective rationality. Tesla's vision of a rotating magnetic field came from within himself, but it didn't just come out of nowhere. Rather, the insight grew out of his ongoing mental engineering and was shaped by a rich stew of ideas, feelings, and impressions he had at the time. Maybe Schumpeter's term—rationality—isn't the best word for it, but Tesla was doing some kind of cognitive processing. But even more important, as we go forward with the story, we will see that what counts with subjective rationality is that inventors like Tesla come to believe in their ideas so strongly that they are willing to rearrange the external world in order to make their ideals into reality. In imposing their ideas on the world, inventors create the revolutionary technology that unleashes the creative destruction of capitalism. But before that can happen in Tesla's case, he had to learn much more about the business of electrical technology.

ALTERNATING CURRENT AT GANZ AND COMPANY

Armed with his insight about using a rotating magnetic field in his motor, Tesla resumed his mental engineering. "For a while," he recalled fondly,

> I gave myself up entirely to the intense enjoyment of picturing machines and devising new forms. It was a mental state of happiness about as complete as I have ever known in life. Ideas came in an uninterrupted stream and the only difficulty I had was to hold them fast. The pieces of apparatus I conceived were to me absolutely real and tangible in every detail, even to the minutest marks and signs of wear. I delighted in imagining the motors constantly running, for in this way they presented to the mind's eye a more fascinating sight. When natural inclination develops into passionate desire, one advances toward his goal in seven-league boots.[1]

As Tesla enjoyed being in the creative flow of visualizing his ideal motor, his efforts were greatly helped by what he learned about alternating current (AC) while working at or visiting the great manufacturing works of Ganz and Company in Budapest in 1882.[2] Founded in

1844 by Abraham Ganz, this company had begun as an iron foundry specializing in wheels for railway cars, cannons, and bullets. After Ganz's death, the firm expanded into the production of water turbines and flour-processing equipment, and in 1878 it expanded again into the new field of electric lighting. Under the direction of Károly Zipernowsky, Ganz began building and installing systems that powered both arc and incandescent lamps. Hence, for a young man intrigued by electricity, the Ganz works would have been an ideal place to work or just hang around.[3]

While Tesla was at the Ganz works, he noticed a broken ring transformer that lay ignored in one corner of the workshop. In all likelihood, this device had been used to power arc lights in an AC series circuit. In a series circuit, if one light failed, then all the lights went out; to overcome this problem, Paul Jablochkoff had ingeniously installed a similar transformer in his lighting system in Paris so that power could be shunted around any defective lamp and keep the other lamps burning. But while Jablochkoff had used a transformer with two coils wound around an iron cylinder, the broken transformer at Ganz consisted of a large iron ring wound with two coils around either side.[4] At some point, Zipernowsky and the other engineers at Ganz began to study this ring transformer to figure out why it was not working properly. Over the next few years, investigations of devices like this ring transformer led Zipernowsky, Ottó Bláthy, and Miksa Déri to develop one of the first AC power systems using transformers to distribute power over a wide area. (For further discussion, see Chapter 4.) Indeed, the first transformers installed in 1885 by Ganz and Company retained the ring shape (Figure 3.1).

But in 1882 Tesla did not know that Zipernowsky, Bláthy, and Déri would pioneer AC transmission. Instead, for Tesla the broken ring transformer was a wonderful device to watch and ponder. While the ring was being powered by an AC generator, Tesla, in a moment of curiosity, placed a metal ball on the wooden surface on the top of the ring transformer. To his delight, the ball began to spin while the current was being applied. As he watched the ball spin, Tesla deduced that because the coils varied in their windings, they produced two different alternating currents.[5] As we saw with Baily's motor in the last chapter, these two currents created a rotating magnetic field that in turn caused the ball to spin. Here was confirmation of the

FIGURE 3.1. First transformers developed by Zipernowsky, Bláthy, and Déri in 1884-1885 in the Museum of Applied Arts, Budapest.
Source: en.wikipedia.org/wiki/File:ZBD.jpg.

hunch Tesla had had while walking in the park with Szigeti: that alternating current could create the rotating magnetic field he wanted for his motor.

To be sure, the ball spinning on the top of the broken ring transformer did not reveal to Tesla how to control several alternating currents so that they created a rotating magnetic field; again, the spinning ball only confirmed that Tesla's ideal of the motor was possible. Tesla would spend the next five years acquiring the knowledge and skill necessary to get electricity to do his bidding. But as we shall see, during this time of learning, the spinning ball and ring transformer became a key way that Tesla represented his ideal. Whenever he thought about or had the chance to experiment with his new motor, he would use a similar ring wound with several coils and in the middle of the ring he would place different metal objects hoping that they, too, would spin in the rotating magnetic field.[6]

JOINING THE EDISON ORGANIZATION IN PARIS

Tesla's musings about spinning balls and rotating magnetic fields, however, came to an abrupt halt when Ferenc Puskás was finally able to hire him to help install the new telephone exchange. Tesla threw himself into improving the exchange and even developed a new telephone repeater or amplifier.[7]

Once the Budapest exchange was up and running, Ferenc Puskás sold it to local businessmen for a profit. While the Budapest exchange was being built Tivadar Puskás had remained in Paris to help introduce Edison's incandescent lighting system. Tivadar now invited Tesla and Szigeti to come to Paris and got them jobs with the Edison organization (Figure 3.2).[8]

Because French law required that any inventions patented in France must also be manufactured there, Edison had sent his closest associate, Charles Batchelor, to France in 1881 to organize a company for manufacturing and installing Edison lighting systems. Imitating how the Edison lighting organization was structured in America, Batchelor established three separate companies in France: the Compagnie Continentale Edison (which controlled the patents); the Société Industrielle & Commerciale (which manufactured equipment); and the Société Electrique Edison (which installed the systems). To manufacture incandescent lamps and dynamos, Batchelor built a factory in Ivry, on the outskirts of Paris.[9] Tesla appears to have been employed primarily by the Société Electrique Edison (SE Edison).[10]

FIGURE 3.2. Tesla when he was in Paris, 1883. From NTM.

Working at the Edison works in Ivry, Tesla acquired a great deal of practical engineering knowledge about dynamos and motors. Up to this point, Tesla had done mostly mental engineering, visualizing in his mind how an AC motor might ideally work. Now Tesla learned firsthand the problems of converting mental inventions into real machines. In order to create a working dynamo or motor, one had to think carefully about the proper proportions for the rotor and stator coils; to secure a particular current output, one had to plan the length and diameter of the coils, the gauge and number of turns of wire, and the speed by which the machine would rotate. In the early 1880s, none of this information had been codified into formulae or design rules; rather, the design of electrical machines was based on trial and error and craft knowledge. Working for the Edison organization, Tesla learned much of what was then known about dynamo and motor design, and this knowledge put him in a position to start thinking about converting his ideal motor into a real machine.

While Tesla acquired engineering know-how from the Edison organization, he also made his own contribution to the company. Most of the Edison men had learned about electrical machines by working either in the telegraph industry or in machine shops, and few had any formal schooling in science or mathematics.[11] In contrast, Tesla had received a thorough education in physics and mathematics at Graz, and the French manager of SE Edison, R. W. Picou, recognized Tesla's ability to apply theory and make calculations. Shortly after joining the company, Tesla was put to work designing dynamos for incandescent lighting installations and was paid three hundred francs a month.[12]

While at the Edison works at Ivry, Tesla continued to think about his ideas for a motor. "We were together almost constantly in Paris in the year 1882," Szigeti later testified, and "Mr. Tesla was much excited over the ideas which he then had of operating motors."[13] One evening, he outlined his plans for an AC motor to Szigeti and four or five Edison men by drawing diagrams in the dirt with a stick. Picking up on his insight in Budapest that several alternating currents should be able to produce a rotating magnetic field, Tesla described to his Edison colleagues an elaborate system in which the generator produced three separate alternating currents that were delivered to the motor over six different wires (see Figure 3.3). In his subsequent patents and lectures, Tesla explained that these three alternating currents would have to be out of phase with each other by 120° in order to create a rotating magnetic field, but there is nothing to indicate that he understood in 1882 the significance of having the currents out of phase. "My idea," Tesla explained, "was that the more wires I used the more perfect would be the action of the motor."[14]

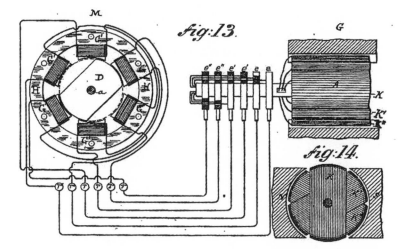

FIGURE 3.3. Diagram from later Tesla patent showing his system in which the generator produced three separate alternating currents that were delivered to the motor over six different wires.

Figure 13 in U.S. Patent 381,968, "Electric Magnetic Motor," (granted 1 May 1888).

Tesla was disappointed that the Edison men were not impressed by his invention. From a commercial standpoint, they were probably not interested because they saw the major opportunities in developing a system for electric lighting and not transmitting power to run electric motors. It was only after 1886 that other electrical pioneers such as Frank Sprague were able to convince central station engineers that they could provide electricity for both lighting and motors.[15]

But from a technical standpoint, Tesla's six-wire scheme probably seemed wrongheaded to these men, not because it used AC but because it would use too much copper in the multiple wires. One of the major concerns faced by the Edison organization in the early 1880s was developing distribution systems that used as little copper as possible. Because the copper wiring was often the largest cost in a new installation, Edison himself devoted substantial effort to evolving more economical wiring schemes. In the early 1880s, Edison introduced his three-wire system to replace his feeder-and-main system. In contrast to Edison's three-wire system, Tesla's proposed six-wire system probably looked uneconomical in terms of the copper wiring it would require. Of course, electrical systems using AC can operate on higher voltages and hence have smaller conductors, but it is not clear that Tesla or the Edison men would have understood this in 1882.

Only one Edison man, David Cunningham, a superintendent of the Edison Lamp Works, showed any interest in Tesla's invention. Edison had sent Cunningham overseas to help Batchelor install equipment at the International Electrical Exhibition in Paris in 1881, and Cunningham stayed on to supervise dynamo building at Ivry. Cunningham, Tesla recalled, "offered to form a stock company. The proposal seemed comical in the extreme. I did not have the faintest conception of what that meant except that it was an American way of doing things." Nothing came of the proposal, and in 1883 Tesla was sent by the company to work as a troubleshooter at different lighting stations in France and Germany.[16]

In between these traveling assignments, Tesla found time to develop an automatic regulator for the Edison dynamos, and his plan impressed Louis Rau, the president of SE Edison.[17] Consequently, when the company needed to send an expert to solve problems at a new station in Strasbourg in Alsace, it chose Tesla.

A MOTOR IN STRASBOURG

At Strasbourg, SE Edison was attempting to install an incandescent lighting system in the new railroad station. During the Franco-Prussian War of 1870–71, Strasbourg had changed from French to German hands. After the war, the German Empire established its presence in Strasbourg by erecting a series of substantial new public buildings, including a new central railroad station.[18] According to Tesla, the German authorities were quite upset with the Edison Company since the wiring in the plant had short-circuited and blown out a large part of a wall during a visit to the train station by Emperor Wilhelm I.[19] To placate the Germans, the company needed to send a German-speaking engineer to finish wiring the new plant. Given his language skills, Tesla was sent to Strasbourg in October 1883 to reinstall the wiring and deal with the upset Germans. To help with this work, Tesla brought Szigeti with him as his assistant.[20]

At Strasbourg, Tesla found that SE Edison was installing a large and ambitious system. The system consisted of four generators that powered twelve hundred lamps. In addition to the Edison equipment, the German electrical manufacturer Siemens & Halske was installing five DC generators and sixty arc lights. The wiring for both the incandescent and arc lights was placed in underground conduits, and since this was a relatively new practice, it was probably the cause of the problems that Tesla had to solve.[21]

Tesla was soon working night and day on the Edison system, but he found time to conduct experiments on his AC motor. Included in the railway station's electrical powerhouse was a Siemens AC generator that had probably been used to power an earlier arc-lighting system with Jablochkoff candles.[22] With Szigeti's help, Tesla constructed a small motor that could be powered by the Siemens AC generator. Anxious to keep the motor a secret, Tesla and Szigeti tested it in a closet where they could tap the AC circuit.[23]

For this motor, Tesla made the stator by winding insulated wire around the outside of an oblong brass ring (see Figure 3.4).[24] The stator's windings were connected to the Siemens generator. For the armature, Szigeti made a five-inch iron disk that was mounted on a horizontal axle.[25] According to Tesla's mental engineering, the AC from the generator should produce a rotating magnetic field in the stator. In turn, the rotating field would induce currents in the disk; the induced currents would be repelled by the rotating field and thus

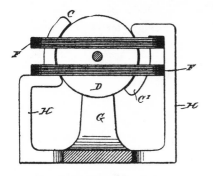

FIGURE 3.4. Diagram of AC motor that Tesla built in Strasbourg in 1882.

This motor consisted of a disk-shaped iron rotor (D) that was mounted on a shaft. The stator (F, F) was two coils of insulated wire mounted on brass rings. Tesla connected the stator to an AC generator and he initially assumed that the AC would produce a rotating magnetic field and induce eddy currents in the rotor. However, since the stator coils were wrapped around brass rings that did not magnetize, Tesla was forced to jam a steel file in the coil [this would have been the equivalent of C or C' in the diagram]. The AC then magnetized the file and induced eddy currents in the disk; since the filed produced by the eddy currents was in the opposite direction to the magnetic field in the file, the disk rotated. Tesla patented a more elaborate version of this motor and this diagram is from that patent.

See, NT, "Electro-Magnetic Motor," US Patent 424,036 (filed 20 May 1889, granted 24 March 1890).

the disk would rotate. "It was," Tesla claimed, "the simplest motor I could conceive of. As you see it had only one circuit, and no windings on the armature or the fields. It was of marvelous simplicity."[26]

As simple as it was, this motor did not work when Tesla first tried it. When he held the stator coil around the disk, the disk did not turn because he had wound the stator coil around a brass core that could not be magnetized.[27] To overcome this difficulty, Tesla jammed a steel file in the coil. Now the alternating current produced a magnetic field in the steel file that in turn induced currents in the iron disk. But still the disk did not rotate and so Tesla tried the file in different positions relative to the disk. Eventually he found a position where the magnetic field in the file and the induced currents in the disk were in the same direction, so that they repelled each other and caused the disk to slowly rotate. Tesla was thrilled to see the disk turn: "I finally had the satisfaction of *seeing rotation effected by alternating currents of different phase, and without sliding contacts or commutator*, as I had conceived a year before. It was an exquisite pleasure but not to compare with the delirium of joy following the first revelation."[28]

The Strasbourg motor was an important turning point for Tesla because this motor tempered his idealized thinking with a strong dose of practicality. Prior to this motor, Tesla had performed only mental engineering and had taken for granted that what he could conjure in his mind's eye could be easily made to work in the real world. In Strasbourg, Tesla realized for the first time that materials count—the core of the stator needed to be made of iron or steel, not brass. Even though he later insisted that he was able to design perfect machines in his head that would then run flawlessly when built, it is clear that, like all inventors, he ran into problems when it came time to convert his ideals into working devices.[29]

While in Strasbourg, Tesla again tried to secure financial backing for his invention. Through his work at the Edison plant, Tesla became acquainted with M. Bauzin, a former mayor of the city. According to Tesla, Bauzin "conceived a great attachment" for Tesla and consequently Tesla revealed to him that he possessed "an invention which would revolutionize the dynamo-machine industry." Bauzin consulted with a wealthy local businessman, Benjamin, but Benjamin declined to invest in Tesla's invention. Bauzin then offered to loan Tesla twenty-five thousand francs that Tesla could repay when he had succeeded in perfecting his motor. Tesla, however, wanted Bauzin to come in as his partner, probably in order to share in the long-term profits that Tesla hoped would come from the invention. Not knowing anything about electricity or inventions, Bauzin declined to join Tesla as a partner, and so Tesla left Strasbourg disappointed.[30]

BACK TO PARIS, ON TO NEW YORK

Tesla returned to Paris in February 1884, expecting to receive a bonus from the Edison Company for solving the problems with the Strasbourg plant. Disappointed when this reward did not materialize, Tesla tried to interest a few Parisians in supporting the development of his motor, but again nothing came of this effort. However, Tesla's work on improving dynamos had caught the eye of Charles Batchelor, who had been head of the French Edison companies. In the spring of 1884, Batchelor was recalled by Edison in order to manage the Edison Machine Works in New York. Intent on improving the dynamos produced at the Edison works, Batchelor requested that Tesla come to America and continue his dynamo work there. To help smooth his entry into the Edison organization in New York, Tesla secured a letter of introduction from Tivadar Puskás addressed to Edison, which stated, "I know two great men and you are one of them; the other is this young man."[31]

Tesla sailed to New York on the *City of Richmond* and arrived on 6 June 1884. As was the case with many immigrants, the customs officer had trouble understanding the nervous young man in front of him, and he recorded Tesla as a native of Sweden when in all likelihood he had told the officer his birthplace was Smiljan. Years later he recalled that process of formally entering the United States consisted of a clerk barking at him, "Kiss the Bible. Twenty cents!"[32]

Having lived in the cosmopolitan cities of Prague, Budapest, and Paris, Tesla was initially shocked by the crudeness and vulgarity of America. As he wrote in his autobiography, "What I had left was beautiful, artistic, and fascinating in every way; what I saw here was machined, rough, and unattractive. A burly policeman was twirling his stick which looked to me as big as a log. I approached him politely, with the request to direct me [to an address]. 'Six blocks down, then to the left,' he said, with murder in his eyes. 'Is this America?' I asked myself in painful surprise. 'It is a century behind Europe in civilization.'"[33]

But Tesla did not dwell on the contrasts between Europe and America, for he was soon busy making a place for himself in the New York Edison organization. Just as he had in Paris, he sought out work as a troubleshooter. The Edison organization had just installed two dynamos in the S.S. *Oregon*, which at the time held the Blue Riband for being the fastest transatlantic passenger ship. Unfortunately the dynamos failed, delaying the ship from departing as scheduled from New York. Drawing on his experience troubleshooting lighting stations in Europe, Tesla volunteered to take a work crew to the *Oregon* and make the needed repairs. Working through the night, Tesla and his crew put the dynamos back in order; the *Oregon* departed New York on 7 June 1884, going on to set a new record for its eastbound run.[34]

Upon returning to Edison's offices in Manhattan at 5 A.M. the next morning, Tesla chanced to meet Edison, Batchelor, and a few other men who were just going home. According to Tesla, Edison said, "Here is our Parisian running around at night." In response, Tesla told Edison that he had just finished repairing the dynamos on the *Oregon*. Edison walked away in silence, but when he thought he was out of Tesla's earshot he remarked, "Batchellor [*sic*], this is a d—n good man." Having impressed Edison, Tesla started work at the Edison Machine Works on 8 June, only two days after arriving in America.[35]

At the Edison Machine Works, Tesla set to work redesigning Edison's long-legged Mary-Ann dynamos, replacing their long magnets with more efficient short-core designs. Tesla claimed that his improved dynamos produced three times as much output using the same amount of iron. Although Tesla worked long hours at the Machine Works, from 10:30 in the morning until 5:00 the next morning, he took time to enjoy good meals and a game of billiards. Not knowing that Tesla had played billiards as a student, Edison's personal secretary, Alfred O. Tate, noted that "He played a beautiful game. He was not a high scorer but his cushion shots displayed skill equal to that of a professional exponent of this art."[36]

While at the Edison Machine Works, Tesla continued to think about his AC motor, but he did not attempt to develop it. Perhaps recalling

FIGURE 3.5. Group of men standing in front of the Edison Machine Works on Goerck Street in New York around the time that Tesla worked there. Tesla is not in the group. From NTM.

how the Edison men in Paris had been indifferent to his ideas, Tesla chose to keep silent. On one occasion, Tesla almost told Edison about his motor. "It was on Coney Island," Tesla recalled, "and just about as I was going to explain it to him, some one came and shook hands with Edison. That evening, when I came home, I had a fever and my resolve rose up again not to speak freely about it to other people."[37]

After working on the design of dynamos, Tesla was next asked to help develop an arc-lighting system. In the mid-1880s, the Edison organization was interested in having its own arc-lighting system in order to compete with its major competitors, the Thomson-Houston Electric Company, the Brush Electric Light Company, and the United States Electric Lighting Company. These rivals had grown up manufacturing and installing arc lighting, and they then expanded their product line by adding incandescent lighting systems. While Edison's incandescent lighting system was suitable for lighting the interiors of homes and offices, it was not particularly effective for outside or street lighting. Consequently, as towns and cities set up new central stations for providing electric lighting for both streets and homes, the Edison organization lost contracts to Thomson-Houston or Brush since these firms could install both arc and incandescent lights.

In response to this competitive need, Edison designed an arc lamp and filed a patent for it in June 1884. Tesla recalled that Edison gave him the basic plan for this arc-lighting system but left it up to him to work out the details.[38] Tesla developed a complete system, and again he expected to be handsomely rewarded for his efforts. However, once the system was completed, it was never put into use.

In all likelihood, Edison and his company shelved the Tesla system for business and technical reasons. At this time, the Edison organization was struggling with the problem of marketing and installing central stations. The difficulty was that most of the new local utility companies that wished to purchase electric lighting systems lacked the capital to purchase the system and the technical expertise to install the equipment; in response, electrical manufacturers experimented with various marketing schemes whereby they could help customers buy their systems while minimizing their financial risk.[39] After supervising the building of power stations through the Thomas A. Edison Construction Department (and losing money in the process), Edison decided in early 1885 to leave the problems of installing his system to others. Consequently, his organization entered into an agreement with Edward H. Goff and the American Electric Manufacturing Company (AEM). Goff had made a name for himself promoting and building arc-lighting stations, and he wished to expand into the incandescent lighting market. The Edison organization and AEM struck a deal whereby when AEM saw an opportunity to install an incandescent lighting system it would sell the local utility an Edison system; in return, when the Edison organization wanted to install an arc-lighting system, it would use the system invented by James J. Wood and owned by AEM.[40] In negotiating with Goff, the Edison organization may have been able to use Tesla's arc system and Edison's arc-lighting patents as bargaining chips in order to negotiate favorable terms. Once this deal went through, however, the Edison organization no longer had any need for the arc system developed by Tesla.

A second reason why the Edison organization did not utilize Tesla's arc-lighting system was that other engineers in the company had developed an incandescent lighting alternative. Called the municipal system, this alternative could be used for street lighting since it used larger incandescent lamps that were placed on a high-voltage series circuit.[41] Thus, when the arc-lighting project was shelved, Tesla went unrewarded and he quit in disgust. As his last notebook entry at Edison, he scrawled, "Good by [sic] to the Edison Machine Works!" Altogether, Tesla worked for the Edison Machine Works in New York for six months.[42]

ARC LIGHTING IN RAHWAY

Once again on his own, Tesla was not without resources. No sooner had Tesla left the Edison organization than he was approached by Benjamin A. Vail from Rahway and Robert Lane, a businessman from East Orange, New Jersey. A descendant of an old Quaker family, Vail had studied at Haverford College and established a law practice in Rahway. Active in the state's Republican Party, Vail served on the Rahway town council in 1875 and was elected to both the New Jersey Assembly and Senate.[43] Excited by the prospects of electric lighting, Vail and Lane were keen to enter this new field. In December 1884, Vail and Lane hired Tesla and they organized the Tesla Electric Light and Manufacturing Company. Although the company could issue stock for up to $300,000, it began with Vail subscribing $1,000 and another $4,000 from other investors in Rahway.[44]

Drawing on what he had learned while working for Edison, Tesla proposed that this company develop its own arc-lighting system. While we tend to assume that the electrical industry grew up around Edison's incandescent lamp, in reality the fastest-growing segment of the electrical industry in the mid-1880s was arc lighting. According to one commentator, the number of arc lights installed doubled every year between 1881 and 1885. Although the industry was dominated by the Brush and Thomson-Houston companies, there were also numerous new, small, start-up companies; by 1886, there were at least forty firms manufacturing arc-lighting systems. Across the country, dozens of businessmen like Vail and Lane were intrigued by the new electrical industry, and they established new companies to manufacture arc-lighting equipment.[45]

To help his new company enter the arc-lighting field, in the spring of 1885 Tesla prepared patent applications covering improvements in generators, arc lamps, and regulators. While his arc lamp and regulator were similar to those invented by Charles Brush and Elihu Thomson, his generator incorporated several improvements that reduced energy losses arising from heat and eddy currents.[46] For help in filing these patents, Tesla turned to Lemuel W. Serrell, Edison's chief patent attorney in New York. While working on these patents applications, Tesla was paid $150 per month. Tesla contemplated trying to convince Vail and Lane that he could develop other electrical inventions (such as his AC motor), but he soon realized that they were only interested in arc lighting (Figure 3.6).

Like other early arc-lighting entrepreneurs, Vail and Lane anticipated that profits could come from both manufacturing equipment and running lighting systems. Consequently, they secured a corporate charter to permit them to do both.[47] Through 1885, Tesla worked to both manufacture his system and run it from a central station. Tesla was probably assisted by Szigeti as well as by a young man, Paul Noyes, whom he recruited from the Gordon Press Works in Rahway.[48]

By 1886, Tesla's system was being used in Rahway to light some of the town's streets and several factories. The company received favorable notice

FIGURE 3.6. Tesla in 1885. From KSP, Smithsonian Institution.

from the New York trade journal *Electrical Review*, which published a front-page feature about the Tesla system in August 1886. In return, the Tesla company ran advertisements in *Electrical Review* announcing "the most perfect Automatic, Self-regulating system of electric arc lighting yet produced."[49]

As the patents for his arc-lighting system were granted, Tesla assigned them to the Tesla Electric Light and Manufacturing Company in return for stock shares. However, once the system was completed, Vail and Lane abandoned Tesla and created a new firm, the Union County Electric Light and Manufacturing Company. Perhaps Vail and Lane decided to exit the manufacturing side of the arc-lighting industry because that side of the business was becoming both highly competitive and capital intensive. By the end of the decade, the manufacture of arc-lighting equipment was dominated by a single firm, Thomson-Houston. Instead, Vail and Lane chose to concentrate on operating as a lighting company for Rahway and the surrounding county. In this situation, Tesla's role as inventor was superfluous since Vail and Lane did not need to improve the system in order to be competitive in the utility business.[50] Having assigned the patents to the company, Tesla was left in a position where he could no longer use his own inventions. All he had to show for his efforts in Rahway was "a beautifully engraved certificate of stock of hypothetical value."[51]

Abandoned by his business patrons in Rahway, Tesla fell on hard times and was unable to find work as an engineer or inventor. After several jobs repairing electrical equipment, he was reduced to working as a day laborer, digging ditches. "I lived through a year of terrible heartaches and bitter tears, my suffering being intensified by material want," Tesla recalled years later, feeling that "my high education in various branches of science, mechanics, and literature were [*sic*] a mockery."[52]

CHAPTER FOUR

A THERMOMAGNETIC MOTOR

In the midst of hardship, Tesla mustered the energy needed to file a patent application for a thermomagnetic motor in March 1886. Just as his arc-lighting inventions saved him after he left Edison, so this new application helped him get back on his feet.

It is likely that Tesla got to thinking about the relationship between magnetism and heat while working at the Edison Machine Works since Edison was then experimenting with a pyromagnetic generator that would directly produce electricity from burning coal. In one dramatic experiment in 1884, Edison heated coal until it incandesced and then introduced a gas that he hoped would be ionized by the glowing coal. Although Edison obtained a very strong current, the gas exploded and blew out his laboratory's windows.[1]

Perhaps taking heed of Edison's disaster with overheated coal, Tesla initially focused on the fact that iron magnets lose their magnetic strength when they are heated. To take advantage of this phenomenon, Tesla designed a small motor consisting of a magnet, iron pivoting arm, spring, Bunsen burner, and flywheel (Figure 4.1). At a normal temperature, the fixed magnet was sufficiently strong to pull the pivot arm and compress the spring. However, when the pivot arm

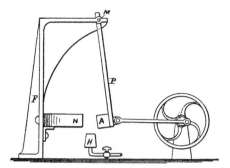

FIGURE 4.1. Tesla's thermoelectric motor from 1886.

Key:
N   fixed magnet
A   moving magnet
P   iron pivoting arm
FM   leaf spring
H   Bunsen burner

From T. C. Martin, *The Inventions, Researches, and Writings of Nikola Tesla,* 2d ed. (1894; reprinted 1995), Fig. 240 on p. 428.

was pulled toward the fixed magnet, it came within the flame of the Bunsen burner. The flame heated the pivot arm and caused it to lose the magnetism induced in it by the fixed magnet. The force of the compressed spring was now greater than the force of the magnetic field, causing the pivot arm to swing away from the fixed magnet. Because the pivot arm was connected by a crank to the flywheel, the motion of the pivot arm caused the flywheel to turn. As the pivot arm swung out of the flame, it cooled and was again attracted to the magnet. Now the strength of the magnetic field was greater than the force of the spring, thus causing the pivot arm to swing back toward the fixed magnet and the flame. In his patent application, Tesla outlined not only the basic principle of this motor but also seven variations.[2]

RESCUE BY PECK AND BROWN

Tesla's thermoelectric motor patent proved to be a turning point in his career because through it he met the men who would become his mentors while he perfected his AC motor. While digging ditches, Tesla told the foreman who hired him about his efforts at invention, and this foreman in turn introduced him to Alfred S. Brown (1836–1906).[3] Brown had entered the telegraph service in 1855 and by 1875 had worked his way up to the position of superintendent of Western Union's New York Metropolitan District.[4] Regarded as a "first class electrician and expert in underground telegraph work," Brown was responsible for supervising the installation of the cables connecting

Western Union's main office to the stock and commodity exchanges located in downtown Manhattan, and so it is entirely possible that a foreman supervising ditchdigging for these underground cables could have introduced Tesla to Brown.[5] As a senior Western Union manager, Brown had watched Edison demonstrate several of his breakthrough inventions, including his duplex (two-message), quadruplex (four-message), and improved telephone.[6] One indication of Brown's prominence in telegraph circles was that he served as a pallbearer at the 1878 funeral of William Orton, the powerful president of Western Union.[7] Based on his Western Union experience, Brown knew well how companies and individuals could use inventions to dramatically reshape an industry.

Sensing an opportunity with Tesla's thermomagnetic motor but realizing that he would need business expertise to turn this invention into a commercial proposition, Brown turned to Charles F. Peck (d. 1890). A lawyer from Englewood, New Jersey, Peck was interested in telegraph and electrical affairs, and he counted another electrical inventor, William Stanley Jr., as a family friend.

Peck had become involved with telegraphy in 1879 when he and John O. Evans investigated setting up a direct telegraph connection between Washington, D.C., and Chicago. In the course of trying to establish this line, Peck discovered that there were banks and merchants who were interested in leasing dedicated wires in order to conduct their business securely. To take advantage of this demand in telegraph leases, he and Evans organized the Mutual Union Telegraph Company in 1880 with capital of $1.2 million to build lines between major cities that could provide this dedicated service. Evans was president of this new company while Peck served as secretary. Mutual Union built a new line between Boston and Washington and then promptly leased individual wires to a number of parties. Peck and Evans realized a handsome profit from the sale of these leases. Together they made a good team; as one historian of the telegraph industry wrote, "Evans was vivacious, quick, adventurous. Mr. Peck was active and cautious."[8]

But Peck and Evans soon realized that there were even greater profits to be made by using Mutual Union to harass Western Union. Ever since it had become the dominant firm in the telegraph industry in the late 1860s, Western Union had been confronted by the threat of a takeover by either the federal government or Wall Street financiers. To fight off these threats, the president of Western Union, William Orton, skillfully used a mix of political lobbying, shrewd rate cutting, erecting lines along major railroad lines, and, above all, encouraging inventors such as Edison and Elisha Gray to develop more efficient telegraph instruments. However, these tactics were not foolproof; if rival financiers could secure either patents for new inventions or new rights-of-way from the railroads, they could easily attack Western Union and attempt a hostile takeover. Jay Gould pursued this strategy twice, first unsuccessfully in 1874–77 and then successfully in 1879–81. "In each raid," noted historian Richard R. John, Gould "mounted a political campaign to whittle away Western Union's legal prerogatives, triggered wild swings in the market price of Western Union shares in which he took advantage of advance information on market trends, and built up a rival telegraph corporation—Atlantic & Pacific in 1874, American Union in 1879—that Western Union found it expedient to buy out."[9]

In 1881, just as Gould had done, Peck and Evans decided to scale up Mutual Union to create their own rival telegraph network. Vowing to acquire "nine-tenths of the profitable telegraph business of the country," Peck and Evans issued stock and bonds for $10 million, convinced Wall Street banker George F. Baker to join the enterprise, and started building new lines. They gained significant strength when the Baltimore & Ohio Railroad leased its telegraph lines to Mutual Union. To supervise operations, Peck hired Brown away from Western Union to serve as Mutual Union's general manager. Anxious to have the latest technology, Mutual Union retained John Wright and John Longstreet as the company's electricians and encouraged them to develop a stock ticker or printing telegraph. All of this was done so energetically that within two years, Mutual Union had over twenty-five thousand miles of lines in twenty-two states. Mutual Union boasted that the annual earning capacity of the network would be $1.5 million and the probable yearly dividends would be 12%.[10]

Gould was not about to have this upstart ruin Western Union, and he counterattacked with the same tactics he had used in his raids on Western Union. Initially Gould purchased 30% of Mutual Union's stock and proposed to Baker that they share control of the company. When Baker refused, Gould retaliated by entangling Mutual Union in a series of lawsuits. Mutual Union's charter limited its capitalization to $1.2 million, and so the $10 million stock-and-bond issue was illegal.

Prodded by Gould, angry investors demanded that the attorney general of New York annul the company's charter. (Disgusted, the attorney general considered canceling the charter of both Mutual Union and Western Union.) Western Union sued Mutual Union for infringing on a patent it held for the telegraph relay invented by Charles G. Page (discussed later in this chapter). Meanwhile, the Chicago city council refused to allow Mutual Union to erect poles in the streets, and Detroit threatened to do likewise. Overwhelmed by these events, Mutual Union's president, Evans, died on Christmas Day 1881.[11]

But Peck knew that such troubles were all part of the game of harassing Western Union; one had to be patient and wait for Western Union to sue for peace. Eventually realizing that he could ill afford to have Mutual Union become the rallying point for his enemies in the telegraph industry, Gould came to terms with Mutual Union in 1885. After much discussion, Western Union agreed to lease Mutual Union's lines. The terms of the lease were that Western Union paid 1.5% per annum on $10 million of Mutual Union's stock and the interest on $5 million of bonds, for which $50,000 was allocated annually to a sinking fund. Under this arrangement, Brown rejoined Western Union as a superintendent.[12] Peck had beaten Gould at his own game and walked away with a fortune.

Based on their experience with Mutual Union, Peck and Brown were ideally suited to mentor Tesla in the world of promoting inventions. Working at the highest levels of the telegraph industry, they had learned how to exploit technological innovation to their advantage. They knew how to create companies, promote new technology, and leverage change. Peck and Brown identified for Tesla key opportunities in the electrical industry and they positioned his inventions so as to receive significant publicity and financial reward. Tesla held both men in high regard, noting that "They were in all their dealings with me the finest and noblest characters I have ever met in my life."[13]

Intrigued by Tesla's thermomagnetic motor and several other ideas, Peck proposed in the fall of 1886 that he and Brown underwrite Tesla's efforts to develop these inventions into practical devices. To permit Tesla to get started on perfecting his inventions, Peck and Brown rented a laboratory for him in lower Manhattan in the fall of 1886. They agreed to share any profits, with Tesla receiving a third, Peck and Brown splitting a third, and a third to be reinvested to develop future inventions. Peck and Brown covered all expenses related to securing patents and paid Tesla a monthly salary of $250. In April 1887, Tesla, Peck, and Brown formed the Tesla Electric Company. And in May 1887, Szigeti came to New York to serve as Tesla's assistant.[14]

Tesla's first laboratory was located in New York's financial district. The laboratory was at 89 Liberty Street, just around the corner from the offices of Mutual Union at 120 Broadway. On the ground floor was the Globe Stationery & Printing Company, and Tesla occupied a room upstairs. The lab was furnished with only a workbench, a stove, and a dynamo manufactured by Edward Weston. To provide power for the dynamo, Peck and Brown struck a deal with the printing company. Because Globe used its steam engine to run the presses during the day, the company could provide power at night to Tesla. As a result, Tesla got into the habit of working on his inventions at night.[15]

In his agreement with Peck and Brown, Tesla promised to develop several different inventions, not just the AC motor that he had long been dreaming about. Consequently, Tesla initially set to work on the problems caused by commutators in motors and dynamos. He had been thinking about commutators for years, and though he preferred to eliminate them from electrical machinery, he nonetheless came up with several improvements, including an AC motor with a short-circuiting commutator and a dynamo commutator that reduced sparking.[16]

THE PYROMAGNETIC GENERATOR

While Tesla duly filed a patent for the dynamo commutator, Peck and Brown were far more intrigued by his ideas about converting the heat from burning coal directly into electricity.[17] They were attracted to this idea because they were broadly interested in energy. Well aware of the growing demand in American industry for cheap power, Peck and Brown had been approached previously by an engineer who proposed to generate steam based on temperature differences in the ocean. In certain situations in the ocean, there can be as much as a 60° difference between the cold water in the depths and the warm water at the surface. One way to take advantage of this temperature differential was to employ the principle embodied in the cryophorus, a device developed by the English scientist W. H. Wollaston. While studying the nature of heat, Wollaston had connected two vessels by a tube and then pumped out all the air. In one vessel he placed water at room temperature while the other vessel was placed in an ice bath. Much to Wollaston's surprise, the temperature differential between the two vessels caused the water in the first vessel to become steam and move through the tube to the second vessel where it condensed. On the basis of this idea, the engineer calculated for Peck and Brown how a large-scale system of pipelines, pumps, engines, boilers, and condensers could be used to generate a seemingly inexhaustible supply of steam from the ocean, which could then be piped to steam engines. Although Peck and Brown found this plan interesting, they were concerned that it would require a huge amount of capital to build the proposed pilot plant. At the same time, they wondered how to distribute all the power that a huge steam plant might generate: how could the power be transmitted to numerous factories, shops, and homes?[18]

Already interested in an ambitious plan like this ocean-steam idea, then, Peck and Brown were naturally drawn to Tesla's ideas about converting the heat from burning coal directly into electricity. The possibility of generating electricity directly from heat was highly appealing to inventors and investors because of the cost and complexity of using steam engines and dynamos. To generate electricity in the 1880s (or even today), one had to burn coal, which heated a boiler and produced steam. The steam was then utilized by an engine that turned the dynamo. At each step in this system, energy was lost as waste heat or friction. If one could eliminate all of these steps and go straight from burning coal to electricity, one would then have an efficient invention that would be even more revolutionary than the dynamo. (Tesla returned to this idea of increasing the efficient generation of electricity a few years later with his mechanical oscillator; see Chapter 10.)

For his pyromagnetic generator, Tesla combined the principle that he had employed in his thermomagnetic motor with Faraday's law of electromagnetic induction. In the thermomagnetic motor, Tesla found that heating a magnet had caused its magnetic field to weaken or change. As Faraday had pointed out, as a magnetic field changes, it induces an electric current in a conductor located within the changing field. Hence, if one placed a conductor in the field of a magnet that

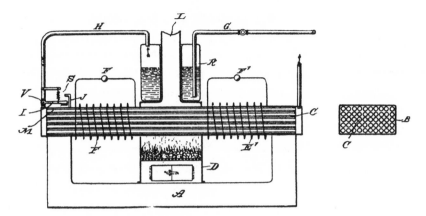

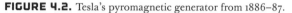

FIGURE 4.2. Tesla's pyromagnetic generator from 1886–87.

Key:
A   horseshoe-shaped magnet
B   thermally insulated metal box
C   hollow iron tubes, inside B
E', F   two coils of wire
D   firebox which heated the iron tubes
K   boiler
H   pipe connecting the boiler to the core so that steam could circulate inside the iron tubes
V   valve controlling steam circulating in core

From TCM, *The Inventions, Researches, and Writings of Nikola Tesla,* 2d ed. (1894; reprinted 1995), Fig. 242 on p. 430.

was being alternately heated and cooled, a current would be induced in the conductor.[19]

To combine these two principles into a practical pyromagnetic generator, Tesla began with a large horseshoe-shaped magnet (Figure 4.2). Across the poles of this magnet, he placed a special core consisting of a thermally insulated metal box holding a number of hollow iron tubes. Because the core sat on the horseshoe magnet, these iron tubes were magnetized. Around the outside of the core were wound two coils of wire. Underneath the center of the core was a firebox, which heated the iron tubes, and above the core's center was a boiler, which was connected by a pipe to the core so that steam could circulate inside the iron tubes. To control when the steam circulated in the core, Tesla placed a valve in the piping between the boiler and the core.[20]

When operating, a coal fire in the firebox heated the iron tubes until they were dull red, about 600°C. At this temperature, the iron tubes would be demagnetized and the changing magnetic field would induce a current in the coils. Next, the valve was opened, and steam (which was 100°C) would circulate inside the tubes and lower their temperature. This cooling process permitted the magnetic field to be restored in the iron tubes, and again, the changing magnetic field would induce another current in the coils. Since heating and cooling would induce currents moving in opposite directions, Tesla's pyromagnetic generator produced alternating current.

Tesla regarded this pyromagnetic generator as a "great invention" and energetically worked on it from the fall of 1886 to the late summer of 1887.[21] In all likelihood, he encountered problems in getting a sufficient temperature differential between heating and cooling. In order to generate a significant amount of electricity, the temperature of the core would have to rise and fall dramatically; if the core retained its latent heat, then little electricity would have been generated. Tesla did apply for a patent for this invention, but it was not granted.

Distressed that he was not able to perfect this invention, Tesla feared that Peck and Brown might abandon him just as Vail and Lane had done in Rahway. However, Peck had ample confidence in Tesla and instead encouraged him to keep inventing. As it became clear that the pyromagnetic generator was not going to work, Tesla recalled, "I met Mr. Peck just at the door of the building in which he had his office, and he spoke to me in a very kind way and said, 'Now do not be discouraged that this great invention of yours is not panning out right; you may bring it to a success after all. Perhaps it would be good if you would switch to some other of your ideas and drop this for a while. I have had an experience that this is a very good plan.' I came back encouraged."[22]

LEARNING TO USE TWO OUT-OF-PHASE CURRENTS

Taking Peck's advice, Tesla shifted his attention from the pyromagnetic generator to electric motors. He now returned to the ideal that had come to him in Budapest five years earlier: a motor with a rotating magnetic field (see Chapter 2). As a first step toward achieving this ideal, Tesla had to test his hunch that several alternating currents could produce a rotating magnetic field. He had thought a great deal about how several alternating currents might be combined, but he had never tried them in practice.

Tesla began by modifying the Weston DC dynamo in the lab so that it could produce two, three, or four separate alternating currents.[23] For his first experiments, he used a large laminated ring for the stator, similar to the one in his Strasbourg motor. Rather than have a single winding around the ring as he did at Strasbourg, Tesla now divided the winding into four separate coils, one in each quadrant. Tesla had the AC generator deliver two separate currents to coils on opposite sides of the ring (Figure 4.3). For the motor's rotor, he balanced a shoe polish tin on a pin in the center of the ring. To Tesla's delight, the rotating magnetic field caused the tin can to spin.[24]

With this motor, Tesla had finally figured out how to combine alternating currents to create a rotating magnetic field in the motor's stator. To do so, the currents delivered to each pair of coils had to be out of phase with one another. In the case of two currents, while one current was at its maximum positive value, the other was at its maximum negative value. If one thinks of the alternating currents as sine waves, then one can say that these two currents are said to be out of phase by 90°. Now understanding the importance of having the currents out of phase, Tesla could build a full-scale electric motor using the rotating magnetic field that he had envisioned in Budapest.

THE RISE OF ALTERNATING CURRENT
IN THE LATE 1880S

Thrilled with this breakthrough motor, Tesla invited Brown, his technically minded patron, to come and see a demonstration in the late summer of 1887. But as Brown watched the tin can spin in this prototype, Tesla now faced the challenge of convincing his patrons that his rotating magnetic field could be used as the basis of a practical, commercial AC motor. Why should they put money into a spinning tin can? While it may seem obvious to us to develop an AC motor, it was not so to electrical experts in 1887. To understand why this was the case, we need to discuss the situation in the electrical industry in the mid-1880s.

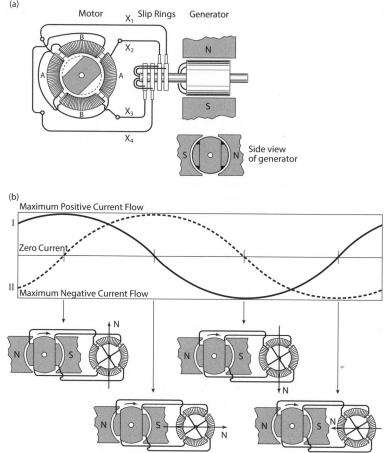

FIGURE 4.3. Tesla's AC motor in 1887.

Like most electrical engineers, Tesla used a combination of stationary electromagnetic coils (called the stator) and rotating electromagnetic coils (called the rotor) to convert motion into electric current and vice versa in his generator and motor. Diagram (a) shows how Tesla used four wires (X_1, X_2, X_3, X_4) to connect his motor to an AC generator. As shown in the side view, the generator's stator consisted of two coils (N, S) and its rotor consisted of two coils mounted at right angles to each other. This generator produced two separate alternating currents that were delivered to the motor via two pairs of slips rings. The two currents traveled to the motor over the four wires and each current energized one pair of coils in the motor's stator (either AA or BB). The rotor in the motor is grey rectangle inside the four coils, but in his 1887 experiments, he used a round shoe-polish tin. Diagram (b) shows how the two separate alternating currents (I, II) were 90 degrees out of phase with each other, meaning that when one was at its maximum, the other was at zero. The images below the current graph show how the magnetic field in the motor's stator rotated as the currents rose and fell over time, with the arrow marked N turning in a clockwise direction. As the magnetic field rotated in the motor, it induced an opposing magnetic force in the rotor, causing the rotor to turn. Note that the rotor is shown in the motor in diagram (a) but is not included in the smaller images in diagram (b), since it would be hard to show both the rotor and the rotating magnetic field.

On the one hand, Peck and Brown were probably comfortable with Tesla investigating motors because of the growing discussion in electrical circles about using motors in central stations. In the mid-1880s, as the number of central stations grew and the utility industry became more competitive, central station operators became interested in expanding their customer base by adding motor service. While they would continue to provide electricity for lighting at night, central station operators saw motors as the means by which they could now sell power to factories and streetcar lines during the day. In response, electrical manufacturing firms added motors to their product lines, and by 1887 there were fifteen firms in the field with a combined output of ten thousand motors.[25] And if central stations could use motors to distribute power to factories, then perhaps a new efficient motor would allow Peck and Brown to distribute power from their ambitious ocean-steam scheme.[26]

On the other hand, Peck and Brown were highly suspicious of Tesla's ideas about developing an AC motor since nearly all of the central stations in the United States in the mid-1880s were using DC, not AC.[27] In the late 1870s, a few electrical inventors in France—as well as Elihu Thomson in America—had experimented with using AC in their arc-lighting systems. Alternating current was attractive to these inventors since it permitted them to use a rudimentary transformer in order to solve the basic problem of how to get a single dynamo to power several arc lights at once; this was what electricians in the 1870s called the "subdivision of the electric light." However, once Charles Brush of Cleveland introduced his DC arc-lighting system with an improved dynamo and regulator, American electricians switched to developing DC systems. Using DC, entrepreneurs were able to establish central stations for arc and incandescent lighting in dozens of American cities.[28]

In Europe, though, AC was not forgotten, and inventors there improved the transformer; by winding two different coils on a single iron core, they found they could raise or lower the voltage of alternating current, and they quickly started using this new device in a variety of ways. For instance, in London in 1883, Lucien Gaulard and John Gibbs used one of the first transformers to connect both arc and different incandescent lights in series to a single large generator.[29] About the same time in Budapest, the engineers Tesla had met at Ganz and

Company—Zipernowski, Bláthy, and Déri (ZBD)—saw AC as a way of developing an incandescent lighting system that could serve a wider area. By having their generator produce high-voltage AC, they found they could distribute power over longer distances using small copper wires. To protect customers from the high voltage, they used a transformer to step down the voltage before the current came into homes and shops. Within a few years, the ZBD system was being used to light several European cities. Both the Gaulard and Gibbs and ZBD systems employed single-phase AC since that was all that was needed to secure the desired voltage change.[30]

The work in Europe on AC transformers was quickly appreciated by astute American electrical entrepreneurs. During a trip abroad in 1885, Charles Coffin of Thomson-Houston learned about the ZBD system and, upon his return, urged Thomson to resume his work on AC. In 1886, Edison agents in Europe warned that they were competing against Ganz and Company for lighting contracts, and they convinced the Edison organization to secure an option on the American patent rights for the ZBD system.[31]

But the American entrepreneur most intrigued by the AC transformer was George Westinghouse (1846–1914). Educated in his father's machine shop in Schenectady, New York, Westinghouse possessed a unique mix of technical genius and business acumen. Not only was Westinghouse able to develop air brakes and improved signal systems for the railroads, he was equally skilled at running the companies needed to manufacture and market these innovations on a large scale. In 1884, Westinghouse became interested in electric lighting, initially as a way of diversifying his Union Switch and Signal Company. As a first step, Westinghouse hired William Stanley Jr., who had patented an incandescent lamp and self-regulating dynamo.

At first, Westinghouse simply intended to develop a DC system similar to that of Edison, but in the spring of 1885 he became intrigued by Gaulard and Gibbs's AC transformer system after reading about it in the journal *Engineering*.[32] Sensing that there would be limited returns from developing one more DC system, Westinghouse decided to strike out in an entirely new direction. In particular, he suspected that AC could be used to establish central stations in municipalities that the Edison organization could not serve. Because of the high costs of its generators and copper distribution network, the Edison organization could sell systems only to towns and cities in which there was a densely populated downtown; for an Edison central station to be profitable, it needed to be located where it could serve dozens of homes and businesses. Westinghouse believed that with AC, one could achieve economies of scale; by employing transformers, one could step up the voltage, distribute power across a wider region, and hence serve more customers. His AC system would be designed to be profitable in towns and cities in which the population was dispersed.

Once he saw the potential for AC, Westinghouse moved decisively. He dispatched an associate, Guido Pantaleoni, to Europe to secure an option on the Gaulard and Gibbs system. In the summer of 1885, Westinghouse ordered several Gaulard and Gibbs transformers shipped to his factory in Pittsburgh, and he asked Stanley to design an AC incandescent lighting system.[33] Working in a small laboratory in Great Barrington, Massachusetts, Stanley developed a practical design for a transformer and confirmed the idea that transformers should be connected to the generator in parallel, not in series as Gaulard and Gibbs had done. To demonstrate the value of his transformer, Stanley strung wires in trees along the streets of Great Barrington to deliver AC to homes and businesses in March 1886.[34] Building on Stanley's demonstration system, Westinghouse installed its first commercial AC system in Buffalo, New York, the following November. Determined to keep up with Westinghouse, Thomson-Houston installed an AC system at the Lynn Electric Light Company in May 1887, and by year's end, Thomson-Houston had installed another twenty-two systems.[35]

The electrical engineering community followed with interest the rapid development of AC lighting systems during 1887. While AC transformers had received only passing mention in *Electrical World*'s annual review of the state of the art in January 1887, the journal regarded the development of lighting systems using transformers as one of the most important breakthroughs in January 1888.[36]

The electrical fraternity was fascinated with AC, not because they were certain that it was the technology of the future but because they saw a serious gap between the ideal and the real. Yes, ideally, AC should permit central stations to distribute power to a larger number of customers, but realistically this had yet to be achieved. As things stood by the end of 1887, AC presented both a commercial opportunity and significant technical problems and risks. Although transformers could step the voltage up and down, engineers at Westinghouse and Thomson-Houston found it difficult to design an efficient transformer. Other critics were concerned about the costs of large AC power stations. Westinghouse claimed that a major advantage of AC was that one could erect a large plant on the outskirts of a city that could generate electricity cheaply. Familiar with the difficulties of raising capital to build stations, both Edison and many central station operators believed that large AC plants would cost too much money to build and the interest charges on the investment would eliminate any operating profits.[37]

Yet another worry was safety. Edison and his associates had spent a great deal of time trying to identify better insulating materials for their low-voltage system, and they simply did not believe that Westinghouse could safely protect people from high-voltage shocks.[38] And finally, several commentators pointed out that the AC systems put forward by Westinghouse and Thomson-Houston were not as convenient or versatile as the DC systems; both companies lacked a meter for measuring how much electricity was used by each individual consumer as well as a motor for providing power to factories and streetcars. After completing a thorough study of the pros and cons of AC, Edison summed up the problems with AC by remarking that it was simply "not worth the attention of practical men."[39]

THE EGG OF COLUMBUS

Peck and Brown were well aware of these trends in the electrical industry. They knew that while there was growing interest in electric motors, no one was sure that the future belonged to AC. Hence, while Peck and Brown encouraged Tesla to investigate electric motors, they were not keen to have him work on an AC motor. AC, for all they knew, might just be a passing fad—interesting, yes, but too difficult to perfect. Perhaps it would be better if Tesla focused on a DC motor for which there was a ready market.

After several discouraging conferences with Peck and Brown to discuss his plans for an AC motor, Tesla realized that he needed a dramatic demonstration. It was not enough to show Brown a shoe polish tin spinning in a rotating magnetic field; Tesla needed to do something that would capture the imagination of his backers.

Consequently, at their next meeting, Tesla asked Peck and Brown if they knew the story of the egg of Columbus. According to legend, Christopher Columbus overcame his critics in the Spanish court of Queen Isabella by challenging them to balance an egg on its end. After the scoffers were unable to get the egg to balance, Columbus made the egg stand upright by lightly cracking one end. Impressed that Columbus had outsmarted his critics, Isabella pawned her jewels to finance Columbus's ships.[40]

When Peck and Brown acknowledged that they had heard the story, Tesla proposed that he could make an egg stand on end without breaking the shell. If he could go one better than Columbus, would Peck and Brown be willing to underwrite his AC experiments? "We have no crown jewels to pawn," replied Peck, "but there are a few ducats in our buckskins and we might help you to an extent."[41]

To gain those ducats, Tesla fastened his four-coil magnet to the underside of a wooden table and secured a copper-plated egg and several balls (Figure 4.4). When Peck and Brown next came to the laboratory, Tesla placed the copper egg on the tabletop and applied two out-of-phase currents to the magnet. To their astonishment, the egg stood on end, but Peck and Brown were stupefied when the egg and

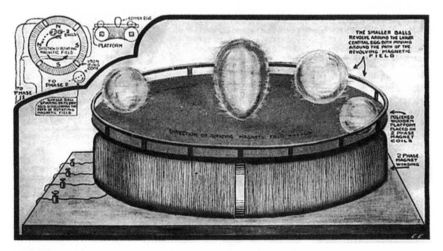

FIGURE 4.4. Tesla's egg of Columbus apparatus, circa 1887.
From "Tesla's Egg of Columbus," *Electrical Experimenter* 6:774–75ff. (March 1919), on p. 774.

balls started spinning by themselves on the tabletop. While it looked like magic, Tesla quickly explained to Peck and Brown that the egg and balls were spinning because of the rotating magnetic field. Highly impressed by this demonstration, Peck and Brown became ardent supporters of Tesla's work on AC motors.

This episode taught Tesla that invention would require a degree of showmanship in order to create the right illusion about his creations. People do not invest in inventions built out of tin cans; they invest in projects that capture their imagination. To draw people in, one often has to draw on metaphors, stories, and themes that have power in a particular culture—that's what Tesla did by invoking the Columbus story. Once he drew them in, Tesla could then get Peck and Brown to think about the commercial potential of his motor.

MAKING POLYPHASE MOTORS

Once his egg of Columbus had convinced Peck and Brown, Tesla pushed ahead with motors that utilized his rotating magnetic field. With Szigeti's help, he fashioned two basic AC motors that he used in nearly all of his experiments in 1887 and 1888. The first was an elaboration of the egg of Columbus apparatus, which consisted of a large laminated ring (the stator) with an iron disk (the rotor) rotating in its center (Figure 4.5).[42] In the second motor, Tesla again used a laminated ring but this time he placed four coils on projections inside the ring (Figure 4.6). In this second motor, Tesla tried several different rotors including both a disk and a drum-wound rotor.[43] He found that both motor designs worked and that when he reversed the electrical connections, the motors changed directions instantaneously. Tesla was enormously pleased with these motors, since they "were exactly as I had imagined them. I made no attempt to improve the design, but merely reproduced the pictures as they appeared to my vision and the operation was always as I expected."[44]

Because these motors used two or more alternating currents that were out of phase with each other, Tesla referred to them as his polyphase motors. Tesla was not the only inventor working on AC motors at the time, but his polyphase motors were significantly different from

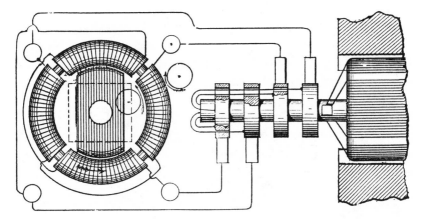

FIGURE 4.5. Tesla's basic experimental arrangement for experimenting with AC motors in the fall of 1887.

    On the left is the motor and on the right is the generator. The generator produced two separate AC currents, as shown by four slip rings on its rotor shaft. The motor consisted of stator in the form of a donut-shaped coil and a rectangular-shaped steel rotor in the center of the donut coil. The stator had four separate coils that were connected in pairs to the generator. When the two currents produced by the generator were out of phase with each other by 90 degrees, they produced a rotating magnetic field in the stator. This field, in turn, caused the rotor to spin on its pivot.

    From TCM, *The Inventions, Researches, and Writings of Nikola Tesla*, 2d ed. (1894; reprinted 1995), Fig. 9 on p. 16.

those developed by Elihu Thomson and other rivals. First, Tesla kept his motor simple by concentrating on finding a way to make the rotor turn by inducing eddy currents in it rather than by delivering currents to it. Second, he developed the motor around a phenomenon that was not necessarily apparent in nature, a rotating magnetic field. And third, unlike any of his contemporaries, Tesla was willing to use several alternating currents to create the magnetic field.

By the end of 1887, Peck and Brown recognized that Tesla had invented a remarkable new AC motor and urged him to patent his ideas. Peck sent Tesla to consult with the law firm of Duncan, Curtis, & Page. Peck had a high opinion of this firm and he assured Tesla that it would secure strong patents for his inventions. At Duncan, Curtis, & Page, Tesla's patent work was taken up by one of the firm's partners, Parker W. Page (1862–1937). Educated at Harvard, Page probably took a special interest in Tesla's motors since his father, Charles Grafton Page, had worked on electric motors—and even a full-scale battery-powered locomotive—in the 1840s and 1850s. Moreover,

FIGURE 4.6. Tesla motor built in 1887–88. From NTM.

patents were central to the Page family. Late in life, Charles Grafton Page secured a special patent from Congress covering the general
form of an induction coil, and after his death, his widow, Priscilla,
convinced Western Union to purchase the rights to this patent for
$25,000 plus royalties paid by licensees. Between his father's work
on electric motors and his mother's success in selling a blockbuster
patent to Western Union, Parker Page was the ideal attorney to work
on Tesla's motor patents.[45]

The offices of Duncan, Curtis, & Page were located at 120 Broadway, in the same building as Peck's office and just around the corner from Tesla's lab on Liberty Street. Tesla regularly visited Page's
office and brought Page sketches and technical descriptions of his
ideas. Tesla carefully prepared his written descriptions in the form of
technical reports not only because he consulted them while he was
experimenting but because he was hoping to write a book titled "The
History of a Thousand and One Alternating Current Motors." According to Page, Tesla emphasized broad principles in his descriptions,

not specific motor designs. Using Tesla's reports and sketches, Page drafted patent applications, which Tesla then checked and revised.[46]

Working together, Page and Tesla now faced a strategic decision: how should he protect his invention? Up to this point, Tesla had followed the custom of most inventors and filed patent applications covering the design of individual components; for instance, with his arc-lighting system, he filed separate applications for his dynamo, lamp, and regulator. However, for his polyphase motor, Page and Tesla decided that a series of applications for individual motor designs did not capture the essence of the invention. Since his student days, Tesla had been thinking about his motor in terms of a system and now he wanted to disclose his invention to the world as a system. Consequently, Page and Tesla chose the bold strategy of filing applications claiming a system for using polyphase motors to transmit power.

Convinced that his motor should be viewed as a complete system, Tesla thus filed a comprehensive patent application on 12 October 1887.[47] In this application, Tesla claimed not only the invention of a new AC motor but also a new system of electrical power transmission. Anticipating that the examiners in the Patent Office might not understand how his new motor worked, Tesla explained his theory as to how the rotating magnetic field caused the rotor to rotate.

During the fall, Tesla and Page followed up the broad application with four more submissions to the Patent Office that covered various ideas for motors dating back to his student days in Prague.[48] All of these applications, however, were regarded as too sweeping by the Patent Office; in particular, the patent examiners did not want Tesla to claim both his motor and his system of power transmission in the same patent. Consequently, in March 1888, Page and Tesla were obliged to split three of the applications with separate submissions for the motor and system designs. As a result, Tesla wound up with seven patents covering his polyphase ideas, and all were issued on 1 May 1888.[49]

A BIT MORE PRACTICAL: SPLIT-PHASE MOTORS

Tesla was quite proud of his polyphase motor and his ideas for power transmission, and he poured a great deal of energy into writing these patent applications. He was captivated by the ideal of developing complete systems for transmitting electric power, systems in which the motor and generator were carefully matched to each other. Tesla felt confident that by developing whole systems he would be able to produce efficient motors and generators in the sense that they would have the greatest possible output for their weight.[50]

But Brown, Tesla's technically minded backer, was not entirely convinced that the systems approach was the best way to develop AC motors. Brown was troubled that Tesla's polyphase schemes required four or even six wires to run between the generator and motor. Yes, the Brush arc-lighting installations in the late 1870s had used four or more wires to connect the lamps and dynamo, but given the high cost of copper wiring, electrical engineers had shifted in the late 1880s to two- and three-wire schemes for electric lighting.[51] Tesla's polyphase system simply did not match the two-wire AC systems currently being built and operated by Westinghouse and Thomson-Houston. Because his polyphase motors were designed to operate on four or six wires, it was not possible to simply connect a Tesla polyphase motor to an existing electrical system. If a central station were to employ Tesla's polyphase system, it would have to start from scratch and install a special generator and wiring network. Hence, Tesla might think his polyphase system was ideal, but in Brown's opinion, it had limited commercial potential.[52]

In September 1887, just after Tesla demonstrated the shoe polish tin motor, Brown asked Tesla if he could design an AC motor that could work on existing single-phase AC circuits—a motor that would need only two wires to connect it to the generator. Because he was caught up in creating his ideal for a polyphase system of power transmission, it had not occurred to Tesla that anyone would want motors that would work on existing AC lines.[53]

However, within a few days, Tesla started showing Brown motors that could run on two wires using single-phase AC. For these motors, Tesla split the incoming AC into two branch circuits and then used several techniques to cause the current in one branch to be out of phase relative to the other. To cause one current to be out of phase, Tesla placed resistance coils, capacitors, and induction coils in one branch of the circuit coming into the motor. Each of these circuits was connected to coils on opposite sides of the stator in the motor, and the out-of-phase currents created a rotating magnetic field. When Brown expressed concern that extra devices such as resistance coils produced waste heat and lowered the efficiency of the motor, Tesla eliminated them by employing two kinds of wire in the stator coils. By using thick wire with a low resistance for one set of stator coils and fine wire with high resistance for the other set, he produced two out-of-phase currents and hence a rotating magnetic field just like the one in his polyphase motors.[54]

Impressed with Tesla's ability to generate a host of split-phase motors, Brown encouraged Tesla to prepare patent applications for all his ideas. With his eye on developments in the electrical utility industry, Brown sensed that patents for AC motors that could be added to existing distribution networks would be valuable. Tesla, however, disagreed with Brown, arguing that he needed to come up with a split-phase design as efficient as his polyphase motors. Clearly Tesla was enamored with his ideal polyphase system and much more interested in how this system could transmit power from one point to another. Tesla was confident that he could raise the efficiency of his split-phase motors, but until he did so, he would file patent applications only for his polyphase inventions. As a result, Tesla decided not to file any applications for the split-phase motors he built during the fall of 1887.[55]

Consequently, for the next few months Tesla continued to prepare his polyphase applications and said nothing to his attorney, Page, about his split-phase motors. In April 1888, though, Page asked Tesla about these motors. While working with Tesla on an application for using a transformer to regulate the phase relationship between two currents, Page happened to inquire if the motor described in this application could run on two wires. When he answered yes, Tesla recalled that "Mr. Page looked at me in amazement and asked me to explain more fully. I remember it very well, because it almost scared me to death." Page couldn't believe that Tesla was unaware of the growing demand for a practical AC motor and that he had not bothered to disclose that he had invented a two-wire motor. Meanwhile, Tesla was worried that if Page knew about his two-wire designs, Page would not take his polyphase system seriously. As Page recalled, Tesla had "purposely kept the knowledge of these [two-wire] motors from me for fear that if I knew that his polyphase motors could be run on a single circuit like any other motors that I would not believe that the invention [i.e., the polyphase system] amounted to anything, and would not, in consequence, draw good claims."[56]

Once he recovered his composure, Page set out to secure strong patents for Tesla's split-phase motors. The situation was complicated because the broad polyphase applications in the Patent Office were ready to issue on 1 May 1888. Anticipating this, Page had been preparing applications for half-a-dozen additional specific polyphase inventions, and he needed to file these remaining applications before the broad patents issued.[57] If he did not do so, the examiners could turn down the specific designs, arguing that these inventions were already covered by the broad patents. In addition, Page had filed patent applications in several foreign countries including England and Germany and was now worried because the foreign polyphase patents had to be based on what was covered in the American patents. If Page revised the polyphase applications to include Tesla's split-phase motors, he ran the risk of significantly delaying the foreign patents. To sort out these complications, Page made a hasty trip to the Patent Office in Washington, and upon his return, he began drafting two-wire motor applications for Tesla to review.[58]

It would be easy to conclude from Tesla's reluctance to patent his split-phase motors that, like other inventors, Tesla tended to have a blind spot about the commercial implications of his work. Elihu Thomson, for example, failed to appreciate fully the importance of developing a single-phase AC system using transformers and only filed patents in 1885 when prodded by his patron, Charles A. Coffin.[59]

However, this episode with the split-phase motors reveals an even stronger characteristic of Tesla's style as an inventor. Tesla was anxious to develop his polyphase system because it embodied an ideal principle. Throughout his AC motor work, he was enthralled by the symmetry of his AC system: just as several alternating currents were produced as the rotor cut through the magnetic field of the generator, so multiphase alternating currents could produce motion in the motor by means of a rotating magnetic field. While Tesla could capture and utilize this ideal symmetry in his polyphase system, he could not do the same with split-phase motors. Yes, he could split the current using a variety of ingenious tricks, but these tricks were not the same as employing a beautiful principle.

As we shall see, over the course of Tesla's career his strength was to identify a grand idea and to seek to develop a system around it. The difficulty with this approach was that it meant that Tesla expected businessmen and consumers to adjust to his systems—based on an ideal—rather than Tesla adjusting his systems to the needs and wishes of society. In the case of his polyphase versus split-phase motors, it meant that Tesla thought that society ought to adopt his beautiful polyphase system even if it meant replacing the existing two-wire, single-phase systems with the more expensive four-wire networks needed for polyphase. Practical considerations and cost meant little to Tesla in comparison to an ideal. Here Tesla was like Steve Jobs, who frequently exhorted his engineers not to worry about costs but instead design "insanely great products" with new capabilities.[60]

The result of this disagreement and confusion with Brown and Page was that Tesla wound up securing two groups of patents: one set covered his ideas for polyphase, multiwire motors and systems, while the second set covered the more practical split-phase or two-wire motors. It was unfortunate that Tesla delayed filing his split-phase applications because the delay weakened his priority claims and led to patent litigation that lasted for the next fifteen years. Yet as we shall see, by having patents covering both broad principles and practical designs, Tesla gave his backers a strong hand to play in negotiating with electrical manufacturers.

CHAPTER FIVE
SELLING THE MOTOR
[1888-1889]

FRAMING A BUSINESS STRATEGY

Prodded by Brown and Page, Tesla spent April and May 1888 working at a feverish pace. He now understood the necessity of testing and
preparing patent applications for as many split-phase motor designs
as possible. "I made daily experiments," he remembered, "and improvised . . . models from pieces of sheet iron and disks and rotors of
various shapes placed in temporary bearings. I may have had possibly
20 finished models that were complete, as nearly as I can recollect." As
his experiments progressed, Tesla gave Page oral reports from which
Page prepared patent applications. Out of the plethora of experimental motors, Page and Tesla decided to concentrate on the most promising method, and they initially filed applications covering a split-phase
motor that had stator coils wound with thick and thin wire (described
in Chapter 4).[1]

While Tesla was busy in the laboratory, Peck and Brown were not
idle. As it became clear that Tesla had indeed come up with several
promising AC motors, they began to think about how they could make
money from Tesla's inventions. As Peck and Brown knew from their
previous ventures, there were three basic strategies they might follow.
First, they could use the patents to create their own new business to

manufacture or use their inventions. Because patents prevented others from manufacturing the product or using the process, the inventor earned a profit from his monopoly position. An example of this strategy is how George Eastman used his patented system of roll film to build up Eastman Kodak beginning in the 1880s.[2]

Second, inventors could grant *licenses* to an established manufacturer. Under the license, the manufacturer might be required to pay inventors royalties for each item manufactured. For instance, after securing a patent for a "road engine" in 1895, George B. Selden collected from automakers a fee of $15 for each automobile manufactured in the United States. Selden was eventually defeated in court in 1911 by Henry Ford.[3]

And third, they could *sell* their patents outright to another entrepreneur or company. The inventor would realize an immediate profit and would avoid the risks of having to manufacture and market his invention. Elmer Sperry, for example, developed an electrolytic process for making white lead in 1904 that he sold to the Hooker Electrochemical Company.[4]

For the most part, historians have concentrated on how inventors in the nineteenth century followed the first strategy, manufacture, largely because that strategy led to creation of long-enduring firms such as General Electric or Eastman Kodak. However, for the average nineteenth-century inventor, this strategy was highly risky, capital intensive, and likely to pay off only over the long run. Moreover, it required that the inventor master the intricacies of manufacturing and marketing, and many inventors lacked these business skills. I suspect that some inventors decided to set up businesses for manufacturing or using their inventions only after they had exhausted the possibilities for selling or licensing their patents. For instance, Bell and his backers initially tried to sell the telephone patent to Western Union in 1876, and it was only after Western Union declined to buy it that they set up the American Bell Telephone Company and began building exchanges.[5]

Given the risks associated with manufacturing, many nineteenth-century inventors preferred to either sell or license their patents. During the 1870s, Munn & Company, the patent agency affiliated with *Scientific American*, urged its inventor clients to pursue a licensing strategy.[6] In particular, licensing was seen as highly profitable since one could grant a large number of licenses to different firms in different territories. With its incandescent lighting system, the Edison Electric Light Company made a handsome profit by granting licenses to power companies in dozens of cities. As a strategy, though, licensing had a downside in that the inventor had to be ever vigilant against competitors infringing his patents, lest the licensees lose their monopoly position. By not aggressively defending its patents in the mid-1880s, the Edison Electric Light Company inadvertently allowed several competitors to spring up, and one of those competitors, the Thomson-Houston Electric Company, eventually took over the Edison Company to form General Electric in 1892.[7]

It was within this context that Peck and Brown framed their business strategy for Tesla's inventions, which can be summed up as patent-promote-sell. As Tesla came up with new electrical inventions, he would patent them. His backers provided the money necessary to cover laboratory expenses and patent fees. Once his inventions were patented, Tesla would then vigorously promote them through interviews, demonstrations, and lectures in order to attract businessmen. To earn a profit on their investment, Peck and Brown then sought to sell or license his patents to either established manufacturers or other investors, who would set up new companies. Hence, the name of the game for Tesla and his backers was not manufacturing his inventions but rather selling or licensing them.

The strategy of selling or licensing patents poses distinct challenges for the inventor and his backers. One has to know people who might be looking for new technology, then one has to generate interest and excitement in the patents that are for sale, and finally one has to negotiate favorable terms. These negotiations involve much bargaining since the seller (i.e., the inventor) asks the highest possible price in order to recover the costs of developing the invention while the buyer seeks to minimize his risk (How much will it cost to convert the invention to a product? Will the product sell?) by keeping the price low. At the same time, the inventor also has to keep in mind that he may not have the only patents for sale and that asking too high a price may send the buyer to other inventors. Hence, to get the best possible price and not drive away the buyer, the inventor and his backers may use all sorts of arguments to persuade the buyer that the invention in question is the best possible version and offers the greatest potential. For the inventor and his backers, then, persuasion is essential for the risky business of selling or licensing patents.[8]

PROMOTING TESLA'S MOTORS

Having chosen patent-promote-sell as their strategy, Peck and Brown now had to promote Tesla's inventions energetically but carefully. The right people—the managers running electrical manufacturing companies—had to learn about his inventions in the right way—scientifically and objectively. In the 1880s, dozens of inventors were turning out hundreds of electrical patents, many of which were of little value. For instance, the Thomson-Houston Company was inundated with offers of patents from inventors, including one for a dubious product called "electric water."[9] Peck and Brown therefore had to chart a course whereby they could capture the attention of electrical manufacturers and convince them of the commercial potential of Tesla's patents. Just as Tesla had captured their imagination with the egg of Columbus, now Peck and Brown had to engage the imagination of the electrical manufacturers.

In framing their promotional efforts, Peck and Brown had to overcome Tesla's obscurity. Since his arrival in America in 1884, Tesla had kept to himself and had not joined any of the recently formed electrical organizations such as the American Institute of Electrical Engineers, the National Electrical Light Association, or the Electrical Club of New York.[10] Aside from the several electricians he had met working in the middle ranks of the Edison organization, Tesla knew few people in the electrical engineering community. Not knowing anything about Tesla, the electrical fraternity might well wonder how a young man of thirty-two from a little-known part of eastern Europe could have developed such a promising AC motor. Was it everything he claimed it was?

To get the right "buzz" going about Tesla's motors, Peck and Brown sought the endorsement of an expert, Professor William Anthony. Educated at Brown University and Yale's Sheffield Scientific School, Anthony was an expert in electricity and optics. From 1872 to 1887, he was Professor of Physics at Cornell University. While at Cornell, Anthony had tested dynamos and established the first electrical engineering program in the United States. Wishing to perfect his own electrical inventions, Anthony left Cornell in 1887 to become the electrician (the chief engineer) at the Mather Electric Company in Manchester, Connecticut. Possessing both academic credentials and commercial experience, Anthony must have seemed to Peck and Brown the ideal person to evaluate Tesla's motors.[11]

In March 1888, Peck and Brown sent Tesla to visit Professor Anthony in Manchester. (Tesla made this trip just as the Great Blizzard of '88 hit, and he was trapped for several days in Manchester.) Tesla prepared two special motors for Anthony to test. Both were polyphase designs, not split-phase machines, since Peck and Brown were worried about revealing too much of what Tesla had accomplished. The tests went well, and Anthony concluded that Tesla's AC motors were as efficient as the DC motors currently available. Anthony was not especially concerned that the polyphase motors would require four wires since he assumed they would be installed in special industrial situations where the need for power would offset the cost of the extra wires.

After conducting the tests, Anthony visited Tesla's laboratory in New York. There Anthony and Tesla discussed specific design problems, such as how to construct a rotor that would respond best to eddy currents and the relationship between the speed of the motor and the number of windings on the rotor. Awed by Anthony's academic credentials, Tesla was reticent with the professor and tried not to disagree with him.[12]

Anthony was highly impressed with Tesla's inventions. As he wrote to Dugald C. Jackson, who was teaching electrical engineering at the University of Wisconsin,

> I [have] seen a system of alternating current motors in N.Y. that promised great things. I was called as an expert and was shown the machines under the pledge of secrecy as applications were still in the Patent Office. . . . I have seen such an armature [i.e., rotor] weighing 12 pounds running at 3,000 [rpm], when one of the (ac) circuits was suddenly reversed, reverse its rotation so suddenly that I could hardly see what did it. In all this you understand there is no commutator. The armatures have no connection with anything outside. . . .
>
> It was a wonderful result to me. Of course, it means two separate circuits from [the] generator and is not applicable to existing systems.

But in the form of motor I first described, there is absolutely nothing like a commutator, the two (ac) chasing each other round the field do it all. There is nothing to wear except the two bearings.[13]

Anthony not only spread news about Tesla's motors among his fellow engineers but also discussed Tesla's achievements in a lecture he gave before the MIT Society of Arts in Boston in May 1888.[14]

After receiving Anthony's favorable evaluation, Peck and Brown contacted the technical press. Knowing that the polyphase patents would issue on 1 May 1888, they invited editors from the electrical weeklies to visit the laboratory. In the closing weeks of April 1888, Tesla demonstrated his polyphase motor to Charles Price of the *Electrical Review* and Thomas Commerford Martin of *Electrical World*. Both Price and Martin were favorably impressed, and Price ran a story about Tesla's motors just after the patents were issued.[15]

THE AIEE LECTURE

The centerpiece of the promotional campaign was Tesla's lecture to the American Institute of Electrical Engineers (AIEE) on 16 May 1888. Since Anthony was a vice president and Martin a former president of the institute, they encouraged Tesla to give a paper on his polyphase inventions. Exhausted and ill from overwork, Tesla initially declined to give the lecture. However, Anthony and Martin persisted, and Tesla wrote the lecture hastily the night before he gave it.

As props for the lecture, Tesla displayed the two motors that Anthony had tested. In order that he could participate in the discussion following the paper, Martin arranged not to chair the session (as he frequently did) but instead asked Francis R. Upton, AIEE vice president and treasurer of the Edison Lamp Company, to preside.

Tesla titled his AIEE lecture "A New System of Alternate Current Motors and Transformers." Though his title was modest, he immediately made bold claims for polyphase AC: "I now have the pleasure of bringing to your notice . . . a novel system of electric distribution and transmission of power by means of alternate currents . . . which I am confident will at once establish the superior adaptability of these currents to the transmission of power." To support his claims, Tesla began by using the step-by-step diagrams from his first polyphase patent to explain how two separate alternating currents could create a rotating magnetic field. To convince his engineering audience that the rotating magnetic field exerted a uniform pull on the motor's rotor, he offered a brief mathematical analysis of the forces involved. Tesla then described his basic polyphase motor, consisting of a ring with four separate coils for the stator and steel disk for the rotor (see Figure 4.5). This motor, he emphasized, could be readily reversed and was also synchronous (i.e., it ran at the same speed as the generator). Anticipating that some might complain that his polyphase motors could not be run on existing AC systems, Tesla argued that it would be relatively simple to change the connections in the rotor coils and the arrangement of the slip rings so that the generators could produce several alternating currents with the appropriate phase relation. Tesla also thought that if power stations would install large multipolar generators (with 64 or 128 coils in the rotor and stator), it would be relatively simple to design motors to run at whatever speed was needed. (As we shall see in Chapter 9, neither of these approaches was especially simple, and both would require a great deal of engineering effort.)[16]

Tesla's lecture was followed by a discussion. To bolster Tesla's presentation, Martin took to the floor and invited Anthony to report on the efficiency of Tesla's motors. Emphasizing that the two motors he tested were of a small experimental type and that small motors tended to be less efficient than large ones, Anthony stated that Tesla's polyphase motors had an efficiency of 50 to 60%.[17]

Anthony was followed by Elihu Thomson, who had been working on an AC motor since 1884 in which he used a commutator to deliver AC to the rotor. By having the commutator cut out different rotor coils at the right moment, Thomson was able to set up a magnetic repulsive force between the rotor and the stator so that the rotor turned. Hoping to stake out the AC motor as his invention, Thomson had given a paper before the AIEE in June 1887 in which he enunciated his principle of induction repulsion.[18] Now faced with Tesla's claims to having developed a practical AC motor, Thomson reminded the audience of his own efforts and promised to report on his motor at a future meeting. In effect, Thomson was politely warning Tesla that he was not the only one working on AC motors and that he could expect competition in the Patent Office and the marketplace. Though Tesla the newcomer was well aware of Thomson's ability as an inventor, he did not back down but rather stood his ground. While acknowledging Thomson as "being foremost in his profession," Tesla responded that he had built a motor like Thomson's but had not pursued it because he was confident that the best AC motor would be one without a commutator.[19]

Thomson's remarks notwithstanding, Tesla carried the day. At the conclusion of the meeting, Upton remarked from the chair, "I believe that this motor—Mr. Tesla can correct me if I am not right—is the first good alternating current motor that has been put before the public anywhere—is that not so, Mr. Tesla?" Upton then announced that Tesla had invited the audience to come see the motors in operation at the Liberty Street laboratory.[20]

SELLING THE TESLA PATENTS

Tesla's ideas captured the imagination of the electrical engineering community and his lecture was reprinted by all of the major engineering journals. In response to his paper, several electrical experts wrote letters to the editor commenting on Tesla's motor, and these, too, were reprinted. With his polyphase motor having been "heralded in the papers as an advance in the art," the stage was now set for Peck and Brown to offer Tesla's patents to the highest bidder.[21]

Tesla entrusted the negotiations related to selling his motor patents to Peck and Brown. Tesla initially hoped that they would sell his patents to the Mather Electric Company since Tesla liked Anthony and thought that he would be able to improve his motor with Anthony's help.[22] Peck and Brown invited Mather to bid on the patents, but at the same time they contacted other electrical manufacturers. In late April 1888, they offered the patents to Thomson-Houston, and Charles A. Coffin asked Thomson to review them. Working on his own AC motor and generally opposed to buying patents from outside inventors, Thomson recommended that Thomson-Houston not purchase the Tesla patents. Thomson considered Tesla's polyphase patents of such little value that they were not worth the required Patent Office fees.[23]

Peck next approached the Westinghouse Electrical Manufacturing Company. As we have seen, George Westinghouse was a latecomer to the electrical industry and had decided to bet on AC rather than DC. Westinghouse and his associates knew that they would only be able to convince central station utilities to purchase their AC equipment if they were able to offer their customers an AC motor. Hence, the Westinghouse Company was a very likely customer for Tesla's patents.

Westinghouse approached Peck about the Tesla patents in late May 1888 because his chief electrician, Oliver B. Shallenberger, had been investigating rotating magnetic fields on his own. In April 1888, Shallenberger had accidentally dropped a small coil spring on top of an electromagnet in an arc light powered by AC. Surprised to see the spring rotating on its own, Shallenberger quickly realized that the changing magnetic field was causing it to turn. Shallenberger recognized that this phenomenon could be used to create a wattmeter and an AC motor. Because the company had a greater need for a meter that could measure power consumed by individual customers, Shallenberger concentrated on developing the wattmeter rather than an AC motor.[24]

While Shallenberger's discovery initially caused jubilation among the Westinghouse engineers, their hopes were soon dashed when they learned that Shallenberger was not the first to discover the rotating magnetic field.[25] Shallenberger had been beaten to this discovery by both Tesla and an Italian physicist, Galileo Ferraris. A few weeks after Shallenberger made his discovery with the spring, Westinghouse learned that Ferraris had also investigated how alternating current could create a rotating magnetic field. As part of his entrepreneurial style, Westinghouse regularly scanned the engineering journals for developments that might enhance his control of key technologies. In the course of his reading, Westinghouse spotted references in May 1888 to an article by Ferraris in the *Proceedings of the Royal Academy of Sciences of Turin*.[26]

A professor of applied physics at the Royal Industrial Museum of Turin, Ferraris had studied optics and took a special interest in mathematically analyzing the behavior of light waves.[27] After testing the AC system of Gaulard and Gibbs at the 1884 International Electrical Exhibition in Turin, Ferraris decided to study transformers.[28] At this time, electrical investigators did not fully understand the relationship between the incoming (primary) current and the outgoing (secondary) current in transformers. Borrowing from his knowledge of mathematical optics, Ferraris theorized that there ought to be a phase difference of 90° between the primary and secondary currents in a transformer. He then hypothesized that if there was such a phase difference, these two currents should produce circular motion, just as two light waves out of phase by 90° created circular interference patterns.[29] To test this hypothesis, in 1885 Ferraris constructed an experimental apparatus consisting of two coils placed at right angles to each other (Figure 5.1). Between the two coils, he placed a small copper cylinder on a pivot, and when he connected the coils to the primary and secondary of a Gaulard and Gibbs transformer, Ferraris found that the cylinder rotated. Ferraris was pleased that his apparatus confirmed that there was a phase difference between the primary and secondary currents in a transformer, and he freely shared his results with other electrical investigators in conversation and correspondence.

Ferraris did not publish the results of his 1885 experiments until 1888 and did so only after reading about Thomson's induction repulsion motor. In his 1888 paper, Ferraris reviewed his findings about the phase difference in transformers, commented on how his ideas could

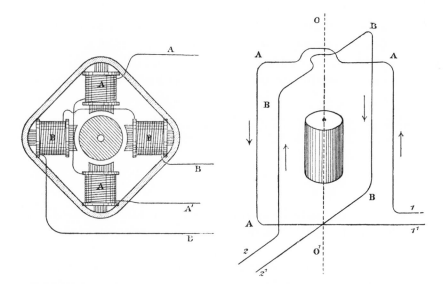

FIGURE 5.1. Ferraris' AC motor circa 1885.
From S. P. Thompson, *Polyphase Electric Currents*, 2d ed., (1900), figs. 332 and 333 on p. 442.

be used to cause Arago's wheel to rotate, and suggested that it might be possible to use his principle to develop a wattmeter. He also reported on how he had created two out-of-phase currents by placing an induction coil and a resistor in two branches of the circuit, the same technique that Tesla had used in his split-phase motors in 1887.

But most important, Ferraris discussed whether a rotating magnetic field could be used to create a practical motor. Ferraris had built a small motor with a copper cylinder for the rotor and had connected it to a dynamometer to measure how much mechanical work the motor performed. In these tests, Ferraris discovered that as the speed of the motor increased, the amount of work decreased. Resorting again to mathematical physics, Ferraris determined that as the motor's speed increased, the induced currents in the copper cylinder created not only a magnetic field but also a great deal of waste heat. According to Ferraris's analysis, when the cylinder reached maximum speed, the induced currents would produce equal amounts of mechanical work and heat, and, as a result, the motor would become inefficient and start to slow down. Based on his tests and mathematical analysis, Ferraris concluded "that an apparatus based upon the principle [of a rotating magnetic field] . . . can have no importance as an industrial motor."[30]

Over the years there has been ample debate over whether Tesla or Ferraris should be recognized as the inventor of the AC induction motor.[31] To some extent, the confusion was created early on by the fact that the first reports in English of Ferraris's 1888 paper left out his analysis of the waste heat produced and thus created the impression that a practical motor would follow from his investigations.[32] But as we have seen, Ferraris drew exactly the opposite conclusion in his paper: he did not think that a practical motor could be developed using a rotating magnetic field. Instead, Ferraris should be credited with being the first to investigate how AC can create rotating magnetic fields. Even more important, Ferraris should be given credit for introducing the notion of phase in discussing alternating current phenomena. Thanks to Ferraris's mathematical analysis, electrical engineers were able to quickly grasp the ideas behind the AC motor and polyphase currents. Nevertheless, it was Tesla who built the first practical AC induction motor.

Meanwhile, we need to get back to Westinghouse. Upon reading reports of Ferraris's paper, Westinghouse decided that it would be best to secure any patent rights that might emanate from Ferraris's work.

Hence, Westinghouse sent his associate, Pantaleoni, to Turin in order to purchase the American rights to Ferraris's ideas for $1,000.[33] Just as he had done a year earlier in securing patent control over AC transformers, Westinghouse purchased the rights to Ferraris's work to gain broad coverage in the area of AC motors.

At the same time that Westinghouse was learning about Ferraris, Shallenberger expressed concern that the Tesla patents might prevent the company from successfully developing an AC motor. In response, Westinghouse dispatched Henry M. Byllesby, vice president of Westinghouse Electric, and Thomas B. Kerr, general counsel, to New York in late May 1888.

Peck asked Tesla to demonstrate his polyphase motors for Byllesby and Kerr at the Liberty Street lab, and Byllesby reported to Westinghouse that the motors seemed to operate satisfactorily. Although Tesla offered an explanation of his motor, Byllesby complained that "his description was not of a nature which I was enabled, entirely, to comprehend." Byllesby noted that Tesla's motors required more than two wires, indicating that Tesla and Peck chose to hold back and not reveal the split-phase designs; after all, why show the customer everything at once? Overall, Byllesby was impressed, and he told Westinghouse that "The motors, as far as I can judge from the examination which I was enabled to make, are a success."

Undoubtedly drawing on his experience with selling Mutual Union to Jay Gould, Peck knew that he would have to bluff in order to get the best possible deal with Westinghouse. Consequently, when Byllesby and Kerr expressed an interest in buying the patents for Westinghouse, Peck informed them that a San Francisco capitalist had offered $200,000 plus a royalty of $2.50 per horsepower for each motor installed. "The terms, of course, are monstrous," Byllesby told Westinghouse, "and I so told them . . . I told them that there was no possibility of our considering the matter seriously. . . . In order to avoid giving the impression that the matter was one which excited my curiosity I made my visit short."[34]

Despite Peck's high price, Byllesby and Kerr nonetheless recommended that Westinghouse buy the Tesla patents in order to secure broad coverage of the principle of using rotating magnetic fields. However, in order to force Peck to accept a lower price, Westinghouse decided to send his star inventors, Shallenberger and William Stanley Jr., to inspect Tesla's work. Perhaps they could persuade Tesla and Peck that Westinghouse was in a stronger technical position and that they should back down. Sending Shallenberger and Stanley to see Tesla was a bit like Steve Jobs's dealings with Xerox's Palo Alto Research Center (PARC) in 1979. Determined to have PARC scientists show him their new graphical user interface, Jobs had arranged for Xerox's venture capital division to invest in his fledging Apple Computer so that PARC would have to cooperate. Much as Xerox capital was the "muscle" needed to force PARC to "open the kimono" and reveal its secrets, so Shallenberger and Stanley were the Westinghouse "muscle."[35]

Shallenberger visited Tesla's lab on 12 June 1888, and Tesla demonstrated his motors operating on four wires. Shallenberger quickly recognized that not only had Tesla discovered the idea of using a rotating magnetic field eight months before he had but Tesla had gone ahead and produced a motor using this principle. Unable to shake Tesla and Peck, Shallenberger returned to Pittsburgh and urged Westinghouse to buy the patents.[36]

Shallenberger's trip was followed by a visit from Stanley on 23 June. As we have seen, Stanley had helped Westinghouse develop single-phase AC electric lighting by designing a practical transformer and confirming the idea that transformers should be connected to the generator in parallel, not in series. While Westinghouse lawyers filed patents for Stanley's transformer design, George Westinghouse decided that the principle of using transformers in parallel should be included in a patent filed on behalf of Gaulard and Gibbs. This decision greatly annoyed Stanley, but he stayed with Westinghouse in order to participate in the development of AC. Regarding Westinghouse as a rascal, Stanley later observed that he took "the advice of my father who said it was better to help out a rascal if your friends were with him than to try to punish him."[37]

Stanley gained access to Tesla because his father was Peck's neighbor in Englewood, New Jersey. Consequently, Peck knew all about the younger Stanley's problems with Westinghouse. Knowing that Stanley Jr. was a pioneer in AC but had quite an ego, Peck was worried that Stanley might well be working on his own AC motor. As Tesla explained, "Mr. Peck thought that Mr. Stanley was a man who would imagine that he had made the invention, and might possibly come in conflict with me."[38] To deal with Stanley, Peck decided to go on the offensive and instructed Tesla to show Stanley both the polyphase and split-phase motors. By doing so, they should be able to counter any claims by Stanley that he had invented a motor better than Tesla's.

Upon arriving at the Liberty Street lab, Stanley promptly announced that the "Westinghouse boys" had developed an AC motor and were ahead of Tesla. Rather than taking the bait, Tesla quietly asked Stanley if he would like to see his motor run on two wires—the motor that Tesla and Peck had not shown to Byllesby and Kerr.[39] Upon seeing this motor, Stanley had to admit that Tesla was indeed ahead of the Westinghouse engineers. "As far as I know every form of motor proposed by Mr. Shallenberger or myself has been tried by Mr. Tesla," reported Stanley to Westinghouse. "Their motor is the best thing of the kind I have seen. I believe it more efficient than most DC motors. I also believe it belongs to them."[40]

Peck kept up the pressure on Westinghouse, telling Stanley that he was just about to sell the patents to another party. With that news, Westinghouse decided that they should not wait any longer, and Kerr, Byllesby, and Shallenberger negotiated an agreement with Peck and Brown.[41] On 7 July 1888, Peck and Brown agreed to sell the Tesla patents to Westinghouse for $25,000 in cash, $50,000 in notes, and a royalty of $2.50 per horsepower for each motor. Westinghouse guaranteed that the royalties would be at least $5,000 in the first year, $10,000 in the second year, and $15,000 in each succeeding year. In addition, Peck and Brown were reimbursed by the Westinghouse Company for all of the expenses they incurred during the development of the motor.[42] In round numbers, the agreement meant that Westinghouse would pay Tesla, Peck, and Brown $200,000 over a ten-year period. Over the life of the patents (seventeen years), Tesla and his backers stood to make at least $315,000. Although it was not specified in the contract, Tesla agreed to come to Pittsburgh and share what he had learned about AC motors with the Westinghouse engineers.

Tesla did not walk away from the Westinghouse deal with $200,000 in his pocket but shared the proceeds with Peck and Brown. Since they had shrewdly handled the business negotiations and assumed all of the financial risk in developing the motors, Tesla gave Peck and Brown five-ninths of the proceeds from the deal while retaining four-ninths for himself. In this way, Tesla acknowledged the essential role Peck and Brown had played in developing the AC motor.[43]

SOJOURN AT WESTINGHOUSE

Tesla moved to Pittsburgh in July 1888 in order to put his AC motor designs into production. While in Pittsburgh, Tesla left Szigeti in New York where he continued to work on several motor patents that Tesla had not assigned to Westinghouse.[44]

During his time in Pittsburgh, Tesla worked closely with Shallenberger and the other Westinghouse boys, and he developed a great admiration for George Westinghouse. "I like to think of George Westinghouse," wrote Tesla,

> as he appeared to me in 1888. . . . A powerful frame, well proportioned with every joint in working order, an eye as clear as a crystal, a quick and springy step—he presented a rare example of health and strength. Like a lion in a forest he breathed deep and with delight the smoky air of his factories. Though just forty then he still had the enthusiasm of youth. Always smiling, affable and polite, he stood in marked contrast to the rough and ready men I met. Not one word which would have been objectionable, not a gesture which might have offended. . . . And yet no fiercer adversary than Westinghouse could have been found when he was crossed. An athlete in ordinary life, he was transformed into a giant when confronted with difficulties which seemed insurmountable. He enjoyed the struggle and never lost confidence. When others would give up in despair, he triumphed. Had he been transferred to another planet with everything against him, he would have worked out his salvation.[45]

Tesla initially worked on improving two polyphase motors that he brought with him from New York, anticipating that Westinghouse would develop a whole new polyphase system using four wires to connect the generators and motors. Since his motors worked best on low frequencies, Tesla set up his motors to run on 50 cycles and he experimented with new transformer designs.[46]

Westinghouse, however, hoped that one of Tesla's motor designs could be used to power streetcars and operate on existing single-phase, two-wire circuits. At that time, the Westinghouse systems were using 133 cycles so that consumers would not complain about their incandescent lamps flickering. Even though he felt his ideal motor was the polyphase version running on 50 cycles, Tesla agreed to work a split-phase version that could be put into production. To adapt his motor for this purpose, Tesla and the Westinghouse boys made several design changes including increasing the amount of copper wire in the rotors as well as replacing the wrought-iron cores of the rotor and stator with soft Bessemer steel. The change to steel cores alone doubled the work that a typical motor could perform and the Westinghouse Company treated this discovery as a trade secret it jealously guarded for years. Tesla also worked with the chief Westinghouse designer, Albert Schmid, to develop a standard frame for the stator that could be easily cast and machined. While working on these changes, Tesla prepared patents for Westinghouse, and in 1889 he filed fifteen applications; in terms of patents, this was the most productive year in his career.[47]

On the basis of these design changes, the Westinghouse Company built between 500 and 1,000 split-phase Tesla motors in early 1889, but it is not clear how many of these motors were actually shipped. Rather than install them in streetcars, Westinghouse initially marketed them for use in mining machinery.[48] The company, moreover, decided to use graphite bearings in these motors, which Tesla thought would overheat and fail. When Westinghouse did not follow his advice and began shipping these motors, Tesla decided that it was time to leave.

Disappointed, Tesla left Westinghouse in August 1889 and traveled to Europe to see the Paris Exposition. To continue the work on his AC motors, Tesla recommended that Westinghouse put the project in the hands of his assistant, Charles F. Scott. An engineering graduate of Ohio State University, Scott had begun simply oiling dynamos for Tesla, but the inventor was impressed by Scott's diligence and quick mind.[49] (We will return to the story of Tesla's motor at Westinghouse in Chapter 9.)

A FEW THOUGHTS ON ILLUSION

Tesla's experience with the AC motor reveals the central role that illusion plays in the process of technological change. Tesla's AC motor was not "automatically" adopted by Westinghouse and the electrical engineering community simply because it was technically superior to other electric motors; in fact, Tesla's motor required several years of serious engineering before it could be used in industry. Rather, Tesla and his backers succeeded in promoting his motor because they created the right sort of illusions about it. Guided by Peck and Brown, Tesla filed the "right" patents, secured the "right" technical endorsement from Professor Anthony, gave the "right" sort of lecture-demonstration before the AIEE, and generated the necessary publicity in the technical press. Once they had the "buzz" going, Peck knew how to play Westinghouse and his associates in order to sell the patents at the highest possible price. Clearly Tesla's motor moved forward not through the disclosure of cold, hard facts but through the careful orchestration of selected information and subtle suggestion.

The story of Tesla and the electric motor suggests that we need to develop more sophisticated ways to think about business decisions and technological choices. In the course of making choices about uncertain technologies, inventors and entrepreneurs must often extrapolate from what they currently know about technology and markets to what may happen in the future. From economists, historians have learned to talk about this situation in terms of bounded rationality and path dependency.[50] Such concepts have done much to guide our efforts to understand how specific contextual factors influence key decisions, yet these concepts fail to take into account the very real ways in which individuals such as Tesla and Peck consciously sought to frame technological decision making. By carefully shaping the discourse about Tesla's motor, they effectively altered the ways in which electrical engineers thought about motors in the utility industry and thus created a "space" for Tesla's invention. Time and time again, it isn't what people say that counts but how they are perceived by others. Illusions play a significant role in guiding our understanding of how the "real" world works.

By August 1889 Tesla had grown restless and was ready to leave West-inghouse. He had uncovered the perfect idea for an AC motor, and it was up to others to work out the details. He was ready to move on to fresh fields.

Living off what he had earned at Westinghouse, Tesla traveled to Europe that summer as a member of the delegation from the American Institute of Electrical Engineers to the Congrès International des Electriciens. This congress was being held in conjunction with the Exposition Universelle in Paris, and so Tesla was able to review the numerous electrical exhibits as well as the unveiling of the Eiffel Tower.[1] While there, Tesla witnessed a demonstration lecture of vibrating diaphragms by the young Norwegian physicist Vilhelm Bjerknes. It is likely that Bjerknes introduced Tesla to the discovery of electromagnetic waves by Heinrich Hertz. In 1887, Hertz reported that he had detected the electromagnetic waves that James Clerk Maxwell had predicted in his theoretical work on electricity and magnetism. Bjerknes had come to Paris to attend Henri Poincaré's lectures on electrodynamics and subsequently spent two years at the University of Bonn as Hertz's assistant. Together, Hertz and Bjerknes studied resonance in oscillatory circuits.[2]

Exhausted from the mundane engineering work at Westing-house, Tesla found Hertz's discovery stimulating, "like ever so many

refreshing berries found on the road by a weary wanderer." Electromagnetic waves seemed to Tesla to be a wide-open field, and as he
waxed poetic in 1899, "The journey is not finished yet, and the wanderer is well-nigh exhausted. He longs for more sweet berries, and
anxiously asks, 'Did anyone pass this road before?'"[3]

During the months that he had been in Pittsburgh, Tesla had paid
the rent for his laboratory so that Szigeti had a place to continue testing devices for Tesla. When Tesla returned to New York, he went to
work in a new laboratory at 175 Grand Street. This lab consisted of one
room divided by partitions; Tesla's backer, Brown, complained that
the space was too small for the work he thought needed to be done.
Along with moving his laboratory, Tesla also changed his residence
to the Astor House on Broadway between Barclay and Vesey streets.
An "old fashioned and conservative establishment," the Astor was the
leading hotel in downtown Manhattan.[4]

To help with the experiments at Grand Street, Tesla assembled a
small team of craftsmen. Along with a German American glassblower,
David Hiergesell, Tesla employed two mechanics, a Hungarian named
Charles Leonhardt and F. W. Clark, who had worked at the Brown &
Sharpe Works. Tesla also hired Paul Noyes, who had helped with the
arc-lighting system in Rahway.[5]

The key person among those working at Grand Street was, of
course, Szigeti. Szigeti had been with Tesla for nine years, having followed him from Budapest to Paris, Strasbourg, and New York. Tesla
valued Szigeti's advice in the laboratory: "He was," Tesla explained,
"a man who had a considerable amount of ingenuity and intelligence,
and [had] been installing electrical apparatus for a long time before he
came to America. He was not exactly a theoretical man, as myself, but
he could understand every idea fully." By now Szigeti was more than
just a trusted employee; as Tesla said later, "So long as he was in my
employ he was, I may say, a very intimate friend of mine, and I treated
him as well as I possibly could."[6]

During this time, Tesla continued to work with the Tesla Electric
Company, the firm organized by Peck and Brown. In March and April
1890, Tesla filed three additional patents on AC motors and assigned
them to this company; these were the last patents he assigned to the
company and all subsequent motor patents were held by Tesla himself.[7] Unfortunately for Tesla, Peck fell ill and moved to Asheville,

North Carolina, perhaps in the hope of regaining his health. Peck died in the summer of 1890.[8] Although Tesla continued to consult Brown over the next few years, Brown could not provide the shrewd business judgment that Peck had contributed to Tesla's early success with the AC motor.

In search of a new field to explore in his Grand Street lab, Tesla took measure of the overall development of electrical science and technology. As he saw it, electrical research could move in three major directions: high voltages, heavy currents, or high frequencies. As he observed, "There were the excessive electrical pressures of millions of volts, which opened up wonderful possibilities if producible in practical ways; there were the currents of many hundreds of thousands of amperes, which appealed to the imagination by their astonishing effects, and most interesting and inviting of all, there were the powerful electrical vibrations with their mysterious actions at a distance." Of these three, Tesla decided that the most promising was the least investigated, namely high-frequency phenomena. There he felt he might make a contribution not only to technology but also to theoretical science. "What better work could one do," asked Tesla, "than inventing methods and devising means for enabling scientific men to push investigation far out into these practically unknown regions?"[9]

In turning to high-frequency phenomena, Tesla could draw on machines that he had already begun to develop. Prior to going to Pittsburgh in 1888, he had started to think about how to run his motors on the existing Westinghouse circuits, which used 133-cycle single-phase AC. At the same time, Tesla wanted to increase the speed of his motors. To address these two issues, Tesla designed a new AC generator. To get higher frequencies, he increased the number of poles in the stator from four to twenty-four. Since his motors were synchronous—meaning they ran at the same speed as the alternator—Tesla had to design the new generator so that it could turn at relatively high speeds. By increasing both the number of poles and the speed in this alternator, Tesla was able to generate currents at 2,000 cycles per second.[10]

But now at Grand Street, Tesla wondered what new inventions could be developed using current with a frequency of 10,000 or 20,000 cycles per second. To reach these frequencies, Tesla built several alternators in 1890 with hundreds of electromagnets in their rotor and stator. The form of these electromagnets had to be carefully planned since the rapid reversal of the current generated unwanted heat in their iron or steel cores; Tesla likened the struggle between maximizing the current while minimizing the heat as "a thoroughly Wagnerian opera" in which the inventor struggled "to get from the Scylla to the Charybdis."[11] In order to run these generators at speeds up to 20,000 rpm, Tesla made some of his rotors in the form of a spoked wheel and he tinkered with the bearings and other mechanical aspects. In this work, Tesla drew on the experience he had acquired in designing dynamos and motors at the Edison Machine Works and Westinghouse.[12]

Tesla used his high-frequency generators to investigate potential applications related to arc lights and power distribution. During the early 1890s, arc lights, which were widely used for street lighting, could only be operated on DC circuits; when operated with AC they created an annoying sizzling sound proportional to the frequency of the alternating current. However, when high-frequency AC was used, the sizzle moved beyond the realm of ordinary hearing and it became possible to operate arc lights on AC circuits. As a result, Tesla patented his first high-frequency generator as a method for operating arc lights.[13]

INVENTING THE TESLA COIL

While developing his generators, Tesla repeated the experiments of Hertz with electromagnetic waves, for in Paris he had "caught the fire of enthusiasm and fairly burned with desire to behold the miracle with my own eyes."[14] This enthusiasm led to one of his most famous inventions, the Tesla coil.

In his classic experiments to generate and detect electromagnetic waves, Hertz used a powerful induction coil connected to a battery, a current interrupter, and a spark gap (Figure 6.1). To appreciate Hertz's experiments, though, we first need to understand how this induction coil worked. Commonly known as a Ruhmkorff coil, the induction coil consisted of two windings—one with thick wire and the other fine—that were carefully insulated from each other using paraffin or gutta-percha and wound around a common iron core. As in a transformer, the thick winding was known as the primary while the fine winding was the secondary. The battery and current interrupter

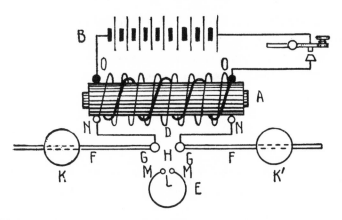

FIGURE 6.1. Diagram of apparatus used by Hertz to study electromagnetic waves.

Key:
A   induction coil with thick and fine windings
B   battery
C   key or current interrupter
H   spark gap
L   -loop of wire with gap that Hertz used to detect waves

From *Hawkins Electrical Guide* (New York: Theo. Audel, 1915), 9:2268, fig. 3104.

were connected to the primary while the spark gap was connected to the secondary.

As with a transformer, it was a change in the current that made the induction coil produce high-voltage sparks. Hence, whenever the current interrupter opened or closed the circuit, the amount of current flowing from the battery into the primary changed and caused the electromagnetic field around the primary coil to expand or contract. As the primary field changed, it induced a current in the secondary coil. Because of the different thicknesses of wire, the secondary coil could have many more turns than the primary coil, thus greatly increasing the voltage of the current induced in the secondary. Because the voltage produced in the secondary was so high, it could ionize the air in the spark gap, allowing a spark to jump between the terminals. Carefully constructed induction coils could produce sparks that could jump a sixteen-inch gap.[15] During the middle of the nineteenth century, physicists used induction coils to generate large amounts of electrical charge in order to study electrostatic effects.

But now let us return to Hertz. Prior to 1887, he conducted several experiments with an induction coil in which he generated a series of sparks in the secondary whenever the current interrupter opened the primary circuit in his apparatus. As the eminent historian of radio Hugh Aitken reminds us, these sparks "represented, of course, a sudden rush of electrical current—precisely the kind of acceleration of current flow, that, according to Maxwell's equations, would generate electromagnetic radiation."[16] Hertz noticed that whenever the sparks were produced at the induction coil, he could also detect sparks elsewhere in his laboratory using a copper loop with a spark gap. By carefully proportioning the diameter of this loop and adjusting the brass balls on either side of the spark gap on the secondary, Hertz was able to show that his apparatus was generating electromagnetic waves that moved through space and were detected by his loop.[17]

In 1890, Tesla repeated Hertz's experiments and may well have been the first researcher in America to do so. Not satisfied with the apparatus Hertz had used, Tesla altered the experimental setup (Figure 6.2).[18] An obvious step was to replace the mechanical current interrupter with his high-frequency generator. Rather than have the apparatus use the few hundred cycles per second produced by the mechanical interrupter, why not use 10,000 to 20,000 cycles from his alternator? Tesla soon discovered that as the frequency increased, so did the amount of heat generated that melted the paraffin or gutta-percha insulation between the primary and secondary inside the induction coil. To address this problem, Tesla made two changes. First, he got rid of the insulation and instead wound his induction coils with an air gap between the primary and the secondary. Second, because the iron core in the induction coil became so hot, he redesigned his version so

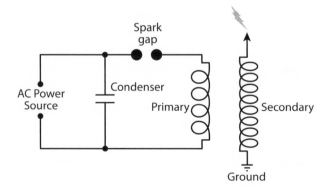

FIGURE 6.2. Diagram of a Tesla coil.

that the iron core could be moved in and out of the primary coil. By moving the core, Tesla found that he could also adjust the inductance of the primary.[19]

Tesla also encountered problems with the condenser or capacitor that was frequently used with induction coils. To increase the strength of the spark produced by the secondary, investigators (beginning with Armand Hippolyte Fizeau in 1853) typically placed a Leyden jar or capacitor in a shunt around the spark gap of the secondary. With the rapid alternations from his high-frequency generator, Tesla found that this capacitor often counteracted the self-inductance of the secondary coil and burned out the coil. In response, Tesla moved the capacitor in his apparatus between the generator and the primary. He also made this capacitor adjustable.[20] Tinkering with the arrangement of the capacitor and coils was quite natural for Tesla; in developing his split-phase motors, he had used combinations of induction coils, resistors, and capacitors to split and give the incoming current different phases.[21]

Tesla now realized that by carefully adjusting the capacitor and induction coil it was possible to boost the frequency to even higher levels. Electrical scientists had initially assumed that when the capacitor discharged, the electricity simply flowed from one plate to the other, much like water running out of a reservoir. However, in 1856, the British physicist Sir William Thomson proved mathematically that capacitor discharge was instead vibratory.[22] Just as a vertical weighted spring bobs up and down when released, the electrical charge surges back and forth between the plates in the capacitor until the stored-up energy is dissipated and flows through the circuit in the form of a high-frequency current.[23]

In order to take full advantage of the vibratory character of the capacitor discharge, Tesla next carefully adjusted the induction coil. Just as the capacitor can be thought of as the electrical equivalent of a weighted spring, an induction coil can be seen as the equivalent of a pendulum. As an alternating current flows through the primary, the current induced in the secondary coil oscillates between a maximum and a minimum value, much like the bob on a mechanical pendulum swings to and fro. Tesla now realized that if he could match each electrostatic discharge or "thrust" so that it coincided with each maximum of the induced current, he could increase the voltage of the current produced by the induction coil. Just as one can keep a mechanical pendulum going longer by giving it a little nudge just as it reaches one end of its swing, Tesla adjusted his capacitor and induction coil so that each "thrust" came just as the current in the induction coil reached its maximum. In doing so, Tesla was taking advantage of the principle of resonance—that one portion of the circuit would reinforce another portion of the circuit and thus boost the output. By creating resonance by adjusting the capacitor and the induction coil, Tesla was soon able to produce a current that alternated up to thirty thousand times per second.[24] Fascinated with how resonance could create such powerful effects, Tesla looked for other places where he could take advantage of resonance, and it quickly became the new ideal guiding his efforts with high-frequency phenomena.

By combining his understanding of the vibratory nature of capacitor discharge with the principle of resonance, Tesla now had an invention that produced currents whose voltage and frequency were higher than those generated by other machines. Tesla called this invention his oscillating transformer, but as it came to be widely employed by other investigators, it became known as the Tesla coil. The oscillating transformer was fundamental for much of Tesla's subsequent work on wireless power, and he felt that it was one of his great discoveries. As he recalled later, "When in 1900 I obtained powerful discharges of 100 feet and flashed a current around the globe, I was reminded of the first tiny spark I observed in my Grand Street laboratory and was thrilled by sensations akin to those I felt when I discovered the *rotating magnetic field*."[25]

In using his high-frequency generator and oscillating transformer together, Tesla soon learned about the physiological effects of high-frequency currents. Early on in his experiments, he accidentally touched the terminals of an oscillating transformer and the high-frequency current passed through his body. Much to his surprise, he was not injured. Tesla realized that, as a consequence of the self-induction of the coil and the high frequency, the current produced in the secondary coil had a high voltage but a small amperage. Moreover, as we know today, currents in the radio frequency range travel along the surface of the human body and do not harm the nerves and internal organs for short exposures. Based on his own experiences, Tesla concluded in February 1891 that "the higher the frequency the greater the amount of electrical energy which may be passed through the body without serious discomfort."[26] This conclusion had potential safety benefits since one way to avoid being electrocuted by high-voltage AC would be to increase the frequency used in the existing distribution systems.[27] Long concerned about the safety of AC, Tesla's rival Elihu Thomson further investigated the physiological effects of high-frequency current.[28] In the meantime, Tesla capitalized on the skin effect in his public demonstrations. It was because of the skin effect that he was able to grasp one terminal of his high-frequency apparatus and take tens of thousands of volts through his body—enough energy to brilliantly illuminate a bulb or tube that he held in his hand.

HERTZIAN WAVES OR ELECTROSTATIC THRUSTS?

As Tesla developed his oscillating transformer, he was confronted with the question of how to utilize this new device. Clearly it lent itself to exploring the waves discovered by Hertz, but how should he direct his investigations? While we now know that Hertzian waves can be used for radio communications, this application was not obvious to Tesla or other early investigators. As with any new phenomenon, there may be different paths by which investigators may exploit it, and the choice of path reflects the knowledge and interests of the investigators.[29]

One path was to concentrate on the theoretical implications of Hertz's discovery. In Britain, a number of physicists and engineers—Oliver Lodge, Oliver Heaviside, and George Francis FitzGerald—studied Hertz's experiments and undertook the mathematical analysis necessary to connect Hertz's results with Maxwell's theory. Calling themselves Maxwellians, this group was primarily concerned with advancing electromagnetic theory.[30] Since Maxwell had emphasized that visible light and the new waves were equivalent—varying only by wavelength—the Maxwellians initially conducted optical experiments to see if the new waves behaved like light. For instance, they tried using lenses made out of different materials (both glass and bitumen) to focus the new waves in the manner of light waves. While one Maxwellian, John Perry, commented in 1890 on the possibility of using Hertzian waves for communication, the Maxwellians assumed that any communication system using the new waves would function more like a system of flashing lights than like an electric telegraph. Indeed, as Sungook Hong has noted, the Maxwellians were not interested in existing telegraph practice in the early 1890s and hence not prepared to convert Hertz's discovery into wireless telegraphy.[31]

Tesla followed the work of Hertz and the Maxwellians by reading the electrical journals, but he possessed neither the inclination nor the expertise to become involved in the complex mathematical discussions of the Maxwellians. (Even Hertz himself was slow to engage in these mathematical discussions, leading Heaviside to remark in 1889 that "I see that Hertz is not a Maxwellian though he is learning to be one.")[32] Instead, Tesla's path forward from Hertz's discovery was marked by two characteristics. First, he decided to focus less on the electromagnetic waves and more on the electrostatic effects created by his apparatus. Second, as an ambitious inventor, Tesla sought to convert this scientific discovery into promising new technology.

When researchers start investigating new phenomena, they often begin by using existing experimental practices and then develop new techniques as they become familiar with the phenomena. Hence, just as the Maxwellians had started by conducting optical experiments, Tesla began studying his new oscillating transformer by repeating the usual demonstrations performed with a Ruhmkorff coil. Previous investigators had used the Ruhmkorff coil to study sparks and the effects of electrical charge or what are called electrostatic effects.

One popular demonstration performed with a Ruhmkorff coil was to use electric sparks to render gases incandescent. To conduct this experiment, investigators used special glass tubes from which most of the air had been evacuated. Known as Geissler tubes, they had two platinum electrodes, and when connected to a Ruhmkorff coil, the high voltages caused the gas to ionize and become luminescent.[33] In his experiments with his oscillating transformer, Tesla connected his coil to a Geissler tube and found that the tube glowed intensely around each terminal while the middle appeared comparatively dark. These dark spaces had been previously studied by the British chemist Sir William Crookes.[34]

Working with Geissler tubes, Tesla then made an important discovery. When he attached the terminals of his oscillating transformer to two spheres, the spark jumped at the point where the gap between the balls was the smallest and then climbed up the sides of the spheres, only to be extinguished at the top and to start again at the closest point. Modern experimenters refer to this demonstration as Jacob's Ladder, and it is often seen in the apparatus used by mad scientists in monster and science-fiction movies. However, what Tesla found startling was that whenever the spark generated by the coil was extinguished—such as at the end of a climb between the spheres—Geissler tubes lying nearby were illuminated and extinguished in unison with the spark. He also noticed that the tubes did not light up when they were at right angles to the terminals of his induction coil; to be illuminated, the tubes needed to be parallel with the terminals and the spark. This suggested to Tesla that the tubes were lit up as a result of the electric field produced by the spark, not by electromagnetic waves; if the waves were causing the tube to glow, he reasoned, then the position would have not mattered. Tesla repeated the experiment with vacuum tubes without any electrodes and was amazed to find that these, too, became illuminated.[35]

These observations convinced Tesla that Hertz and the Maxwellians had been misled. While the Ruhmkorff coil built up large amounts of charge and generated electromagnetic waves, in Tesla's view, most of the energy in the apparatus went into electrostatic effects and not electromagnetic waves. As far as he was concerned, the buildup of charge created an electrical field in the space surrounding the coil, and when the spark was extinguished, the voltage of this field soared, causing the Geissler tubes to glow. According to Tesla, electrostatic "thrusts," not Hertzian waves, caused the tubes to glow.

In his May 1891 lecture on high frequency phenomena (described in the next chapter), Tesla made clear that he disagreed with Hertz and Lodge. "Many have been carried away by the enthusiasm and passion to discover, but in their zeal to reach results some have been misled," he wrote. "Starting with the idea of producing electro-magnetic waves, they turned their attention, perhaps too much to the study of electro-magnetic effects, and neglected the study of electrostatic phenomena. . . . It has thus been thought—and I believe asserted— that in such cases most of the energy is radiated into space. In the light of the experiments, which I have described above, it will now not be thought so. I feel safe in asserting, that . . . the amount of the directly radiated energy is very small."[36]

Given that we now accept the Maxwellian interpretation of electromagnetic waves to be true—it is found in all physics and electrical engineering textbooks—it may seem ridiculous that Tesla chose to challenge this interpretation. However, we should keep in mind several points. First, Tesla was following a well-established principle in science: one does not propose a new theory when the old theory seems to explain most of what is going on. Second, much of the power of the interpretation put forward by the Maxwellians lay in their ability to mathematically represent electromagnetic phenomena and to subsequently use mathematics to predict new phenomena. Tesla was not interested in these mathematical endeavors, and, even more important, he was not persuaded by them. What counted for him were the phenomena he was able to produce and observe in his laboratory.

Indeed, having observed how his oscillating transmitter caused the Geissler tubes to glow, Tesla's next step was to convert this phenomenon into an exciting demonstration. During one all-night work session, he sent his men out at 3:00 AM to get something to eat. When they returned, they found him standing in the middle of the laboratory, holding a long glass tube in each hand and with no connection to his high-frequency coil. "If my theory is correct," Tesla told them, "when the switch is thrown in[,] these tubes will become swords of fire." He then ordered the room darkened, the switch thrown, and instantly the glass tubes lit up brilliantly.

"Under the influence of great exultation," Tesla recalled, "I waved them in circles round and around my head. My men were actually scared, so new and wonderful was the spectacle. They had not known of my wireless light theory, and for a moment they thought I was some kind of a magician or hypnotizer." With that experiment, Tesla knew wireless lighting was a reality and that he had a demonstration that could capture the imagination of audiences and attract new investors.[37]

CHAPTER SEVEN
A VERITABLE MAGICIAN
[1891]

At one bound, [Tesla has] placed himself abreast of such men as Edison, Brush, Elihu Thomson, and Alexander Graham Bell. Yet only four or five years ago, after a period of struggle in France, this stripling from the dim mountain border-land of Austro-Hungary landed on our shores, entirely unknown, and poor in everything save genius and training, and courage.

JOSEPH WETZLER in *Harper's Weekly*, July 1891

BROKEN CONTRACTS, BROKEN HEARTS

During the winter of 1890–91, Tesla was probably thinking about how to generate public interest and financial support for his inventions because his major patron, Westinghouse, was in deep financial trouble. Thanks to the company's innovative AC product line, annual sales at Westinghouse had grown from $800,000 in 1887 to $4.7 million in 1890.[1] As sales boomed, though, Westinghouse had to develop an engineering staff and enlarge his factories. At the same time, Westinghouse joined Edison General Electric and Thomson-Houston in buying out smaller companies and engaging in vigorous patent litigation. Westinghouse partly financed this expansion by advancing the company $1.2 million of his own money, but he also borrowed heavily. By mid-1890, the firm was carrying an additional $3 million in short-term liabilities when its total assets were about $11 million and current assets were $2.5 million.

Disaster struck in November 1890 when the failure of a major London brokerage house, Baring Brothers, set off a financial panic and prompted Westinghouse's creditors to call in their loans. The Westinghouse Company was forced into receivership, and George Westinghouse struggled over the next two years to save the company. After failing to gain support from Pittsburgh bankers, Westinghouse turned to Wall Street banker August Belmont, who organized a committee of powerful investors to reorganize the company.[2]

According to John O'Neill—Tesla's biographer in the 1940s—the investors who backed this financial reorganization insisted that if Westinghouse wanted to retain control of his company, he would have to terminate Tesla's contract, which called for royalty payments of $2.50 per horsepower for each motor installed. O'Neill claimed that the investors insisted on terminating the contract to avoid paying Tesla millions of dollars in royalties and that this money would have supported much of his subsequent research.[3]

However, viewed from the perspective of early 1891, it is not likely that Tesla's royalty payments would not have been a major cost confronting the reorganized company. Based on the terms of the 1888 contract with Tesla, Peck, and Brown, Westinghouse would have paid $105,000 by 1891, with Tesla receiving about $47,000. Because there were only a handful of AC power systems that could employ Tesla's motors, Westinghouse had sold very few motors and had probably not paid any significant royalties before 1891. Moreover, given that Westinghouse engineers had not solved the technical difficulties related to Tesla's motor designs (see Chapter 10), neither Westinghouse nor the bankers had any reason to worry that Tesla's royalty payments might amount to millions of dollars.[4] Tesla's motor did prove to be a commercial success in the late 1890s, but there was no way to anticipate this in early 1891.

Instead, it is more likely that investors insisted that Westinghouse terminate the contract with Tesla because they felt that Westinghouse had spent too much money and energy on developing new technology. As one Pittsburgh banker complained, "Mr. Westinghouse wastes so much on experimentation, and pays so liberally for whatever he wishes in the way of service and patent rights, that we are taking a pretty large risk if we give him a free hand with the fund he has asked us to raise. We ought to at least know what he is doing with our money."[5] At the same time, the investment committee organized by Belmont wanted more say in the affairs of the reorganized Westinghouse Company. Viewing Westinghouse as "a bright & fertile mechanic" who lacked tact and an understanding of high finance, the bankers sought to circumscribe his power.[6] Hence, the request to terminate his contract with Tesla probably grew more out of a desire on the part of the bankers to rein in Westinghouse and less out of fear that Tesla's royalties would run into the millions.

Reluctantly, then, Westinghouse went to Tesla and asked him to give up the contract and help him retain control of the company. According to O'Neill, Tesla then tore up the contract in a grand flourish, demonstrating his loyalty to Westinghouse.[7] Yet at the same time, Tesla may well have been thinking about his own future in terms of who would control his patents. If Tesla retained the contract, then he would be negotiating with the investors instead of Westinghouse, and they might not be so inclined to spend money to develop or promote his inventions. O'Neill suggested that Tesla preferred to continue to deal with Westinghouse on an informal basis and trust that the Pittsburgh magnate would continue to support him in some way. (For instance, see Chapter 14.) For Tesla, personal loyalty counted more than a legal contract.[8]

Around the time that he tore up his contract with Westinghouse, Tesla was confronted with an equally great personal disappointment. After having been with Tesla for nine years, Szigeti left sometime in 1890 to develop what he thought would be his own great invention: a new compass for steering ships. When Szigeti returned five or six months later, Tesla told him that his compass invention had already been developed by Sir William Thomson, and this prompted Szigeti to leave for a second time in 1891. Tesla thought that Szigeti went south this second time, perhaps to South America, to pursue another invention scheme. Deeply hurt that Szigeti had left him, Tesla testified twenty years later how "I would have much desired to see him, because I would have wanted him."[9] With Szigeti, we see that Tesla was attracted to men and sought to form intimate friendships with them; we will further explore this facet of Tesla's life in Chapter 12.

NEW LAMPS FOR THE WORLD

In the aftermath of losing Szigeti and tearing up his contract with Westinghouse, Tesla began working harder to develop and publicize what he had learned about the new realm of high-frequency phenomena. Tesla could no longer count on royalty income from Westinghouse and now had to generate interest in his new inventions in order to attract investors. Pursuing the business strategy he had learned from Peck of patent-promote-sell, Tesla filed patent applications, published several articles in the electrical journals, and gave a second major lecture during the first half of 1891.

As he experimented with his oscillating transformer, Tesla applied his discoveries about electrostatic "thrusts" to developing new forms of electric lighting. Somehow the electrostatic thrusts were transmitting power through space; how might he use these thrusts to create new inventions? Knowing as we do that Marconi converted Hertz's discoveries into wireless telegraphy, it may seem puzzling at first that Tesla would choose to concentrate on lighting and not communications. Yet this choice made sense on several levels. For years, scientists had been fascinated by how Geissler tubes converted electricity into light without heat. Prior to the introduction of electric lighting, artificial illumination—from candles, oil lamps, or gas lights—had involved a flame and the production of heat; how was it that the Geissler tube avoided producing heat? At the same time, Maxwell had emphasized in his theory that light and electricity were related; why not follow up this idea and look for ways to convert electricity directly into light?

Tesla's first step in this direction was to follow up on the experiment in which a thin insulated wire wriggled and glowed brilliantly when attached to one terminal of his oscillating coil. For Tesla, the rapid movement and streams of light in the wire were the result of electrostatic thrusts causing intense vibrations in the molecules of the wire. To better capture this intense action, Tesla placed a thin platinum wire inside an evacuated bulb and found that it spun around, creating a cone of illumination.

Tesla knew that the platinum wire became incandescent not because the metal itself had a high resistance but because he was using a very thin wire. However, suspecting that he would get even better results using a high-resistance material, Tesla did what Edison had done in developing his incandescent lamp in 1879 and substituted carbon for the platinum wire. Rather than use a filament (as Edison did in his lamp), Tesla fashioned the carbon into a small spherical button that he placed at the end of one wire; when connected to one terminal of his oscillating transformer, the high-frequency, high-voltage current brought the button to incandescence and it threw off a bright light.[10] To focus the light from the incandescent button, Tesla added a metal reflector outside the lamp. Since this lamp needed to be connected to the power source by only one wire—normally incandescent lamps required two wires—Tesla immediately saw that this carbon button lamp had commercial potential since it would reduce the wiring needed for electric lighting by half, and he proceeded to patent several variations.[11]

THE COLUMBIA COLLEGE LECTURE

By the spring of 1891, Tesla realized that all of his new devices—his high-frequency alternator, oscillating transformer, and new lamps—constituted a technology platform from which he could make a series of bold claims: that Hertz and the Maxwellians were paying too much attention to electromagnetic waves, that high-frequency AC could be readily converted into light, and that he was on the threshold of revolutionizing the electrical industry with his new lamps.

Tesla reported his initial findings in the *Electrical Engineer* in February 1891 and was immediately challenged in print by Elihu Thomson. He, too, was working with high-frequency currents, but his experiments were under 10,000 cycles and so Thomson did not always observe the same effects as Tesla did. Not willing to relinquish even the smallest amount of ground to the other, Tesla and Thomson sparred with each other in a string of articles published in the electrical journals through March and April 1891.[12]

This sparring with Thomson must have revealed to Tesla that he would need to make a dramatic move if he were going to establish himself as the leading expert investigating high-frequency phenomena. Perhaps because Thomson had given a keynote lecture on AC phenomena at the spring 1890 meeting of the American Institute of Electrical Engineers (AIEE), Tesla decided to lecture again before the institute; just as he had introduced his rotating-field motor by presenting it to this group three years earlier, now Tesla would launch his ideas about high-frequency AC by speaking before the institute. In all likelihood, he was able to get on the spring 1891 program since William Anthony was serving as institute president and Tesla's acquaintance T. C. Martin was serving as the chair of the institute's Committee on Papers and Meetings.[13]

As he had done in 1888, Tesla made sure that he had protected his inventions by filing applications prior to the lecture. In late April and early May, he filed two U.S. patent applications for high-frequency incandescent lighting, and on the day before the lecture, he submitted patent applications for protection in Britain, France, Germany, and Belgium.[14]

Tesla spoke on the evening of 20 May 1891 at Columbia College in New York in the lecture hall of Theodore W. Dwight, Dean of the Law School. Although an electrical engineering department had been established in 1889 in the Columbia School of Mines, the department lacked its own classroom, but the two electrical engineering professors, Francis B. Crocker and Michael Pupin, were probably anxious to attract attention to their new department by hosting Tesla's lecture. To provide power, Tesla installed his high-frequency alternator in the college's electrical workshop (a modest brick building nicknamed the "cowshed") and powered it by an electric motor; using a switch onstage, Tesla could adjust the speed of the motor and hence control the frequency produced by his alternator (Figure 7.1).[15]

Speaking in "pure, nervous English" to a large and eager audience, Tesla began by remarking that modern science had been able to make rapid progress by recognizing the ether as the medium in which invisible waves travel but that the exact nature of electricity was still unknown. Tesla proposed that electrostatic phenomena might be considered as the ether under strain while dynamic electricity or currents should be seen as "phenomena of ether in motion." Alluding to

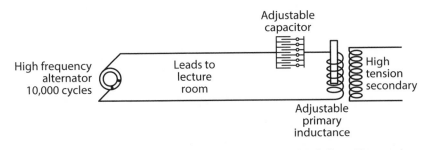

FIGURE 7.1. Circuit used by Tesla for his 1891 lecture at Columbia College. The capacitor, inductance, and transformer on the right constituted his oscillating transformer or Tesla Coil. Redrawn from NT, "The True Wireless," *Electrical Experimenter,* May 1919, 28–30 ff., on p. 29.

the work of Hertz and Lodge, Tesla informed the audience that the luminous effects in Geissler tubes were caused not by electromagnetic waves but electrostatic "thrusts."[16]

To support these claims, Tesla provided a series of demonstrations. As he increased the number of cycles delivered to his oscillating transformer, the arc began to "sing"—to emit a high-pitched note. Tesla then demonstrated how his coil could generate a variety of streamers, sparks, and electric flames. Next he showed how high-frequency current could be used to light up Geissler tubes and his new lamps.[17] "Here," reported the *Electrical Review*, "Mr. Tesla seemed to act the part of a veritable magician. It seemed to make little difference whether the lamps were lying on the table or whether they were connected by one terminal to one pole of the coil, or whether the lecturer took a lamp in each hand and held one to each pole of the coil, . . . in each and every case the filaments were brought to incandescence, to the supreme delight of the spectators."[18]

To help the audience appreciate the full potential of high-frequency AC for electric lighting, Tesla offered a breathtaking demonstration (Figure 7.2). Two large zinc sheets were suspended from the ceiling about fifteen feet from each other and connected to the oscillating transformer. With the auditorium lights dimmed, Tesla took a long gas-filled tube in each hand and stepped between the two sheets. As he waved the slender tubes, they glowed, charged by an electrostatic field set up between the plates. As Tesla explained, high-frequency current now made it possible to have electric lighting without wires, to have lamps that could be moved freely around a room.[19]

FIGURE 7.2. Tesla demonstrating his wireless lamps before the American Institute of Electrical Engineers in May 1891.

From "Experiments with Alternate Currents of Very High Frequency and Their Application to Methods of Artificial Illumination," *Electrical World*, 11 July 1891, pp. 18–19 (TC 3:86–87).

This demonstration created a sensation and was featured in all of the articles published about Tesla's Columbia lecture. Enthralled with the idea of illumination without heat or flames, Joseph Wetzler predicted in *Harper's Weekly* that Tesla's lamps would "bring a fairy-land within our homes." "It is difficult to appreciate what those strange phenomena meant at that time," Tesla later recalled. "When my tubes were first publicly exhibited they were viewed with [an] amazement impossible to describe."[20]

Lest the audience be concerned about the safety of the high-frequency currents, Tesla followed the wireless lighting demonstration with a physiological experiment. Holding a brass ball to one terminal of the oscillating transformer, Tesla adjusted the potential of the coil so that a stream of light came out of the other terminal of the coil. Estimating the potential across the terminals to be 250,000 volts, Tesla then brought a second brass ball to the other terminal of the coil and let the full current pass through him. Thanks to the skin effect, the current remained on the surface of his body and he was unharmed.[21] Tesla concluded the lecture—which had lasted for three hours—by remarking that he had conducted additional interesting experiments in his laboratory but regretted that he could not show them since he had run out of time. Ever the showman, Tesla knew the importance of always keeping the audience hungry by promising them more.

Like his 1888 lecture, the Columbia lecture was a great success. "All who attended Mr. Tesla's brilliant lecture on Wednesday evening," noted the *Electrical Review*, "will remember that occasion as one of the scientific treats of their lives."[22] The lecture was reported in the technical press as well as the New York newspapers. The written version of the lecture—prepared in the weeks following the presentation—was widely reprinted and an excerpt appeared in *Literary Digest*.[23] For the most part, the press was excited not only by the sensational demonstrations but also by the commercial potential of Tesla's wireless electric lights. His experiments with high frequencies seemed to show that AC was "the El Dorado of the electrician" whereby it would be possible to produce light efficiently and without losses from heat or flames.[24] "It is impossible to read Mr. Tesla's epoch making paper without admiration at the clearness of view and ingenuity of mind exhibited throughout," remarked the *Telegraphic Journal and Electrical Review*. "It would seem that we are at last fairly progressing towards a means of transforming energy into any form we wish without such a disastrous loss of availability as is now inevitable, and a large measure of the credit must be given to Mr. Tesla for helping so much forward to this great end."[25] Supremely confident of Tesla's ability to move from theory to practice, the *Electrical Engineer* chimed in, "With the method now clearly pointed out, we believe that it will take but a comparatively short time to work out and present to the public the practical details necessary for the general application of such a system."[26]

Although the press was impressed with Tesla's creative accomplishments, not everyone in the electrical world was captivated by the way Tesla came across in print. In particular, the English journal *Industries* took the inventor to task: "We think, however, that anyone who has read many of Mr. Tesla's articles must have difficulty in understanding the frequent vague and idiomatic statements with which they abound. We do not think it too much to ask an electrician occupying such a prominent position as Mr. Tesla has gained for himself in America to omit passages that may detract from his reputation, and to allow us to admire him even more. If Mr. Tesla could keep phantom ideas about the electromagnetic theory of light and Hertz and Dr. Lodge out of his work, we feel sure he would make his interesting experiments more clear."[27]

Nevertheless, the Columbia lecture firmly established Tesla as one of the leading electrical inventors in America, and he had done so in a few short years after landing in New York. "At one bound," exulted Wetzler, Tesla "placed himself abreast of such men as Edison, Brush, Elihu Thomson, and Alexander Graham Bell. Yet only four or five years ago, after a period of struggle in France, this stripling from the dim mountain border-land of Austro-Hungary landed on our shores, entirely unknown, and poor in everything save genius and training, and courage."[28]

Basking in the role of the poor immigrant who made good, Tesla decided it was time for him to become an American citizen. In July 1891, he filed an application for naturalization in the Common Pleas Court of New York. As his former nationality he listed "Austrian," but as his occupation he listed "civil engineer," reflecting what he had studied in Graz.[29] Tesla had come a long way from his student days in Austria.

BRINGING THE EARTH INTO HIS CIRCUITS

While Tesla enjoyed the publicity following his Columbia lecture, he was nonetheless determined to keep shaping his ideas into practical inventions. He was especially keen to follow up on the demonstration in which gas-filled tubes glowed when placed between two electrified plates. "That was an experiment which carried the whole world by storm," Tesla recalled, "but to me it was the first evidence that I was conveying energy to a distance, and it was a tremendous spur to my imagination."[30]

Through the summer and fall of 1891, Tesla began to scale up his demonstration apparatus. Onstage he had used his high-frequency alternator and oscillating transformer to transmit energy between two plates separated by fifteen feet; how much farther might he be able to convey energy without wires? To find out, Tesla substituted a large metal can for one plate and placed it on the roof of his Grand Street

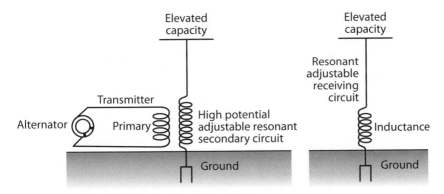

FIGURE 7.3. Diagram showing how Tesla grounded his transmitter and receiver.
In his 1916 testimony, Tesla reported that he used arrangements like this as early as 1891.
Tesla used a large metal can for the elevated capacity in the transmitter.
Redrawn from NT, "The True Wireless," *Electrical Experimenter,* May 1919, 28–30 ff., on p. 29.

lab (Figure 7.3). In lieu of the other plate, Tesla connected his apparatus to the building's water pipe system to ground it. In configuring his
system in this manner, Tesla may have been influenced by a circuit patented in 1886 by Amos E. Dolbear of Tufts College; by connecting one
terminal of an induction coil to a large capacitor and grounding the
other terminal, Dolbear found that he was able to transmit telephone
signals from the college laboratory to his home nearby.[31] In grounding
the circuit, Dolbear was using a technique already in common practice in telegraphy and telephony. First discovered by Carl August von
Steinheil in 1838, electricians had discovered that it was possible to
operate a telegraph circuit with one wire running between the transmitter and receiver and then connecting both devices to plates buried
in the ground since the current could move through the Earth and
hence complete the circuit.[32] With the can and Earth connected to
the terminals of his high-frequency apparatus, Tesla's idea "was that I
would disturb the electrical equilibrium in the nearby portions of the
earth, and the equilibrium being disturbed, this could then be utilized
to bring into operation in any way some instrument."[33]

Just as he had carefully adjusted the components in his oscillating
transformer in order to boost the voltage and frequency of the output
(see Chapter 6), Tesla found that the elevated can did not provide sufficient capacity to match the frequency of the generator. To remedy
this, he introduced an adjustable induction coil.[34] However, he did not

at this point think about adjusting the induction coils in the transmitter and the receiver so that they would resonate at the same frequency and hence be tuned (see Chapter 10). Instead, Tesla adjusted the induction coil and capacitor so as to create a maximum electrical current in the Earth and hence potentially illuminate as many wireless lamps as possible.

On the receiving end, Tesla tried a variety of lamps. As we have seen with the Columbia lecture, some lamps needed no wires and simply glowed, charged by the electric field set up between the Earth and the can. Tesla also got good results connecting the lamps to a plate or to the ground. These results gave him hope that with the right lamp design, it should be possible to develop a wireless lighting system that could compete with Edison's incandescent lighting system. Consequently, Tesla kept his glassblower busy as he experimented with a wide range of lamps—some with filaments and others with carbon buttons.[35]

But as the inventor of an AC motor, he was also intent on transmitting power. Toward this end, Tesla returned to a motor reminiscent of his experiments in Strasbourg, one consisting of a copper disk placed next to the iron core of an induction coil (see Figure 9.2). When powered by AC, the coil created a changing magnetic field that induced eddy currents in the disk; since the eddy currents were opposed to the magnetic field, they caused the disk to rotate.[36] Tesla now found that it was possible to provide power to this motor using a single wire connected to his oscillating transformer and a suspended plate. Here the circuit was completed by the electrical field set up between the metal can on the transmitter side and the plate on the motor side. Knowing that the motor needed to be connected to a large capacitor and that the human body offered a large capacitance, Tesla found that he could remove the plate and make the motor run simply by holding a wire connected to the motor. Pleased that he could make the motor run on a single wire, he experimented further with a "no wire" motor in which the motor was connected to a plate and to the ground. Although Tesla could make the motor turn, the wireless arrangement did not deliver as much energy as the single-wire connection.[37]

These experiments suggested to Tesla that he could transmit electricity for light and power over a distance and that he might be able to eliminate all the copper wires used in electric lighting, telegraph, and telephone systems. Thrilled with this possibility, Tesla sought to convey his vision to his assistants. "They had seen me run the wire up the building, they had seen me operate continuously with those machines," Tesla testified later. "I had shown them wonderful results, and had told them all the time that I was going to transmit energy without wire[s]—telephone, telegraph, run [street]cars and lights at any distance—and that these were the primary steps toward this end. How much these men could tell . . . I am unable to say; but, certainly, I had plenty of witnesses to follow my work, and to know what I had been doing."[38]

These experiments from 1891 may look suspiciously like modern radio and may suggest that Tesla invented radio before Marconi; indeed that was the argument Tesla tried to make years later through testimony and publications.[39] Clearly Tesla was the first investigator of electromagnetic waves to appreciate the importance of grounding the transmitter and receiver, a fundamental insight that Marconi came to in 1895.[40] Moreover, Tesla devised novel circuits using capacitors and induction coils, and his circuits were subsequently used and modified by Marconi and other early radio researchers to perfect wireless telegraphy.

But while he understood the significance of grounding for the exploitation of electromagnetic waves and devised several key circuits, we should note that, even at this early stage, Tesla was making choices that would lead him away from what we commonly think of as radio. First and foremost, Tesla was not particularly interested in creating a communications system. For him, the big opportunity was not to imitate telegraph systems but develop the next generation of technology for delivering light and power; as we will see, it was Marconi who wanted to utilize electromagnetic waves to create a wireless alternative to the telegraph. Second, though Tesla knew he was generating waves that radiated through space, he was becoming much more curious about the current that passed through the ground; he was fascinated by having the Earth in his circuits. And third, though one can adjust either the capacitance or inductance of the circuits, we see that Tesla was already concentrating on changing the inductance.

Hence, rather than thinking about the history of radio as a race to a specific goal, we should realize that a new discovery (like the existence of electromagnetic waves) does not have to lead to a single new technology (like wireless telegraphy). Instead, what makes the history of a technology like radio interesting is that the same discovery can prompt different investigators to pursue different paths. All too often, by focusing on Marconi's commercial success with wireless telegraphy, we overlook the diversity of approaches pursued by rival inventors. In the chapters that follow we will see that Tesla's personality, skills, and insights led him to shape his 1891 experiments into a technology that was distinctly different than the wireless telegraphy pursued by Marconi. As the poet Robert Frost said, "Two roads diverged in a wood, and I— / I took the one less traveled by, / And that has made all the difference."

In the months following the Columbia lecture, Tesla tried to ignore the public acclaim and concentrate on his high-frequency experiments. "When my tubes were first publicly exhibited they were viewed with [an] amazement impossible to describe," Tesla recalled. "From all parts of the world I received urgent invitations and numerous honors and other flattering inducements were offered to me, which I declined." With cans on the rooftop and grounded circuits, he was getting promising results and felt little desire to interrupt his labors. As the *Electrical World* observed in January 1892, "In his skillful hands the experiments have extended far beyond their merely theoretical importance in the direction of important practical applications. . . . Many of the practical difficulties that at first appeared have been overcome and we may ere long see the results in commercial work."[1]

Yet developments in Europe soon drew Tesla out of the laboratory and back to the lecture hall. For several years, the electrical journals in Britain had periodically raised the question of whether Ferraris had developed a motor with a rotating field, and Tesla continued to insist that he had filed patents months before Ferraris had published his results (see Chapter 5). Meanwhile, a German engineer, F. A. Haselwander, claimed that he had invented a three-phase motor that delivered ten horsepower in the summer of 1887. Haselwander did not get his motor to actually work until 12 October 1887, which was a month after Tesla's successful demonstration of his rotating field using the shoe-polish tin. Equally, while Tesla promptly filed patent applications in October 1887, Haselwander did not file an application for his design until July 1888.[2]

But even more worrisome for Tesla were events at the Electrotechnical Exhibition in Frankfurt, Germany, in August and September 1891. Desiring to establish a municipal electric power system and unable to determine the best system for its needs, the city of Frankfurt commissioned an electrical engineer, Oskar von Miller, to organize an exhibition so that experts could study the state of the art.[3] Hoping to secure the Frankfurt contract, the leading electrical manufacturing firms exhibited at Frankfurt, and many companies emphasized their AC equipment in their displays.

But in addition to the exhibits, von Miller also arranged for a spectacular demonstration of the potential of polyphase AC for transmitting power over long distances. Using a hydroelectric station he had set up at a cement plant in Lauffen on the Neckar River, von Miller convinced the Imperial German Postal Authority to build a high-voltage line to carry power 110 miles (175 kilometers) from Lauffen to Frankfurt. At Lauffen, the generators and transformers were designed by Charles E. L. Brown of the Swiss firm Oerlikon. At the Frankfurt end of the line, von Miller commissioned Michael von Dolivo-Dobrowolsky, the Russian chief engineer of Allgemeine Elektricitats-Gesellschaft of Berlin, to build the motors. Drawing on British patents he had filed in 1890 and 1891, Dolivo-Dobrowolsky used three-phase current but reduced the number of wires needed for his system. While Tesla's most promising three-phase system required six wires to run between the generator and motor, Dolivo-Dobrowolsky employed a Y-connection (also known today as wye connection) that joined the three wires coming out of the generator and transformers to a common grounded return and thus reduced the number of wires needed in his system. To differentiate his ideas from the existing single-phase and polyphase schemes, Dolivo-Dobrowolsky called his system *drehstrom*, which means "rotary current" in German.[4] Skeptical engineers expected the Lauffen-Frankfurt system to be able to transmit only 50% of the power generated at Lauffen; they were astounded when the system operated with an efficiency of 75%. Based on the careful engineering work of von Miller, Brown, and Dolivo-Dobrowolsky, the Lauffen-Frankfurt line demonstrated for the first time the full commercial potential of polyphase AC.[5]

Although the Lauffen-Frankfurt line confirmed his original ideas about the value of polyphase current, Tesla was disturbed to see reports in the electrical journals that gave credit to Brown and Dolivo-Dobrowolsky for the idea of using three-phase current. While Brown bluntly stated "The three-phase current as applied at Frankfurt is due to the labors of Mr. Tesla, and will be found clearly specified in his patents," the patent situation in Europe was far from clear.[6] In the course of developing the AC motor, Tesla had filed patent applications in several foreign countries including England and Germany, but he had neither issued any licenses to European manufacturers nor enforced them by taking legal action against infringers.[7] "There is some ill feeling here," reported Carl Hering from the Frankfurt Exhibition, "as to who the inventor is of this system [i.e., three-phase AC] and who has the right to use it, but it is quite likely that it originated in the United States and is public property here."[8]

Worried that he would not be recognized as the inventor of polyphase AC and anxious to solidify his patent position in Europe, Tesla decided to travel to Europe to lecture on his high-frequency research and look after his foreign interests. Given that Westinghouse was no longer paying him any royalties, Tesla also needed to generate income by licensing European electrical companies to manufacture his motor. Sir William Crookes, the president of the Institution of Electrical Engineers, had invited him to lecture in London, and Tesla had also received an invitation to speak in Paris before the Société de Physique and the Société International des Electriciens.[9]

After Paris, Tesla planned to visit his family in Croatia and Serbia. He was especially anxious to see his mother. As he recalled in his autobiography, he missed her terribly but had found it too hard to break away from the laboratory in order to travel home to see her. Now, however, "a consuming desire to see her again gradually took possession of me. This feeling grew so strong that I resolved to drop all work and satisfy my longing."[10]

Tesla sailed from New York on the *Umbria* on 16 January 1892 and arrived in England ten days later. In London, Sir William Preece, a distinguished electrician and head of the telegraph department of the British Post Office, invited Tesla to stay at his home.[11] Determined to "change the attitude of scientific men and engineers very considerably, both as regards the utilisation of rotary-current motors and as to the credit which should be given to this most interesting discovery," Tesla met at once with a reporter from the London *Electrical Engineer*. Three days after his arrival, the journal ran a profile of Tesla that detailed how his research on AC motors preceded the work of Ferraris, Haselwander, and Dolivo-Dobrowolsky.[12]

To help set the stage for Tesla's London lecture, Crookes published a highly speculative piece on electricity in the *Fortnightly Review*. Along with discussing how electricity might improve harvests, kill parasites, purify sewage, and control the weather, Crookes introduced general readers to the latest discoveries made by Hertz, Lodge, and Tesla about electromagnetic waves. Like other British electrical scientists, Crookes likened Hertzian waves to light and assumed that they would be manipulated using lenses. At the same time, he speculated about how these waves might be used for communications:

> Rays of light will not pierce through a wall, nor, as we know only too well, through a London fog. But the electrical vibrations of a yard or more in wave-length of which I have spoken will easily pierce such mediums, which to them will be transparent. Here, then, is revealed the bewildering possibility of telegraphy without wires, posts, cables, or any of our present costly appliances. Granted a few reasonable postulates, the whole thing comes well within the realms of possible fulfillment. At the present time experimentalists are able to generate electrical waves of any desired wave-length from a few feet upwards, and to keep up a succession of such waves radiating into space in all directions. . . . Also an experimentalist at a distance can receive some, if not all, of these rays on a properly-constituted instrument, and by concerted signals messages in the Morse code can thus pass from one operator to another.

Crookes went into particular detail about Tesla's experiments using high-frequency AC to power lamps without using any wires, promising that homes might soon be lighted with brilliant wireless lamps.[13]

THE LONDON LECTURES

The stage thus set, Tesla lectured before the Institution of Electrical Engineers on 3 February 1892. Anticipating a large turnout, the electrical engineers decided to hold the lecture not in their usual meeting place, the Institution of Civil Engineers (which held four hundred), but the Royal Institution (which could accommodate eight hundred). In return for this favor, the managers of the Royal Institution asked that Tesla repeat his lecture the following night for its members.[14]

At first Tesla was reluctant to repeat the lecture, and it fell upon James Dewar, the Fullerian Professor of Chemistry at the Royal Institution, to convince him to do so. "I was a man of firm resolve but succumbed easily to the forceful arguments of the great Scotchman," recalled Tesla. "He pushed me into a chair and poured out half a glass of a wonderful brown fluid which sparkled in all sorts of iridescent colors and tasted like nectar. 'Now,' said he, 'you are sitting in Faraday's chair and you are enjoying whiskey he used to drink.'"[15] Honored, Tesla agreed to a second performance.

It was not lost on Tesla that he would be lecturing on the same stage where in the 1830s Faraday had introduced the fundamental principles of electromagnetic induction.[16] But as thrilling as it was, it must have also been intimidating. The audience at Royal Institution lectures was an erudite group, and the meetings were both social as well as scientific occasions. Evening dress was worn, and a substantial number of ladies often attended. The lecture hall was an amphitheater, with the seats rising steeply in front of the stage. Traditionally the lectures were expected to last only one hour, with no lengthy introductions or votes of thanks.[17]

Before a packed audience that included the leading British electrical engineers and scientists in the front row, Tesla began by praising Crookes, telling the audience that "what I have to tell you and to show you this evening concerns, in a large measure, that same vague world which Professor Crookes has so ably explored." Tesla recalled that while in college he had read a paper in which Crookes described his early experiments with radiant matter and that these experiments had made a deep impression on him.[18]

Having invoked Crookes, Tesla proceeded with several brilliant demonstrations. Holding a long, evacuated glass tube in one hand, Tesla grasped one terminal of his oscillating transformer and the tube "glowed with a brilliant lambent flame from end to end, and recalled to every one the idea of the magician's enchanted wand." Standing on an insulated platform, he brought his body into contact with one terminal of his oscillating transformer and streams of light burst forth from the other terminal. Turning to the audience, Tesla asked, "Is there anything more fascinating than the study of alternating currents?"[19]

Although the British journal *Engineering* grumbled that it was "a breach of the dramatic canons to start with an experiment of such brilliancy, and then to descend to others of less importance," the audience loved it and broke into applause. Now fired up, Tesla used his coil to perform more wonders: six-inch sparks jumped between balls; two long wires, one foot apart and stretched across the well of the theater, glowed blue along their entire length; and between two wire circles he created "a palpitating purple disk of great beauty." In honor of Lord Kelvin, the prominent British physicist, Tesla used his coil to illuminate a sign that spelled out his common name, William Thomson (see Figure 8.1).[20]

As Tesla "showed wonder after wonder," reported a commentator in *Nature*, "the interest of [the] audience deepened into enthusiasm." Captivated by his modesty and charm, the audience ignored his "broken English and imperfect explanations did not detract from his success. His marvelous skill as an experimentalist was evident and unmistakable."[21]

Tesla then showed the audience what he had observed about electric brush phenomena in evacuated bulbs with a single electrode inside. When powered by his high-frequency coil, a glowing discharge—called the brush—could be seen between the electrode and the inside wall of the bulb. (Today we regard the brush to be a stream of electrons.) Tesla reported that he had noticed that the brush could be manipulated by a magnet and that it rotated in a clockwise direction as a result of the earth's magnetic field. Impressed with how the brush inside the bulb responded to slight electrical and magnetic changes, Tesla speculated that it might "find practical applications in telegraphy. With such a brush it would be possible to send dispatches across the Atlantic, for instance, with any speed, since its sensitiveness may be so great that the slightest changes will affect it."[22] Here Tesla was

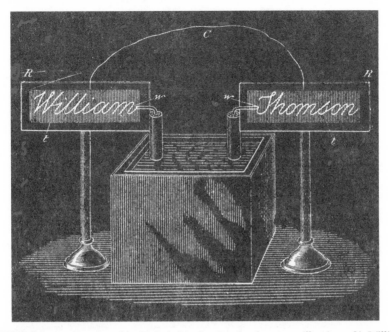

FIGURE 8.1. Apparatus used by Tesla in his 1892 London lecture to illuminate Sir William Thomson's name.

From NT, *Experiments with Alternate Currents of High Potential and High Frequency* (New York: McGraw Publishing Co., 1904; rep. Hollywood, Calif.: Angriff Press, 1986), fig. 9 on p. 27.

anticipating the first electronic vacuum tubes of Lee de Forest and J. A. Fleming that were used fifteen years later to detect and amplify weak radio signals. However, to make a practical radio tube, de Forest and Fleming discovered that it was necessary to use several electrodes to manipulate and control the electron stream inside the tube.

But in the lecture, Tesla did not elaborate on this speculation and moved to a topic that intrigued him much more. "A most curious feature of alternate currents of high frequencies and potentials," he told the audience, "is that they enable us to perform many experiments by the use of one wire only." Tesla then demonstrated how his disk motor could be operated with one wire connected to the transformer and another connected to a suspended plate, and he boldly hypothesized how this motor could even be run without any wires, simply drawing power from the atmosphere charged with electricity (Figure 8.2).[23]

Tesla next showed a variety of single-wire lamps. These lamps consisted of a tiny button of a high-resistance material such as carbon

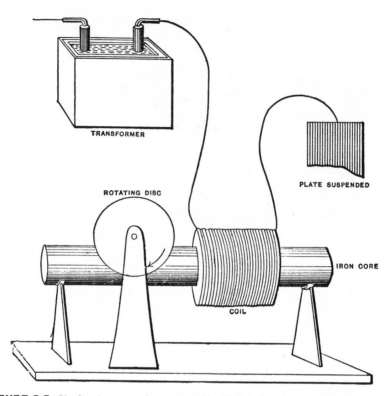

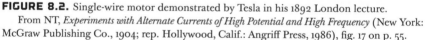

FIGURE 8.2. Single-wire motor demonstrated by Tesla in his 1892 London lecture.
From NT, *Experiments with Alternate Currents of High Potential and High Frequency* (New York: McGraw Publishing Co., 1904; rep. Hollywood, Calif.: Angriff Press, 1986), fig. 17 on p. 55.

or carborundum that was rendered incandescent when powered by a high-frequency power current. One observer estimated that the lamps produced about five candlepower.[24] In showing his lamps, Tesla theorized about the causes of incandescence and phosphorescence, and he discussed Crookes's notion of radiant matter, but this theorizing was not central to the performance. As A. P. Trotter, editor of *The Electrician*, reminisced, "[Tesla] did not write and read a paper, nor did he give a lecture, and he was so occupied in waving long glowing electrodeless tubes in the air, and lighting up of ordinary incandescent lamps by a current taken through his body, that he had no time to explain 'how it was done.' Nor, I think, could he."[25]

Tesla then repeated his celebrated demonstration of placing a long tube between two plates and invited the audience to imagine similar wireless lamps glowing in their homes.[26] For his finale, Tesla topped this demonstration by introducing a new tube, which, like a Crookes radiometer, contained a tiny fan with mica blades. However, while the fan in a radiometer turns as a result of light hitting the blades, Tesla made his fan turn when placed in the electrostatic field between the two suspended plates. More than a glowing lamp, the tiny fan demonstrated to the audience the power being derived from the electrostatic field. When shown the fan spinning as a result of the invisible field, the audience was amazed. "The scientists," recalled Tesla, "simply did not know where they were when they saw it."[27]

"For a full two hours," noted the *Electrical Engineer*,

> Mr. Tesla kept his audience spellbound, with easy confidence and the most modest manner possible displaying his experiments, and suggesting, one after another, outlooks for the practical application of his researches. . . . Even at the end, Mr. Tesla tantalisingly informed his listeners that he had shown them but one-third of what he was prepared to do, and the whole audience . . . remained in their seats unwilling to disperse, insisting upon more, and Mr. Tesla had to deliver a supplementary lecture.[28]

Although it was not customary, at the conclusion of the second performance at the Royal Institution Lord Rayleigh, a prominent British physicist, insisted on giving a vote of thanks to Tesla. Praising Tesla, Rayleigh remarked that "Mr. Tesla has not worked blindly or at random, but has been guided by the proper use of a scientific imagination. Without the use of such a guide we can scarcely hope to do anything of real service. Mr. Tesla has the genius of a discoverer, and we may look forward to a long career of discovery for him."[29]

Tesla took Rayleigh's remarks as a great compliment and a source of inspiration. "Up to that time," Tesla said, "I never realized that I possessed any particular gift of discovery but Lord Rayleigh, whom I always considered as an ideal man of science, had said so." Tesla interpreted Rayleigh's compliment in a particular way; if he was indeed destined not just to invent but to discover, Tesla felt that henceforth "I should concentrate on some big idea."[30]

In the week following Tesla's lectures, the London press "was full of thrilling accounts of this wizard who defied scientific explanation."

Anxious to know more about the man behind the magic, Trotter and several engineers organized an informal dinner for Tesla. "We were all young and eager to know more of Tesla's attractive personality," Trotter recalled. Over dinner, Tesla delighted his British hosts with humorous stories about life in America, including the following: "I heard a noise, one morning, under the window of my room at the Westinghouse Works. I looked into the yard and found two boys quarreling. 'I told you.' 'You didn't. You're a liar.' 'I ain't.' 'You're a stinking little liar, you know you never said it.' 'Yes, I did, you'll find it in my British Association paper of last year!'"[31]

Tesla's lectures inspired one British engineer, J. A. Fleming, to photograph the sparks produced by an induction coil in order to determine whether the sparks did indeed oscillate. Fleming invited Tesla to view the resulting photographs and congratulated him on his lectures. Calling the performances "a grand success," Fleming told Tesla "that no [one] can doubt your qualifications as a magician of the first order, viz the 'Order of the flaming sword.'"[32] (We will meet Fleming again in 1901 when he designed the transmitter that Marconi used for his transatlantic tests.)

During his stay in London, Tesla also spent time with Crookes. They conducted experiments together, and Tesla wound a coil for him. They discussed the future of electricity as well as Crookes's interest in the occult and psychical phenomena. Having read widely in "Spiritualism, Demonology, Witchcraft, Animal Magnetism, Spiritual Theology, Magic, and Medical Psychology," Crookes had investigated séances and had come to believe there was some basis for the claims made by mediums of being able to contact the dead. Up to this time Tesla had given little thought to such matters, but he was deeply impressed that a man of science such as Crookes took spiritualism so seriously.[33]

BREAKDOWN ON THE CONTINENT

From London, Tesla traveled on to Paris and booked a room in the Hotel de la Paix. On 19 February he gave a lecture before the Société de Physique and the Société International des Electriciens (Figure 8.3).[34] Finding his demonstrations highly persuasive, the French electrician

FIGURE 8.3. "Paris—Mr. Tesla Lecturing before the French Physical Society and the International Society of Electricians."
From "Mr. Tesla's Experiments of Alternating Currents of Great Frequency," *Scientific American,* 26 March 1892, p. 195.

Édouard Hospitalier observed, "The young scientist is . . . almost as a prophet. He introduces so much warmth and sincerity into his explanations and experiments that faith wins us, and despite ourselves, we believe that we are witnesses of the dawn of a nearby revolution in the present processes of illumination." Just as in London, Tesla's performance generated a great deal of excitement and praise. "The French papers this week are full of Mr. Tesla and his brilliant experiments," reported the *Electrical Review.* "No man in our age has achieved such a universal scientific reputation in a single stride as this gifted young electrical engineer."[35]

During his stay in Paris, Tesla encountered several dignitaries, including the physicist André Blondel and Prince Albert of Belgium, who was interested in improving the electrical power systems in his country. Intent on generating some income from his foreign patents, Tesla met with representatives of Schneider & Co. of Creusot, France, and the Helios Company of Cologne, Germany, and licensed these companies to manufacture his motors in France and Germany.[36]

Tesla found all these activities—giving lectures, meeting important people, negotiating with businessmen—exciting yet stressful. While he was still in London, Crookes noticed that Tesla was nearing exhaustion; worried, he wrote Tesla in Paris that "I hope you will get away to the mountains of your native land as soon as you can. You are suffering from over work, and if you do not take care of yourself you will break down. Don't answer this letter or see any one but take the first train."[37]

But Crookes's letter probably arrived too late, and Tesla was overtaken by exhaustion and depression. As often happened with his depressive bouts, Tesla retired to his hotel room to sleep. When he awoke, he was greeted by terrible news about his mother, Djuka; as Tesla remembered, he was "just coming to from one of my peculiar sleeping spells, which had been caused by prolonged exertion of the brain. Imagine the pain and distress I felt when it flashed upon my mind that a dispatch was handed to me at that very moment bearing the sad news that my mother was dying."[38]

Tesla rushed from Paris to his family home in Gospić without stopping. His fear that he would not arrive in time to see Djuka alive caused a patch of hair on the right side of his head to turn white overnight. (Within a month it was restored to its original jet-black color.) When Tesla arrived at his mother's side, she murmured to her only son, "You've arrived, Nidzo, my dear."[39]

Over the next several weeks Tesla stood vigil by his mother's bedside, only to suffer a breakdown. As he recalled:

> I had become completely exhausted by pain and long vigilance, and one night was carried to a building about two blocks from our home. As I lay helpless there, I thought that if my mother died while I was away from her bedside she would surely give me a sign. . . . [F]or my mother was a woman of genius and particularly excelling in the powers of intuition. During the whole night every fiber in my brain was strained in expectancy, but nothing happened until early in the morning, when I fell in a sleep, or perhaps a swoon, and saw a cloud carrying angelic figures of marvelous beauty, one of whom gazed upon me lovingly and gradually assumed the features of my mother. The appearance slowly floated across the room and vanished, and I was awakened by an indescribably sweet song of many voices. In that instant a certitude, which no words can express, came upon me that my mother had just died.[40]

Deeply disturbed by this clairvoyant dream, Tesla immediately wrote Crookes about it, as it seemed to confirm Crookes's ideas about spiritualism. Tesla puzzled over this dream for years and eventually concluded that the music he had heard was based on hearing a nearby church celebrating Easter mass on the morning his mother died. The angels were inspired by a painting by Arnold Bocklin depicting one of the seasons and showing a group of allegorical figures on a cloud; Tesla had seen this painting during a visit to Munich and it had made a deep impression on him as the figures seemed to be floating in the air. Thus he was able to explain "everything satisfactorily in conformity with scientific facts."[41]

Djuka was buried on Easter Sunday, beside her husband, in the Jasikovac cemetery in Divoselo. As a sign of the deep involvement of the Tesla and Mandic families in the Serbian Orthodox Church, six priests officiated at the funeral. Tesla arranged for white obelisk headstones over the graves of his mother and father.[42]

Tesla remained in Gospić for the next six weeks, grieving with his family. "I don't have to tell you that I am very sad and holding myself with restraint," he wrote Uncle Pajo in April 1892. "I was afraid of this event a while ago, but, the blow was heavy."[43]

As Tesla regained his strength, he journeyed across Croatia to Plaski to visit his sister Marica, to Varazdin to see Uncle Pajo, and to Zagreb where he lectured at the university. From Zagreb, Tesla traveled to Budapest to meet with representatives of the electrical manufacturing firm Ganz and Company.[44] Along with learning about the firm's current efforts to build a 1,000-horsepower alternator, Tesla negotiated a license so that Ganz could manufacture his motors. Overall Tesla was quite pleased with how his patent negotiations went, and he reported to Westinghouse that "The patents are in the hands of three of the most powerful companies who will cooperate and they are earnest in their intention to push the manufacture. The introduction of the motor in Europe on an extensive scale will, no doubt, have a very favorable influence upon the value of my American patents which are owned by your Company."[45]

In May, Tesla went to Belgrade, the capital of Serbia, where he received a hero's welcome. King Alexander I conferred on him the title of Grand Officer of the Order of St. Sava. The Serbian poet Jovan Jovanović Zmaj composed a poem, "Pozdrav Nikoli Tesli," which he read at a ceremony honoring Tesla. In response to these honors from his fellow Serbs, Tesla thanked the audience, expressing both his ambition and his national pride: "[I]f I were to be sufficiently fortunate to bring about at least some of my ideas it would be for the benefit of all humanity. If these hopes become one day a reality, my greatest joy would spring from the fact that this work would be the work of a Serb."[46]

In returning to America from Serbia, Tesla traveled through Germany, stopping to see the physicist Hermann von Helmholtz in Berlin and then on to Bonn to confer with Hertz.[47] Tesla had repeated Hertz's original experiments using his oscillating transformer, and while he felt that Hertz had been correct in showing that electromagnetic waves were being propagated in space, Tesla disagreed with Hertz about the form of the waves. In his experiments, Hertz had found the waves were transverse, meaning that the disturbances are at right angles to the direction of propagation. (A familiar example of transverse waves is the waves in the ocean.) To demonstrate this, Hertz had set up tests that showed that the waves could be reflected and interfere with each other, thus revealing that electromagnetic waves were like light. In replicating Hertz's experiments, Tesla concluded that the waves he observed were not transverse but longitudinal, in that the displacement is parallel to the direction of propagation. (A simple example of a longitudinal wave would be how a train moves backward—as the locomotive reverses, each car bumps the next so that a pulse moves through the string of cars.) For Tesla, electromagnetic waves were more like sound waves than light waves. If the new waves were not transverse like light waves, then this meant that Hertz had not provided the experimental proof for Maxwell's theory. Needless to say, Tesla's claims would have been very troubling to Hertz, and as Tesla recalled, "He seemed disappointed to such a degree that I regretted my trip and parted from him sorrowfully." It is perhaps not surprising that there is no mention in Hertz's diary of meeting with Tesla.[48]

Though the death of his mother made the latter portion of his trip to Europe "a most painful ordeal," Tesla returned from Europe with a major insight. As we have seen, he had left London challenged by Lord Rayleigh's praise to focus his efforts on one big idea, and this idea came to him while walking in the mountains of his homeland. During this hike, a thunderstorm came up suddenly, but Tesla was able to find shelter before rain began. As he described in his autobiography:

[S]omehow the rain was delayed until all of a sudden, there was a lightning flash and a few moments after a deluge. This observation set me
thinking. It was manifest that the two phenomena were closely related,
as cause and effect, and a little reflection led me to the conclusion that
the electrical energy involved in the precipitation was inconsiderable,
the function of the lightning being much like that of a sensitive trigger.
Here was a stupendous possibility of achievement. If we could produce
electric effects of the required quality, this whole planet and the conditions of existence on it could be transformed. . . . The consummation
[of this idea] depended on our ability to develop electric forces of the
order of those in nature. It seemed a hopeless undertaking, but I made
up my mind to try it and immediately on my return to the United States
in the summer of 1892, work was begun which was to me all the more
attractive, because a means of the same kind was necessary for the successful transmission of energy without wires.

Watching how the lightning seemed to cause the rain to start, Tesla
was fascinated by the notion of a "sensitive trigger": that a small force,
properly applied, could be used to harness tremendous forces in the
earth. Recalling his experiments with grounding his oscillating transformer from the previous fall, Tesla now realized that if he could scale
up his transformer, he might very well have the trigger needed to harness the Earth and "provide motive power in unlimited amounts."[49] To
Tesla, here was a challenge worthy of his talent and genius.

Tesla sailed from Hamburg on the *Augusta Victoria* and arrived back in New York on 27 August 1892.[1] Upon his return, he changed both his laboratory and residence. He enlarged his laboratory by relocating from Grand Street to 33–35 South Fifth Avenue (today LaGuardia Place) where he occupied the fourth floor in a nondescript factory building. Located just south of Washington Square, his new laboratory was "in the heart of that picturesque neighborhood known as the French Quarter, teeming with cheap restaurants, wine shops, and weather-beaten tenements." In late September, Tesla switched from the Astor House to the Gerlach Hotel on 27th Street, between Broadway and Sixth Avenue. Built in 1888 at a cost of $1 million, the Gerlach was an imposing eleven-story fireproof building that featured elevators, electric lights, and several sumptuous dining rooms.[2]

With his return to New York, Tesla was keen to follow up on his new vision for his high-frequency inventions, but he also felt the need to improve his polyphase motors and do everything he could to convince Westinghouse to promote them. Since Tesla had torn up his contract with Westinghouse, the company was under no obligation to work with him but Tesla was anxious to ensure that his polyphase system was not ignored in the United States. Having talked with the engineers at Ganz and Company and other electrical firms in Europe,

Tesla was well aware that the Europeans were moving forward quickly with developing systems for transmitting power using two- or three-phase current.

When Tesla left Westinghouse in 1889, his former assistant, Charles Scott, was given the task of continuing to develop marketable motors based on Tesla's patents. However, before Scott and other engineers at Westinghouse could do so, the company went into receivership, and it took George Westinghouse the better part of 1890 and 1891 to secure new financing for the company (see Chapter 7).

While they were waiting for the company's financial situation to be resolved, Scott and his associates made several decisions about the frequency and phase of future polyphase systems. Over the short term, they decided to build two-phase systems using 60-cycle alternating current. By doing so, they would be able to combine motor and lighting loads since they could split the two-phase current into two separate single-phase currents for lighting circuits and 60 cycles would not produce a noticeable flicker in incandescent lamps. They planned to build three-phase, 30-cycle power systems later on that would be better suited for industrial applications. In particular, Scott found that it was possible to connect two-phase generators with three-phase motors by using his special "T" transformer connection. As a result, it became possible to use three-phase, 60-cycle AC to service both lighting and power loads in a single network.[3]

Such was the technical situation by the start of 1892 when, having finally stabilized the company, George Westinghouse was able to start thinking about what his company should do with AC. Although Westinghouse owned the American rights to Tesla's patents for using polyphase AC, he was not, for most of 1892, particularly concerned with developing polyphase. Instead, he was much more interested in pursuing single-phase AC since there was a ready market for lighting systems using single-phase at 133 cycles.[4]

The Westinghouse Company employed single-phase AC and Tesla's split-phase motors to make its first power transmission installation at the Telluride gold mine in Colorado. Unable to secure power nearby, the mine owners had asked Westinghouse to install a turbine at a stream four miles away and erect a 3,000-volt AC power line over the rugged terrain to a 100-horsepower motor at the mine. Reporting that the Telluride system was delivering power with an efficiency of

83.5% at full load, Scott proudly boasted that "work in this field is fast passing from experimental investigation into practical engineering."[5] But Telluride was only an isolated plant using single-phase AC to transmit power a few miles; compared to the Lauffen-Frankfurt line and other work going on in Europe, Telluride was small potatoes.

Instead of pursuing polyphase AC—which must have seemed promising but unproven—Westinghouse decided in the spring of 1892 to concentrate on promoting AC by going after the contract to provide electric lighting for the 1893 Chicago World's Fair. Westinghouse did so because he needed to do something dramatic to regain visibility as a major electrical manufacturer. Many people had assumed that Westinghouse, having nearly gone under, would be a much smaller player in electrical manufacturing. At the same time, Westinghouse now faced an even larger adversary since Edison General Electric and the Thomson-Houston Electric Company had merged in February 1892 to form General Electric (GE). In May 1892 Westinghouse won the contract to provide electric lighting for the fair by substantially underbidding GE. Since the fair buildings were to be decorated with two hundred thousand incandescent lights, the fair was an ideal opportunity for Westinghouse to demonstrate how AC could be used to power an entire city.[6]

But Westinghouse had made such a low bid that his engineers were forced to design larger alternators and operate at higher voltages than previously used; as one history of the fair recounted,

> The Westinghouse Electric and Manufacturing Co., having secured the contract to furnish this immense service at a figure far below the cost, as such work had always been done, it became necessary to devise a system more economical and at the same time more flexible. This was done. They devised and constructed in less than six months larger machines than had ever been built for this work before, and on radically different lines, embodying the principles of the alternating system of transmission. By this system hundreds of thousands of dollars' worth of copper wire were saved, as it was possible to send the current under high pressure [i.e., voltage] to its destination on small wires, and then transform it down to the point of utility.[7]

Not only did Westinghouse have to worry about designing new equipment for the fair, he also had to come up with a new incandescent lamp. In October 1892, after a long legal battle, the courts upheld Edison's original lamp patent and ruled in favor of GE. In response, Westinghouse and his engineers designed a new stopper lamp that avoided the Edison patent. Though less efficient than the Edison lamp, this new design allowed Westinghouse to complete the installation at the Chicago World's Fair.

Consequently, when he arrived back in New York in late August 1892, Tesla found that Westinghouse was not giving much thought to promoting his motors or polyphase system. Westinghouse was not opposed to polyphase, but it was not the most pressing technology to pursue at that moment. Concentrating on designing the generating equipment and lamps needed to satisfy the lighting contract for the fair, Westinghouse was not even thinking about an exhibit for the fair in which Tesla motors could be displayed.[8] Yet Tesla was anxious to make sure that the best possible version of his motor was available for the fair. As Tesla told Westinghouse in mid-September:

> I intend myself to go to Pittsburgh this evening if I can manage to find time. I must consult with Mr. Schmid [the general superintendent] about the speedy carrying out of some improvements on my . . . motors. It is necessary to bring the motor to high perfection before the exhibition as this is of prime importance. . . . Please [ask] your staff to aid me in all they can. My conviction is that a motor without brushes and commutator is the only form which is capable of permanent success. To introduce other forms I consider a . . . waste of time and money.[9]

Knowing that the Telluride installation used only single-phase, Tesla was concerned with improving on his polyphase designs since he asked a few weeks later to borrow several transformers as well as an alternator that produced either a two- or three-phase current.[10] Tesla probably investigated how to patent around the Y-connection used by Dolivo-Dobrowolsky in the generators and motors at Frankfurt. At the same time, Tesla also studied the trade-offs between using two- or three-phase current. Although he had emphasized three-phase current in his patents, he had heard enough in discussions with European engineers to realize that two-phase might be better in some situations for transmitting power.

As he fretted over his polyphase motor and his relationship with Westinghouse, Tesla was approached by Henry Villard, a German financier who had helped funnel German capital into American railroads in the 1880s. He had been the driving force behind the consolidation of various Edison companies into Edison General Electric in 1889. But in the negotiations leading to the formation of GE in 1892, Villard was outmaneuvered by Charles Coffin of Thomson-Houston, and Coffin became president of GE.[11] Rebuffed but undaunted, Villard was still determined to play a role in the electrical industry. In the fall of 1892, Villard approached Tesla with some sort of scheme. Although the correspondence does not tell us, Villard's scheme could have involved electric street railways, the promotion of Tesla's polyphase system, or even a consolidation of other electrical firms built around the Westinghouse Company—all were ideas that Villard had considered over the previous few years.

Whatever the scheme, it intrigued Tesla but it meant convincing Westinghouse to come onboard. Tesla, however, was unable to persuade Westinghouse; as he explained to Villard in October 1892:

> I have approached Mr. Westinghouse in a number of ways and endeavoured to get to an understanding with him in the sense of our last conversation. The results so far have not been very promising and have impressed me that to follow up on this matter would require more time than I am able to spare at present.
>
> Realizing this, and also considering carefully the chances and probabilities of success, I have concluded that I cannot associate myself with the undertaking you contemplate. For the time being I am working on an invention which, if only in a measure successful would radically transform the present system of electric lighting, and the subject demands the concentration of all my energies.[12]

THE RACE FOR THE NIAGARA CONTRACT

Westinghouse was indifferent to Villard's scheme because as the fall of 1892 progressed, he could see even greater opportunities taking shape. Along with lighting the upcoming World's Fair in Chicago,

Westinghouse decided to make another big move: to pursue the contract for the power equipment that would be used to harness Niagara Falls. As we shall see, the successful development of power at Niagara proved to be the turning point for Tesla's polyphase inventions.

Thanks to geography and population, Niagara Falls was an ideal place to pursue the development of power transmission. Connecting Lake Erie with Lake Ontario, the Niagara River carries the full flow from the upper Great Lakes as the water makes it journey to the Atlantic Ocean via the St. Lawrence River. The falls occur where the bedrock beneath the river suddenly changes from hard to soft, and the river drops dramatically 160 feet. But rather than being isolated in the wilderness, Niagara Falls was located near much of the industrial population of the United States and Canada. In 1890, about one-fifth of all Americans lived within four hundred miles of Niagara Falls, and Buffalo, a city of 250,000, was twenty miles to the south.[13] To the north, across the Niagara River, lay much of the population and industry of Canada in the province of Ontario.

But while Niagara Falls held the promise of generating vast amounts of power, the scenic beauty of the cataract posed a challenge to those wishing to harness that power. In 1885, local industrialists on the American side dug a canal to provide waterpower to several factories at the foot of the falls. Concerned, however, that such industrial development would spoil the beauty of the falls, the state of New York declared the remaining land near the falls to be a special nature reservation. The effect of the reservation was to remove permanently from use the land that would have been ideal for a major industrial district. Unable to build factories right at the falls, industrialists now had to get around the reservation.

In response, a civil engineer, Thomas Evershed, unveiled a plan in 1886 using canals, shafts, and a tunnel to divert water around the reservation. A broad canal located a mile above the falls would bring water to a series of branch canals that would power 238 separate waterwheels. After passing through a waterwheel, the water would then plunge down a 150-foot shaft to a 2.5-mile-long tailrace tunnel. Sloping under the town of Niagara Falls, this tunnel would carry the water to the lower portion of the river.

Although the tunnel would have to be cut through solid limestone, Evershed's scheme captured the imagination of both local investors and William Birch Rankine, a prominent New York City attorney. As a young man Rankine had clerked for a lawyer in Niagara Falls and became fascinated with the possibility of harnessing the cataract.[14] Recognizing that Evershed's plan would cost millions, Rankine brought the idea to J. P. Morgan. Although Morgan was willing to invest in the scheme, he told Rankine that the project would need a strong leader to serve as its promoter. Since the project would involve both finance and engineering, Morgan suggested a fellow Wall Street banker, Edward Dean Adams (1846–1931). "If you can get him," Morgan told Rankine, "I'll join you."[15]

A Boston Brahmin and an indirect descendant of two presidents, Adams had studied engineering at Norwich University and MIT. He came to Wall Street in 1878, joining the investment house of Winslow, Lanier & Company. As one of his first projects, Adams helped organize the Northern Pacific Terminal Company and the St. Paul & Northern Pacific Railway Company, and through this work he met Villard. Villard was the American representative of the Deutsche Bank, and when Villard stepped down from this post in 1893 he was replaced by Adams; over the next several decades, Adams was responsible for directing millions of German marks into American railroads and industrial enterprises. Adams further made a name for himself by streamlining the finances of railroads and manufacturing companies. Impressed with his ability, Morgan frequently asked Adams to participate in industrial reorganizations. In 1896, Adams particularly endeared himself to Morgan by getting the Deutsche Bank to underwrite a quarter of the $100 million that Morgan loaned to the U.S. Treasury in order to save the dollar from collapsing. His biographer summed up Adams's character in the following manner: "In a cynical and materialistic age it is refreshing to find a business man of this type—analytical, indefatigable and industrious, yet with all the grace and refined culture of the aristocracies of old."[16]

Long interested in the commercial potential of electricity, Adams had been a stockholder in the Edison Electric Light Company since 1878. Enthusiastic about the possibilities for Niagara, Adams organized a syndicate of Wall Street capitalists who put up $2.63 million. Adams then formed the Cataract Construction Company to develop the power potential of the falls.[17]

As president of the Cataract Company, Adams made a key decision early on. Rather than utilize the power generated in new factories in the small town of Niagara Falls, Adams thought the real opportunity lay in transmitting power to factories in Buffalo and other cities. At that time, Buffalo factories were using steam engines to generate 50,000 horsepower daily, so there was clearly a ready demand for power in that city. Moreover, by utilizing the power away from Niagara Falls, Adams would avoid the expense of the branch canals and numerous vertical shafts needed to connect the individual waterwheels with the tailrace tunnel. As exciting as this idea was, it meant that Cataract would need to find a way to transmit large amounts of power over the twenty miles between Niagara and Buffalo.

To determine the most efficient way to transmit power to Buffalo, Adams consulted first with Edison, who, not surprisingly, suggested using direct current. Adams turned next to Westinghouse. Because of the low price of coal in Buffalo, Westinghouse was dubious that electric power could compete with steam power in that city. Also aware that factory owners might balk at the cost of replacing their existing steam engines with electric motors, Westinghouse recommended transmitting power via a compressed-air pipeline. Westinghouse was knowledgeable about compressed air since he used it in his railway air brakes, and he suggested to Adams that compressed air could power existing steam engines. In general, Westinghouse was concerned that Adams was underestimating the problems of finding enough customers for all the power generated by the falls.[18]

Faced with a medley of opinions about power transmission, Adams consulted with engineers in England, Germany, and Switzerland. In June 1890, Adams brought together the leading experts and established the International Niagara Commission. The commission announced a contest to determine the best method for generating and transmitting power at Niagara and invited twenty-eight European and American engineering firms to submit proposals. The commission offered $20,000 in prizes, with the top award being $3,000. Upon hearing of the competition, a Westinghouse engineer, Lewis B. Stillwell, was keen to enter and approached his boss. Annoyed that the commission was fishing for major information but offering such paltry rewards, Westinghouse refused to participate. "These people are trying to secure $100,000 worth of information by offering prizes, the largest of which is $3000," he growled. "When they are ready to do business, we will show them how to do it."[19]

Westinghouse was right to be suspicious because although the commission received fourteen proposals, it deemed none satisfactory and awarded no first prize. Instead the commission mined the proposals for technical information and forwarded a series of recommendations to Adams. Using the tunnel that Cataract had started digging in October 1890, the commission recommended that the company place several 5,000-horsepower turbines in the tunnel and connect them via 150-foot-long shafts to power generators located at the surface in a central power house. The commission could not decide between compressed air or electricity for transmission, but Adams chose to go with electricity because of the efficiencies demonstrated on the Lauffen-Frankfurt line. Determined to move ahead, in December 1891 Adams invited six electrical companies—Westinghouse, Thomson-Houston, Edison GE, and three Swiss firms—to provide estimates on the electrical equipment needed at Niagara.

In response to Cataract's call for bids, the newly formed General Electric offered a plan in the fall of 1892 that would provide DC for local industry in Niagara and AC transmission to Buffalo. Meanwhile, the Swiss firms—which specialized in designing hydroelectric plants—offered other schemes. Cataract rejected the Swiss proposals since American tariffs of 40% on imported machinery made their plans prohibitively expensive. In addition, as Tesla pointed out to Westinghouse, the foreign firms could not bring polyphase equipment into the United States without infringing on his patents.[20]

Having resisted getting drawn into the Niagara race for two years, Westinghouse now plunged into the fray. In December 1892, with Adams's call for bids having come and gone, Westinghouse boldly announced that his company was ready to provide polyphase equipment for Niagara. What gave Westinghouse confidence were developments coming out of his engineering department. Continuing the work begun by Charles Scott to improve Tesla's polyphase motors, Benjamin G. Lamme had come up with new arrangements for the coils in the stator so that Tesla's designs now worked as well as Dolivo-Dobrolowsky's *drehstrom* motors. In addition, tests of these new motors suggested a more efficient way of winding the rotor, leading to what came to be the standard rotor design: the squirrel cage. And Lamme designed a new rotary converter; consisting of an electric motor and generator on a single shaft, this machine could convert polyphase AC into either single-phase

AC or DC. Using a rotary converter, a power company could now use polyphase AC to transmit power over long distances and then convert the power so that customers could use it with their existing single-phase AC or DC equipment. Westinghouse was quick to see that the rotary converter meant power companies would be able to find customers for all the power that they could generate and transmit.[21]

These engineering developments meant that, for the first time, Westinghouse could make full use of Tesla's AC polyphase motor, and the company began emphasizing its ownership of the Tesla patents. In January 1893 the Westinghouse Company issued a pamphlet that included the twenty-nine Tesla patents it owned. In part, the pamphlet highlighted the opportunity to use polyphase AC to transmit power from waterfalls to cities, but at the same time Westinghouse warned customers not to buy polyphase equipment from other manufacturers since they could be sued by Westinghouse for infringement.[22]

Armed with these engineering developments and the Tesla patents, Westinghouse now energetically pursued the Niagara contract. In January 1893, Adams and his associates from Cataract visited the Westinghouse plant in Pittsburgh where they were shown the latest equipment and preliminary plans. The following month, Adams made a similar tour of GE's factories.

SELLING ADAMS ON POLYPHASE

But Tesla was not satisfied to leave the choice of transmission system to be negotiated between Westinghouse and Adams. Recalling his childhood dream of harnessing Niagara, Tesla was determined that his polyphase system should be used to transmit power. As he explained in 1917, "When I heard that such authorities as Lord Kelvin and Prof. W. C. Unwin had recommended—one the direct-current system and the other the compressed air—for the transmission of power from Niagara Falls to Buffalo, I thought it dangerous to let the matter go further, and I went to see Mr. Adams."[23] Tesla met with Adams and corresponded with him during the first few months of 1893.

After meeting with Adams, Tesla reviewed plans for the powerhouse and suggested that Cataract should not run the turbines and generators at 150 rpm (as suggested by Unwin) but at 250 rpm (as recommended by Schmid at Westinghouse). Tesla argued against the lower speed since it would spoil the splendid view of the dynamos inside the powerhouse: "If you would reduce the speed it would necessitate . . . that the diameter of the Dynamo be made considerably larger, which would leave still a smaller space between the wall and the machine, and this I think, with such magnificent machinery which will be viewed by Monarchs, would be decidedly bad."[24]

Even though Adams was a close observer of the electrical industry, he quizzed Tesla on developments in the field. For instance, Adams was puzzled by why European electrical firms had suddenly switched from pushing polyphase to promoting single-phase AC. Had not Oerlikon and Allgemeine Elektricitats-Gesellschaft used polyphase at Frankfurt? For Tesla, the answer was that his patents were now being used by the Helios Company in Germany to sue infringers. "I have not the slightest doubt," Tesla wrote Adams in February 1893, "that all companies except Helios, who have acquired rights from my Company, will have to stop the manufacture of [poly] phase motors. Proceedings against infringers have been taken in the most energetic way by the Helios Co. It is for this reason that our enemies are driven to the single phase system and rapid changes of opinion."[25]

Finding Tesla's responses useful, Adams asked him to review various articles on AC from electrical engineering journals. Tesla dismissed the plans proposed by others as inefficient or impractical but regularly emphasized the virtues of his motor. As he explained to Adams:

> [I]t is much easier to run [my] machine which has no commutator or brushes . . . than a direct current machine. . . . Not to speak of the advantages of the ideal simplicity of these machines affords ov[er] the long run.
>
> Under frequent conditions in practice it would be entirely impossible for any system to compete [with] this ideally simple one. . . . This is the third form of my motor. This form has never been criticized by the adversaries of my system and for a good reason, because it is the most efficient form of electric machine that has been produced to this day. I have shown that in such machines under favorable conditions an efficiency of 97% can be obtained.[26]

While Tesla saw his letters to Adams as an opportunity to advance his polyphase system, Adams viewed the correspondence as a way of getting inside information. By March 1893, Adams was growing especially concerned about the patent situation. Although the Westinghouse Company was claiming that the Tesla patents gave it exclusive control over polyphase AC, the situation was not necessarily that clear. GE had been developing its own polyphase technology, drawing on Elihu Thomson's research and by purchasing AC patents from several inventors including Charles S. Bradley. The ambiguity of the patent situation alarmed the chief patent attorney for Cataract, Frederick H. Betts, and he warned Adams in March 1893 that if he used the Tesla patents, Adams might find himself embroiled in patent litigation with GE.[27]

To gain some perspective on the patent situation, Adams turned to Tesla. The day after Betts had warned Adams, Tesla provided Adams with his evaluation of the Thomson and Bradley patents held by GE:

> The patent of Thomson . . . has absolutely nothing to do with my discovery of the rotating magnetic field and the radically novel features of my system of transmission of power disclosed in my foundation patents of 1888. . . .
>
> As to the Bradley patents . . . I would think it fair to all concerned that a thorough examination be made to their history and bearing. Such [an] examination, which has been made by myself without the slightest prejudice, will convince you that in the earliest patent there is not the least hint of a method of transmission of power which would be novel.[28]

While Adams was probably suspicious of Tesla's claims to have reviewed Bradley's patents "without the slightest prejudice," the letter signaled to Adams that Tesla and Westinghouse believed that they had a strong legal position and that they would actively defend their patents.

Adams was also wondering why Westinghouse was advocating a two-phase system and GE a three-phase system in plans submitted by each company in March 1893. In a two-phase system, the generators produce two currents that are 90° out of phase, while in a three-phase system the generators produce three currents that are 60° out of phase. In either system, these currents would be transmitted over separate circuits and then combined to run a motor at the point of use. In response to Adams's query, Tesla recommended two-phase. In his patents and publications, Tesla had previously emphasized three-phase, largely because he had found that three currents had produced a more uniform rotating magnetic field in his motors than two currents did. Because of his patents, Tesla had a personal interest in advocating three-phase. Now, however, he told Adams that there were practical advantages to using two-phase. As Tesla pointed out, a key advantage for two-phase was that Westinghouse had found that each of the two currents could be separated and used to power single-phase incandescent lights. For Cataract, this meant that there would be yet another way of selling the enormous output of the new Niagara plant, and as a result, the company decided ultimately to go with two-phase current for local distribution and three-phase for long-distance transmission.[29]

But as he pumped Tesla for information about matters such as two-versus three-phase systems, Adams was keeping his options open. As the official history of the Niagara project noted, "[T]he type of current to be generated was left until the last moment in order to have the guidance of the latest foreign and home experience." Much to Tesla's dismay, Adams was still considering the possibility of using DC since at 10,000–20,000 volts it is possible to transmit power efficiently over a distance. Such a plan, which Adams sent to Tesla, was being put forward by a "prominent advocate." Adams may well have been swayed to take such a DC plan seriously since commission member Lord Kelvin was continuing to push for DC and Kelvin cabled Adams in May 1893, "Trust you avoid gigantic mistake of adoption alternating current."[30]

For Tesla, it would be an even bigger mistake if Cataract adopted DC. To persuade Adams of this, Tesla first argued that the generation and transmission of electric power were fundamentally alternating:

> I hope that you are not seriously considering the statement of the "prominent advocate." To gain an idea of the infinite inexperience of the man and the absurdity of his view you need only to realise that all power transmissions are alternating. The process in the direct system is this. We generate in the machine alternate currents—(this is true for any machine in use)—which we reclaim by means of the commutator and brushes. The direct currents which go over the line can not drive a motor, they must be again made alternating by the commutator and

brushes in the motor. Now what my system has offered was to do away with the commutator and brushes both in the generator and motor, otherwise the action is the same. This renders the system simpler, cheaper and more efficient in general. But these are only incidental advantages. The chief gain consists in these features: absolutely constant speed, facility of insulating for high voltage, easy conversion to any voltage, and facility of [unclear word] to the wires on all points along the route where energy is needed. These features are practically unattainable in the direct system, especially when transmissions at a great distance are contemplated. In fact I think that such a scheme if ever carried out could not help being a commercial and perhaps also a technical failure. Of course with a sufficient outlay of capital any scheme, however absurd, can be carried out but the question here is to achieve a practical commercial success and the best and safest appliances should be employed by all means.

Next, drawing on his own practical experience, Tesla argued that a high-voltage DC system would encounter serious problems: it would require a great deal of additional insulation, it would be difficult to avoid current variations, and extra equipment (probably motor-generator sets) would be needed to adapt the current to different lighting and power applications:

I suppose very few engineers have wound direct current machines for 10,000 volts. My experience with such machines which I constructed for certain experimental purposes was that they invariably break down. The reason is that owing to the presence of the commutator it is very difficult to insulate. It has been found impracticable to operate successfully arc light machines (direct beyond 4000 volts and with 20,000 volts the difficulties are—figuratively speaking—25 times as great). When a very thick insulation is employed, the machine looses [*sic*] efficiency, and what is the worst, it becomes poorly regulating and unfit for the purpose. The power you would furnish with such machines would be unavailable for many uses, for instance, electric lighting. There might be—and very likely would be—variations of 20%, and you can not afford 2–3%, for this would give you an unsatisfactory light and produce considerable variations in smaller machines. Suppose all difficulties of such Nature [were] overcome, one would still be very far from commercial success. You would not be able to carry out your plan of making connections along the route,—at least it would be very difficult—and in such case you would have to use two machines in every place, for you could not expect to have two windings in one machine, it would be impracticable and dangerous. The cost of maintenance would be an important item. I think it fair to estimate that you could not possibly operate such a system with adequate safety with double the outlay of capital—not to speak of some of the mentioned insuperable difficulties. Let the "prominent advocate" carry out such a system and he will receive the pay he deserves.

In closing, Tesla played to Adams's dream of transmitting Niagara power not only to Buffalo but throughout New York State, even to New York City; Tesla reminded him that he was aware of Adams's "intended plans for the utilization of the energy at much greater distance than Buffalo—and [it is] only that case I have been considering. With the alternate system you will have absolute, unquestionable success."[31]

In May 1893, Adams and the Cataract Company announced their decisions about which technology to use at Niagara. Swayed by Tesla's arguments, Adams declared that Niagara would use two-phase AC.[32] And although he had concluded that Westinghouse was better prepared to build the large-scale equipment needed, Adams once again rejected both of the plans submitted by GE and Westinghouse. In part, this rejection was probably caused by the fact that blueprints for Westinghouse designs had been discovered in GE's engineering offices; with accusations of industrial espionage flying, Adams may have wanted to steer clear of both companies. But another reason for rejecting the plans was that another member of the Niagara Commission, Professor George Forbes, had been formulating his own design for the generators, and he was agitating for Cataract to follow his plan.

Still determined to land the Niagara contract, the Westinghouse Company pulled out all the stops at the Chicago World's Fair during the summer of 1893 and demonstrated a fully integrated AC system. At the fair, Tesla had his own personal display that highlighted the magic and potential of AC by showing his early motors, egg of Columbus apparatus, oscillating transformer, and an array of his new lamps.[33] Tesla's personal exhibit, though, helped draw visitors' attention to all of the Westinghouse equipment being used to power the fair.

To supply power for all of the incandescent lighting at the fair, Westinghouse installed twenty-four 500-horsepower single-phase, 60-cycle generators. These generators were installed in pairs on single shafts so as to provide two-phase AC for the Tesla motor circuits. To provide power for their electric railway, Westinghouse engineers used a rotary converter to change the AC into 500-volt DC. Transformers were also used in the network to step the voltage up or down, as needed for different applications using Tesla motors. The first installation to use alternating current to service both electric lighting and power applications, the Westinghouse display at the World's Fair convinced electrical engineers from both America and Europe that AC was here to stay.[34]

At the same time, Tesla continued his campaign with Adams on behalf of polyphase AC; the result was that Adams awarded Westinghouse the contract for building the generators in October 1893. Once Forbes's plans arrived at Westinghouse, they were substantially revised by Lamme. To ensure that he could draw on both major electrical manufacturers in the future, Adams awarded another contract to GE for building the twenty-mile transmission line from Niagara to Buffalo to GE. Even though they had to share the Niagara business with GE, it was not lost on Westinghouse executives that Tesla had helped them secure the business; as one Westinghouse manager congratulated Tesla in November 1893, "It must certainly be gratifying to you to think the largest water power in the world is to be utilized by a system which your ingenuity originated. Your successes are gradually pushing to the front. . . . Let the good work go on."[35]

From 1893 to 1896, Adams and Rankine were deeply absorbed in supervising the construction of a powerhouse that would ultimately hold ten Westinghouse generators, each rated at 5,000 horsepower. To design the building for the powerhouse as well as several dozen houses for employees, Adams hired the prominent architect Stanford White. Since the new powerhouse would deliver four times the amount of electricity than any previous power station, Adams and Rankine began looking ahead to using polyphase current to distribute power over a wider region, first to other cities besides Buffalo in New York State, and then even further afield. As Rankine proclaimed: "If it be practicable to transmit power at a commercial profit in these moderate quantities to Albany, the courage of the practical man will not halt there, but inclined to following the daring promise of Nikola Tesla, would be disposed to place 100,000 horsepower on a wire and send it 450 miles to New York in one direction, and 500 miles in the other to Chicago—and supply the wants of these great communities."[36] As we will see, Adams and Rankine were sufficiently impressed with Tesla's technical prowess that they helped him set up a company for the promotion of his wireless-power inventions in 1895 (see Chapter 11).

The Niagara powerhouse began transmitting power to Buffalo in November 1896, and over the next decade Niagara power was running machines across New York State. Excited by the possibilities revealed by the Cataract Company, Rankine launched a second company to build a similar power plant on the Canadian side of the falls. As a result of the success of the Niagara Falls power plant, American and European utilities shifted to polyphase AC; it now forms the standard current distributed in most parts of the world today.[37]

Fascinated by the idea that the natural sublime of Niagara Falls had been supplanted by the technological wonder of AC, American newspapers waxed eloquently about Niagara power and Tesla.[38] Quite understandably people came to think that Tesla, working with the Westinghouse Company, had designed this new system. Though Tesla did not design the system used at Niagara, he nonetheless played a profound, but subtle, role in harnessing the falls. Having articulated the ideal of using polyphase AC to transmit significant amounts of power, Tesla took up the task of shaping the thinking of the key decision maker, Edward Dean Adams. Through his letters and meetings with Adams, Tesla not only provided technical data but mobilized the beliefs and values necessary for Adams to favor AC. Through his correspondence and conversations with Adams, Tesla played a decisive role in getting AC used at Niagara and hence the world.

Though the journalists did not necessarily know how hard Tesla had worked behind the scenes to convince Adams to use polyphase AC, they did recognize that he had introduced into electrical engineering practice the fundamental idea of using polyphase AC to send large amounts of power over distances. Regarding the harnessing of Niagara as "the unrivalled engineering triumph of the nineteenth century," the *New York Times* commented in July 1895 that

[p]erhaps the most romantic part of the story of this great enterprise would be the history of the career of the man above all men who made

it possible . . . a man of humble birth, who has risen almost before he reached the fullness of manhood to a place in the first rank of the world's great scientists and discoverers—Nikola Tesla. . . .

Even now the world is more apt to think of him as a producer of weird experimental effects than as a practical and useful inventor. Not so the scientific public or the business men. By the latter classes Tesla is properly appreciated, honored, perhaps even envied. For he has given to the world a complete solution of the problem which has taxed the brains and occupied the time of the greatest electro-scientists of the last two decades—namely, the successful adaption of electrical power transmitted over long distances.[39]

Thus the success of AC at Niagara played a major role in establishing Tesla's reputation as one of America's leading inventors. Building on his celebrity from Niagara, the Wizard was now ready to introduce an even more remarkable power distribution system.

Through the winter of 1892–93, while he was corresponding with Adams about Niagara, Tesla was also working on his high-frequency apparatus. In doing so, several strands from his recent trip to Europe came together. Lord Rayleigh had told him that he was destined to discover great things, Sir William Crookes had suggested the possibility of using electromagnetic waves to transmit messages, and he had had a moment of insight during the thunderstorm that the forces of the Earth might somehow be harnessed. Weaving these strands together, Tesla decided to see if he could discover a way of using the Earth to transmit both messages and power.

LECTURING IN PHILADELPHIA AND ST. LOUIS

But before he could get too far along with new experiments, Tesla agreed to lecture again, first before the Franklin Institute in Philadelphia on 25 February 1893 and again the following week at the National Electric Light Association in St. Louis. In this lecture, Tesla followed a strategy similar to the one he employed in his performances in London and Paris, offering American audiences both his philosophical musings on the relationship between electricity and light along with sensational demonstrations.[1]

In St. Louis, Tesla lectured in the Exhibition Theater, which seated four thousand, but the hall was packed to suffocation as another several thousand people crowded in, most of whom came hoping to see Tesla's spectacular demonstrations. The demand for seats was so great that tickets were being scalped outside the hall for three to five dollars.[2]

Tesla did not disappoint this huge crowd. In his first demonstrations he allowed 200,000 volts to pass through his body; as he described in the published lecture:

> I now set the coil to work and approach the free terminal with a metallic object [most likely a ball] held in my hand, this simply to avoid burns. As I approach the metallic object to a distance of eight or ten inches, a torrent of furious sparks breaks forth from the end of the secondary wire, which passes through the rubber column. The sparks cease when the metal in my hand touches the wire. My arm is now traversed by a powerful electric current, vibrating at about the rate of one million times a second. All around me the electrostatic force makes itself felt, and the air molecules and particles of dust flying about are acted upon and are hammering violently against my body. So great is this agitation of particles, that when the lights are turned out you may see streams of feeble light appear on some parts of my body. When such a streamer breaks out on any part of the body, it produces a sensation like the pricking of a needle. Were the potentials sufficiently high and the frequency of the vibration rather low, the skin would probably be ruptured under the tremendous strain, and the blood would rush out with great force in the form of fine spray or jet so thin as to be invisible. . . .
>
> I can now make these streams of light visible to all, by touching with the metallic object one of the terminals as before, and approaching my free hand to the brass sphere [connected to the coil's other terminal]. . . . [T]he air . . . is more violently agitated, and you see streams of light now break forth from my fingertips and from the whole hand. . . . The streamers offer no particular inconvenience, except that in the ends of the finger tips a burning sensation is felt.[3]

In the rest of the lecture, Tesla reviewed systematically the different means by which electricity could produce light using effects based on electrostatics, impedance, resonance, and high frequencies. Waving differently shaped tubes in the strong electromagnetic field created by his oscillating transformer, Tesla produced "wonderfully beautiful effects . . . the light of the whirled tube being made to look like the white spokes of a wheel of glowing moonbeams." Near the end of the performance, Tesla held up in his hand one of his phosphorescent bulbs and announced that he would illuminate this lamp by touching his other hand to his oscillating transformer. When this lamp burst into light, recalled Tesla, the audience was so startled that "there was a stampede in the two upper galleries and they all rushed out. They thought it was some part of the devil's work, and ran out. That was the way my experiments were received."[4]

After the lecture, Tesla was mobbed in the lobby as several hundred of St. Louis's leading citizens rushed up to greet him and vigorously shake his hand. Never a fan of crowds, Tesla found the whole episode overwhelming. As the *New York Times* reported, in St. Louis he "had expected a little gathering of expert electricians, and though he went through the ordeal bravely, no power on earth would induce him to try anything like it again."[5]

EXPERIMENTING WITH WIRELESS TRANSMISSION

Although Tesla's 1893 lecture covered many of the same topics as his previous lectures, what was new was that Tesla outlined for the first time his hopes for wireless transmission, remarking that

> I would say a few words on a subject which constantly fills my thoughts and which concerns the welfare of all. I mean the transmission of intelligible signals or perhaps even power to any distance without the use of wires. . . . My conviction has grown so strong, that I no longer look upon this plan of energy or intelligence transmission as a mere theoretical possibility, but as a serious problem in electrical engineering, which must be carried out some day. . . . Some enthusiasts have expressed their belief that telephony to any distance by induction through the air is possible. I cannot stretch my imagination so far, but I do firmly believe that it is practicable to disturb by means of powerful machines

the electrostatic condition of the earth, and thus transmit intelligible signals or perhaps power. In fact, what is there against the carrying out of such a scheme? We now know that electric vibration may be transmitted through a single conductor. Why then not try to avail ourselves of the earth for this purpose?[6]

For Tesla, wireless transmission did not mean using the waves discovered by Hertz but sending power through the earth. As we have seen, he had conducted experiments that revealed how high-frequency currents could be used to power lamps and motors that were connected to his oscillating transformer by only one wire. Since the Earth is a conductor, wondered Tesla, why not ground both the transformer and the lamps and send the current through the earth? In doing so, he would be able to eliminate all of the expensive copper wiring that was being used in the existing electrical networks.

Tesla had intended to say much more about his ideas for harnessing the Earth and had gone so far as to write up an extensive description of his plans for the wireless transmission of power and messages, boldly speculating on its future potential. However, at the last moment he cut his remarks, fearing that he might scare off potential investors. As Tesla later explained, "I had prepared an elaborate chapter on my wireless system, dwelling on its various instrumentalities and future prospects, [but] Mr. Joseph Wetzler and other friends of mine emphatically protested against its publication on the ground that such idle and far-fetched speculations would injure me in the opinion of conservative business men. So it came that only a small part of what I had intended to say was embodied in my address."[7]

While he was careful about what he said in public about transmitting messages and power by disturbing the electrical condition of the earth, Tesla nevertheless got busy in 1893 pursuing this dream. "A point of great importance," wrote Tesla, "would be first to know what is the [electrical] capacity of the earth? and what charge does it contain if electrified?"[8]

To answer these questions, Tesla turned to his new ideal of resonance. Just as it was possible to produce a sound wave at exactly the right frequency so that the waves cause a glass goblet to resonate and then break, Tesla had found it possible to generate electromagnetic waves at a particular frequency and to then create a receiving circuit that would respond—that is, resonate—with that frequency. To create tuned circuits, Tesla tinkered with various arrangements of coils and capacitors, matching the inductance and capacitance in the transmitter and receiver.[9]

Using resonance, Tesla now set out to study how high-frequency currents traveled through the ground, and he returned to the apparatus he had put together in the fall of 1891 (see Figure 7.3). Again, his high-frequency generator and oscillating transformer served as the transmitter. With one terminal of his oscillating transformer grounded to the water mains, he connected the other terminal to "an insulated body of large surface" (what we would call an antenna) on the roof of his laboratory downtown on South Fifth Avenue. The receiver consisted of several capacitors and an electromagnetic relay. When the capacitors were adjusted to the frequency of the transmitted signal, the relay caused a tightly stretched wire to vibrate and produce an audible hum (Figure 10.1). Tesla arranged these components into a wooden box so that he could carry the receiver tucked under his arm.[10]

With the transmitter running in his lab, Tesla carried the receiver throughout Manhattan, stopping periodically to ground the receiver and see if it would hum and detect the oscillating current produced by the transmitter. He would often take the receiver uptown to the Gerlach Hotel and found that he could detect the current there, about 1.3 miles (2.09 kilometers) from his lab.

FIGURE 10.1. Receiver used by Tesla to detect electromagnetic waves in the mid-1890s. From NTM.

However, in transmitting between his lab and the Gerlach, Tesla became frustrated because he often detected no signal at his hotel even though the generator was running just fine at the lab. Tesla found that the generator did not produce waves at a single frequency but instead at several frequencies. In particular, it did not produce oscillations with the same time period, and this made it difficult to tune the receiver to the right frequency. This variation in frequency was caused by slight changes in the speed of the steam engine that drove the alternator.[11]

THE OSCILLATOR

To overcome this problem, Tesla designed a new AC generator. Rather than generate a current by having coils rotate through a magnetic field (as in typical electric generators), Tesla instead built a generator using the reciprocating (or to-and-fro motion) of a piston. His inspiration for this new generator dated back to 1884. Shortly after arriving in America, Tesla had visited the International Electrical Exhibition organized by the Franklin Institute in Philadelphia. At the exhibition, he had played with a thick copper washer with handles so that visitors could move the washer within a strong magnetic field; because the field would induce eddy currents in the washer as it was moved, the visitors would feel mechanical resistance to the way they were manipulating the washer. In moving the washer through the field, Tesla realized that it would be possible to create a generator just by moving a conductor using reciprocating motion.[12]

To do so, Tesla combined a piston engine with generating coils and a magnetic field. As steam or compressed air drove the piston back and forth, a shaft connected to the piston moved the generating coils through the magnetic field (see Figure 10.2). By using high pressure and keeping the stroke of the piston short, Tesla was able to move the coils far more quickly than in a traditional rotating generator and hence produce currents with higher frequencies than were previously possible. Moreover, the oscillations produced were completely isochronous, so much so that Tesla boasted that they could be used to run a clock.[13] Tesla called this new machine his oscillator and he filed patent applications covering

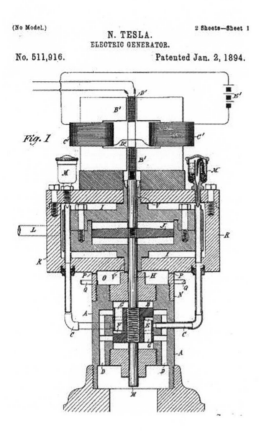

(No Model.)

N. TESLA.
ELECTRIC GENERATOR.

No. 511,916.

2 Sheets—Sheet 1

Patented Jan. 2, 1894.

Fig. 1

FIGURE 10.2. Tesla's oscillator, or combination steam engine and generator.

This device consisted of three units: a generator at the top, an air spring in the middle, and a steam engine at the bottom. All three units were connected to the shaft that runs up the middle. When steam (or compressed air) was introduced into the cylinder of the steam engine, a piston on the main shaft moved upward and pushed a second wider piston, constituting an air spring. The air compressed behind this wider piston created a cushion which eventually pushed back on both pistons, thus reversing the motion of main shaft (center). As this shaft went up and down, it moved the coils of the generator in and out of an electromagnetic field, thus producing a current.

From NT, "Electrical Generator," US Patent 511,916 (filed 19 Aug. 1893, granted 2 Jan. 1894).

several versions in August and December 1893. He unveiled this new invention in a lecture he gave at the Chicago World's Fair.[14]

Tesla found that this new invention provided the precise oscillations he needed for his high-frequency experiments, and he installed one in his South Fifth Avenue laboratory that ran on 350 pounds of pressure. With this oscillator, Tesla could power fifty incandescent lamps, several arc lights, and a variety of motors, and he regularly showed it to visitors to the lab.[15]

Tesla, however, soon realized that his oscillator could also be promoted as a solution to the amount of energy wasted in electrical generating stations. He estimated that barely 5% of the potential energy in coal was actually delivered in the form of light or power to the consumer—the remaining 95% was lost due to the thermal inefficiency of boilers and steam engines, mechanical losses arising using belts to connect engines and generators, and electrical losses on transformers and distribution lines. While other inventors might work on raising the efficiency of each component in an electrical generating system, Tesla preferred to go to the heart of the matter and use as few components as possible to convert steam into electricity. Hence in his oscillator, he strove to eliminate all the extra parts typically found in a steam engine—flywheels, control valves, and governors—so that the oscillator was "bared to the skin like a prize-fighter, every ounce counting." Thanks to his oscillator, Tesla predicted that "we are going to have very shortly a means at hand of producing twice as much electricity from coal as we can produce at the present time."[16]

In making such a claim for his oscillator, Tesla was clearly hoping that he had another major invention that he could sell—that an entrepreneur would buy the patents and develop the oscillator as a product in its own right. However, the engineering community was not necessarily impressed with Tesla's oscillator since the most promising replacement for the reciprocating steam engine in electrical generating plants were the steam turbines already being developed by Charles A. Parsons in England and Gustaf de Laval in Sweden.[17] As rotating machines, these turbines could be directly coupled to existing electric generators, they were probably more efficient than Tesla's oscillator, and, most important, they could be scaled up to deliver power to larger and larger generators. Although Tesla continued to promote his oscillator over the next few years, he might well have heeded the advice of a prominent engineer at the Chicago World's Fair; after hearing Tesla's lecture on the wonders of the oscillator, the engineer said to him, "Well, don't you work on steam engines. You have done some work in electricity. If you stick to it you will do some good work, but if you work on steam engines you are bound to fail."[18]

SPARK GAPS, STOMACH AILMENTS, AND ARTIFICIAL EARTHQUAKES

At the same time that he was working on the oscillator, Tesla conducted another set of experiments related to developing a lighting system. Prior to going to Europe, he had focused much of his high-frequency work on developing new bulbs to replace Edison's inefficient incandescent lamps. To power these new lamps, Tesla planned to replace traditional AC transformers with his new oscillating transformer; in fact, the first patent showing his oscillating transformer was for a system of electric lighting.[19]

But to create a functional lighting system, Tesla had to do something about the spark gap connected to the capacitors. In his oscillating transformer, the spark gap played the vital role of serving as the release mechanism for the capacitors. Initially the spark gap consisted of two polished brass balls placed close to one another. At the start of a cycle of charging and discharging the capacitors, no current could jump the gap between the balls and so electrical charge accumulated in the capacitors. Once the charge was sufficiently high, the built-up charge would ionize the air between the brass balls and a spark would jump across the gap. As the current rushed across the gap in the form of a spark, electromagnetic waves radiated from the circuit. Once the charge was depleted from the capacitors, the spark would be extinguished, and the cycle would begin again. Of course, to produce waves with high frequency, this charge-discharge cycle would occur thousands of times per second.

As he studied the ordinary spark gap—like the one used by Hertz—Tesla realized that the train of electromagnetic waves produced by the transmitter was erratic because at times the air between the brass balls would remain ionized and a current would pass continuously in the form of an electric arc. This arcing was undesirable because it meant that the current was flowing through the circuit instead of building up in the capacitors. Hence, to create a more regular train of waves—and even increase the frequency or number of waves produced—it was necessary to carefully control the conditions in the spark gap.[20]

To get a more regular train of electromagnetic waves from his transmitter, Tesla tried a variety of devices in place of the spark gap. Since the spark could be extinguished when a strong permanent magnet was brought near it, he fashioned a controller in which the spark gap was at right angles to a horseshoe magnet. Tesla next tried using several gaps between adjustable wheels (later known as a quenched spark gap). He experimented with replacing the air in the spark gap with a gas such as hydrogen that would ionize easier and hence allow sparks to jump more frequently; later patented by the Danish wireless pioneer, Valdemar Poulsen, this design became known as the Poulsen arc.[21]

Tesla also tried to regulate the waves produced by his transmitter by using a mechanical oscillator that was similar to the combination steam-engine-and-electric-generator described above. In this device, he initially employed very strong steel springs that required several tons of force to compress, and he carefully delivered that force using a piston driven by steam or compressed air. However, as he increased the steam or air pressure in order to get high rates of vibration, Tesla found that the steel springs broke and so he replaced them with an air spring in which the piston was driven back when a column of air was compressed and then released.[22] (See Figure 11.2 for a more complete description of how the air spring worked.)

Although this mechanical oscillator was not particularly suited for regulating the transmitter in Tesla's wireless lighting system, it nonetheless fascinated him. As he recalled in the 1930s,

> I had installed . . . one of my mechanical oscillators with the object of using it in the exact determination of various physical constants. The machine was bolted in a vertical position to a platform supported on elastic cushions and, when operated by compressed air, performed minute oscillations absolutely isochronous, that is to say, consuming rigorously equal intervals of time. . . . One day, as I was making some observations, I stepped on the platform and the vibrations imparted to it by the machine were transmitted to my body. The sensation experienced was as strange as agreeable, and I asked my assistants to try. They did so and were mystified and pleased like myself.

Suddenly, however, Tesla and his assistants had to respond to a call from nature: "[S]ome of us, who had stayed longer on the platform, felt an unspeakable and pressing necessity which had to be promptly satisfied," namely that they needed to rush to the bathroom and have a bowel movement. However, never one to miss a new opportunity, Tesla saw an application for his invention here, as he realized that the rapid oscillations were helping move food more quickly through the intestines and that the vibrating platform might serve to cure digestive ailments. "A stupendous truth dawned upon me" and as a result, "When I began to practice with my assistants MECHANICAL

THERAPY we used to finish our meals quickly and rush back to the laboratory. We suffered from dyspepsia and various stomach troubles, biliousness, constipation, flatulence and other disturbances, all natural results of such irregular habit. But after only a week of application, during which I improved the technique and my assistants learned how to take the treatment to their best advantage, all those forms of sickness disappeared as by enchantment and for nearly four years, while the machine was in use, we were all in excellent health."[23]

Along with his assistants, Tesla also invited visitors to the laboratory to try his mechanical therapy, including Mark Twain. Tesla had read Twain's books growing up in Serbia, and he probably met Twain either dining at Delmonico's or at his club, The Players (see Chapters 11 and 12).

For his part, Twain was interested in Tesla as a result of his involvement with an inventor named James W. Paige, who developed an automatic typesetting machine. Twain was fascinated by the possibility of a machine that could set the type for books and newspapers and believed that the machine would be worth millions. He had first heard about Paige's invention in 1880, and he promptly invested $5,000 to develop it. By 1887 he had invested a total of $50,000 and was pouring about $3,000 a month into the typesetter. But along with duplicating the actions of a human typesetter, Paige intended that his machine would be powered by an electric motor, and he convinced Twain in 1887 to invest $1,000 specifically in developing a motor. "We tried a direct current—& failed," noted Twain. "We wanted to try an alternating current, but we lacked the apparatus." Hence Twain was quite excited to learn that Tesla had perfected an AC motor, and he had noted in his journal in November 1888, "I have just seen the drawings & description of an electrical machine lately patented by a Mr. Tesla, & sold to the Westinghouse Company, which will revolutionize the whole electric business of the world. It is the most valuable patent since the telephone. The drawings & description show that this is the *very* machine, in every detail which Paige invented nearly four years ago."[24]

Sometime in the early 1890s, Twain became a regular visitor to Tesla's lab, where he tried Tesla's mechanical cure. "He came to the laboratory in the worst shape suffering from a variety of distressing and dangerous ailments," recalled Tesla, "but in less than two months he regained his old vigor and ability of enjoying life to the fullest extent."[25]

Tesla conducted another experiment with his mechanical oscillator a few years later at his laboratory on Houston Street in order to understand resonance, and the result was an artificial earthquake. By now he had developed a much smaller version that "you could put in your overcoat pocket." "I was experimenting with vibrations," he explained, and

> I had one of my machines going and I wanted to see if I could get it in tune with the vibration of the building. I put it up notch after notch. There was a peculiar cracking sound.
>
> I asked my assistants where did the sound come from. They did not know. I put the machine up a few more notches. There was a louder cracking sound. I knew I was approaching the vibration of the steel building. I pushed the machine a little higher.
>
> Suddenly all the heavy machinery in the place was flying around. I grabbed a hammer and broke the machine. The building would have been down about our ears in another few minutes. Outside in the street there was pandemonium. The police and ambulances arrived. I told my assistants to say nothing. We told the police it must have been an earthquake. That's all they ever knew about it.[26]

A SYSTEM FOR WIRELESS LIGHTING

But while it was fun to play with the mechanical oscillator, it was not the solution that Tesla needed in order to perfect his wireless lighting system. Consequently, in 1893, he devised yet another substitute for the spark gap by inserting a fan or turbine spinning between the terminals of the spark gap; sparks would jump between the stationary terminals and the turbine blades, and the sparks would be shorter and quicker as a result of the turbine spinning rapidly. To minimize any chance of arcing between the blades and the terminals, Tesla immersed the entire spark gap in an oil tank; by pumping oil through the tank, the flow of the oil turned the turbine blades that interrupted the sparks. Using this circuit controller, Tesla was able to produce frequencies in the range of 30,000 to 80,000 cycles per second.[27]

Tesla was quite proud of this new circuit controller, and he filed a patent application for it in August 1893. However, he did not appreciate its full potential for transmitting power and messages without wires until he demonstrated it for the great German physicist Hermann von Helmholtz at the Chicago World's Fair. After showing him this invention and describing his hopes for using it for wireless transmission, Tesla asked Helmholtz, "Excellency, do you think that my plan is realizable?" to which the distinguished scientist replied, "Why, certainly it is, but first you must produce the apparatus."[28]

Encouraged by Helmholtz, Tesla redoubled his efforts to understand what was occurring as the capacitors in his oscillating transformer were rapidly charged and discharged by the circuit controller. As he studied this phenomenon, Tesla realized that they functioned as the electrical analog to a pile driver. Just as more energy is stored as the heavy weight in the pile driver is raised higher and higher, one can control the capacitors in order to store more energy with each charge-and-discharge cycle. And just as one releases the pile driver suddenly so that it delivers all its stored-up energy with one downward blow, one can discharge a capacitor in a very short interval and deliver a tremendous amount of electrical energy. "For instance," explained Tesla, "if the engine [used to power the AC generator] is of 200 horsepower, I take the energy out for a minute interval of time, at a rate of 5,000 to 6,000 horsepower, then I store [it] in a condenser and discharge the same at the rate of several millions of horsepower."[29]

To convey to visitors how much energy could be concentrated by the capacitors in his oscillating transformer, Tesla would sometimes pass through his apparatus

> energy at a rate of several thousand horsepower, put a piece of thick tinfoil on a stick, and approach it to that coil. The tinfoil would melt, and would not only melt, but while it was still in that form, it would be evaporated and the whole process took place in so small an interval of time that it was like a cannon shot. Instantly, I put it there, there was an explosion. That was a striking experiment. It simply showed the power of the condenser [i.e., capacitor], and at that time I was so reckless that in order to demonstrate to my visitors that my theories were correct, I would stick my head into that coil and I was not hurt; but I would not do it now.[30]

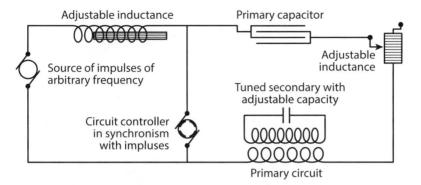

FIGURE 10.3. Circuit used by Tesla to deliver wireless power to his lamps in his South Fifth Avenue laboratory circa 1894.

In practice, the primary shown above was a thick cable around the perimeter of the main room and the secondary was the moveable coil shown in Figure 10.4.

Redrawn from NT, "True History of the Wireless," p. 29.

As Tesla came to appreciate how he could use his oscillating transformer to concentrate power, he realized that he could now modify his distribution circuits in order to transmit power across a room and light lamps without any wires. In his previous patents, Tesla had connected his oscillating transformer to a second transformer that in turn distributed the power to his lights. But now thanks to the increased concentration of power, he could separate the two windings in this second transformer and transmit power between them even if they were separated by ten or twenty feet (Figure 10.3).

On the transmitting side of this new wireless lighting system, Tesla now used his oscillator to charge a bank of capacitors that in turn were connected to a large cable that ran around the perimeter of the main hall (which was 40 × 80 feet) of the lab. Effectively a coil with just one turn of wire, this large cable took the place of the primary winding of the old second transformer. On the receiving end, Tesla used a three-foot-tall coil in place of the secondary winding of the old second transformer (Figure 10.4). Mounted on casters, this receiving coil could be rolled about the main hall to see where it worked best. Even more important, it could be adjusted so as to resonate to the frequency generated by the transmitter. Instead of the energy being transmitted between two plates—as in his lectures at Columbia and the Royal Institution—the energy now moved wirelessly between two coils.[31]

FIGURE 10.4. The receiving coil for Tesla's resonant transformer, as used in the South Fifth Avenue laboratory circa 1894.
From TCM, "Tesla's Oscillator and Other Inventions," *The Century Magazine* 49:916–33 (April 1895), Fig. 10.

Tesla had this system perfected by February 1894 and showed it off in spectacular private demonstrations to friends, select professional groups, and a few journalists.[32] One of these lucky reporters described his experience in the following way:

> I glanced about in some bewilderment after hearing Mr. Tesla say that he had a surprise in store for me. Promptly suiting the action to the word, he called in several employees from the workshop and issued a succession of hurried orders which I followed but vaguely. Presently, however, the doors were shut and the curtains drawn until every chink or crevice for the admission of light was concealed, and the laboratory was bathed in absolutely impenetrable gloom. I awaited developments with intense interest.
>
> The next minute exquisitely beautiful luminous signs and devices of mystic origin began to flash about me with startling frequency.

Sometimes they seemed iridescent, while again a dazzling white light prevailed.

"Take hold," said a voice, and I felt a sort of handle thrust into my hand. Then I was gently led forward and told to wave it. On complying, I spelled the word "Welcome" flaming before my eyes. Unfortunately, I was totally unable at the time to appreciate the kindly sentiment implied.

A hand approached mine ere I had quite recovered, and I felt the tips of my fingers lightly brushed. Fancy my dire dismay when I immediately experienced an acute tingling sensation, accompanied by a brief pyrotechnic display that was surprising to say the least. When the daylight as well as my equanimity was in a measure restored, I learned something of the meaning of these wondrous experiments, which may be said to foreshadow in a way the electric light of the future.[33]

As the reporter subsequently learned, the "devices of mystic origin" he saw lit up in the dark were but a sampling of the many lamps that Tesla had devised: some were tubes with gases at a low pressure and some had phosphorescent coatings (like modern fluorescent tubes), but none had filaments (Figure 10.5). Tesla believed that the oscillations—or, as he often said, the electrostatic "thrusts"—emanating from his transmitter loaded energy onto the ether between the two coils. Because there were relatively few gas molecules inside the tubes, the molecules were easily agitated by the "thrusts" and caused to glow. Since neither the transmitter nor the receiving coil was grounded, the energy not through the Earth but rather by means of the "thrusts" or electromagnetic waves was moving through the ether.

Lest one think that his wireless lighting scheme would work only with these new tubes, Tesla also demonstrated how ordinary incandescent lamps could be powered by his apparatus. To do so, he connected a standard sixteen candlepower lamp to the terminals of the resonating coil in the center of the room, and it was powered wirelessly by energy generated by the transmitter (see Figure 10.4).

In these increasingly elaborate demonstrations, Tesla was becoming more and more confident about the potential of his wireless lighting system and that it could compete with the reigning Edison

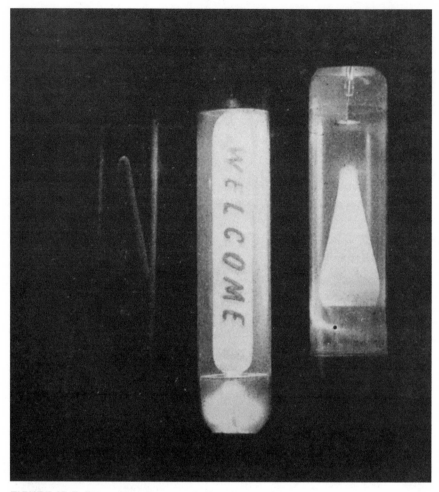

FIGURE 10.5. "Three phosphorescent bulbs under test for actinic value, photographed by their own light."

From TCM, "Tesla's Oscillator and Other Inventions," *The Century Magazine* 49:916–33 (April 1895), Fig. 5.

incandescent system. Indeed, Tesla might well have been thinking about his wireless lighting system when he wrote to his uncle in November 1893, "I have just completed a new invention that is great! Success is wonderful to me in every aspect except money. That will come soon."[34]

CHAPTER ELEVEN
EFFORTS AT PROMOTION
[1894–1895]

It is difficult to give you an idea [of] how I am respected
here in the scientific community. I received many letters
from some of the greatest minds proposing that I stay
the course. They say that there are enough educated men
but few with ideas. They inspire me instead of taking me
away from my work. I [have] received many awards and
there will be more. Think how things are that I recently
received a photograph from Edison with the inscription,
"To Tesla from Edison."

TESLA TO HIS UNCLE PETAR MANDIC, DECEMBER 1893

By 1894, based on his investigation of high frequencies, Tesla had
worked out a new system of electric lighting as well as a new steam-
powered oscillator. It was now time to start promoting these inventions in order to attract new customers for the patents he had secured.
Just as he had done with Peck and Brown five years earlier with the
AC motor, Tesla set out to shape the expectations—the illusions—that
people would have for these new inventions. To do so, Tesla decided

that he needed to establish his reputation as a leading electrical inventor
(Figure 11.1).

It is easy to assume that reputations in technology and science grow
automatically out of an individual's accomplishments. However, reputations,
like everything else in technology and
science, are actively constructed by
individuals and groups. In particular,
individuals draw on the resources of
their time and place—their culture—to
construct a reputation that makes them
credible.[1]

FIGURE 11.1. Tesla circa 1894–95.
From NTM.

During the last quarter of the nineteenth century, it was especially
difficult to construct a credible reputation as an inventor or scientist in
America. In the decades following the American Civil War, there was a
tremendous sense of uncertainty about professions and social roles—
that anyone who wanted to be a lawyer or doctor or engineer could
simply announce that they were a practitioner in one of these fields
and set up a business. No one was quite sure what the rules were: Did
one need a college degree or other credential? Should one belong to
a professional organization? Did one need to publish papers or demonstrate their expertise in some other way? What did it mean to be an
authority and what did it mean to be famous?[2]

To be sure, organizations in a variety of fields moved quickly to
establish standards and clarify the role of the professional in their discipline, but it took from the 1870s to the 1910s for this work to be
completed.[3] In the meantime, in these fluid decades, individuals were
free to experiment with how they shaped their professional personae,
drawing on different elements of American culture.

In this chapter we will look at how Tesla, with the help of his
friends, shaped his reputation. He now cultivated an image of being
a brilliant, even eccentric, genius. Tesla delighted in showing off his
wireless lamps, and after dinners at Delmonico's he would invite ce-
lebrities to late-night demonstrations in his laboratory. Just as newspaper reporters had covered Edison's exploits at Menlo Park in the

1870s, they flocked to Tesla's laboratory in the 1890s to cover his sensational discoveries. Like Edison, Tesla delighted in telling lively stories and promising great results for his inventions.

T. C. MARTIN AND THE BOOK

Tesla's efforts at promotion were strongly shaped by his friendship with Thomas Commerford Martin (1856–1924), the editor of *Electrical Engineer*, one of the leading weekly electrical journals. Martin functioned as Tesla's publicity manager in the mid-1890s and did more than anyone else to help Tesla establish his reputation.

Born in England, Martin spent part of his boyhood traveling aboard the massive steamship *Great Eastern* while his father helped lay the transatlantic telegraph cable. After studying theology, Martin immigrated to the United States to work with Edison at Menlo Park. Noticing that Martin had a gift for writing, Edison encouraged the young Englishman to publish stories about the telephone and phonograph in the New York newspapers. In 1882 he became an editor at the telegraph journal *The Operator*, which was soon renamed *Electrical World*. Along with his editorial work, Martin helped found the American Institute of Electrical Engineering in 1884 and served as the institute's president in 1887–88.[4]

As we have seen, Martin first became acquainted with Tesla's work in April 1888 when he was invited to see a demonstration of Tesla's AC motor in the Liberty Street laboratory (see Chapter 5). A few weeks later, he arranged for Tesla to lecture before the AIEE. Knowing that AC motors represented a breakthrough for the utility industry, Martin and his fellow editor Joseph Wetzler published a book in 1889 on the development of electric motors.[5]

Martin watched with interest the growing public excitement stimulated by Tesla's lectures on wireless lighting. In particular, Martin must have noticed that prior to Tesla's lecture before the National Electric Light Association (NELA) in St. Louis, a young entrepreneur had printed up extra copies of the NELA bulletin containing a biographical article on Tesla and that he sold four thousand copies on the street before Tesla spoke. As the *New York Herald* reported, this was "something unprecedented in the history of electrical journalism."[6] If a pamphlet had sold so well, why not a book about Tesla?

Consequently, in the spring of 1893, Martin and Tesla started planning a book that would bring together Tesla's lectures, a description of his inventions, an outline of his work on high-frequency lighting, and a biographical sketch. After a nasty contract dispute with the publisher of *Electrical World*, Martin had left in 1890 to edit the rival *Electrical Engineer*, and he undoubtedly hoped that the Tesla book would help draw readers away from *Electrical World* and over to his paper.[7] Although Tesla could rely on Martin as "one of the best writers in the technical field," he found the process of composing a book in English to be difficult; nevertheless, he understood that the book was essential to establishing his reputation. As he explained to a cousin back home in Serbia: "In addition to all my work, I am going to write a book. It is going to be unusually evil [Tesla probably meant difficult or demanding]. I intend to announce various apparatus and experiments that are going through my head after some years of work. I have compiled everything from what I have read in magazines and what is new. It can hurt me or possibly help me. My ambition is to come out not as a technician but as an inventor."[8]

Published under the title *The Inventions, Researches, and Writings of Nikola Tesla*, the book appeared in January 1894. At Martin's suggestion, it was dedicated to Tesla's fellow countrymen in eastern Europe. In its review, the *New York Times* noted that the assembly of the materials for the volume must have been "by no means [an] easy task" but that Tesla and Martin had successfully accomplished it.[9]

Thrilled to have a book summarizing his work, Tesla eagerly sent copies to his family in Serbia, friends, and his former associates at Westinghouse. Hoping to sell the book rather than give it way, Martin told Tesla that "[y]our request [for more free copies] is just too hard. It seems to me that the Pittsburgh boys, if they love you, ought to be willing to blow a little money of their own on the book." Since Tesla was so keen to give out copies, Martin suggested that "[p]erhaps you would like to make us a bid on the whole edition." Hoping to make up the losses by selling Tesla's signature, Martin closed by saying in jest, "When you write me, make it autographic as often as you can. People are beginning to deplete my stock."[10]

The Inventions, Researches, and Writings proved popular. The first edition sold out within a month, a second edition by the end of 1894, and a third edition was released in February 1895. It was favorably reviewed in Europe as well as America, and a German edition appeared in 1895. While the book earned a tidy sum for Martin, Tesla convinced his editor to lend him the proceeds from the book but he never repaid Martin. As a result, grumbled Martin decades later, "Two years of work went for nothing."[11]

Although Martin complained later, in 1894 he regarded Tesla as an up-and-coming scientific celebrity and did everything he could to help promote this new wizard. In February he escorted Tesla to the swanky New York apartment of Gianni Bettini, an early audiophile who had made several improvements to Edison's cylinder phonograph. Drawing on his social connections, Bettini recorded the voices of opera singers, presidents, and popes, and Martin was anxious to have Tesla's voice in Bettini's collection.[12] Martin also arranged a sitting for Tesla before a sculptor named Wolff, and he put Tesla in contact with S. S. McClure, who was looking for contributors for the eponymous magazine he had just started. Tesla and McClure dined together, and as Martin reported back to his protégé, McClure "knows now personally . . . that you are a great man and a nice fellow." Although McClure was keen for a contribution, Tesla declined: "Much as I would like to comply with your amiable request, I find it impossible at present, as every moment of my time is taken up by work which I must finish without delay."[13]

"THE FILIPOVS": ROBERT AND KATHARINE JOHNSON

But the most important thing that Martin did for Tesla was introduce him to Robert Underwood Johnson (1853–1937) and his wife, Katharine (d. 1924) (Figures 11.2 and 11.3), in the fall of 1893. Socially prominent and gracious, Robert and Katharine became the inventor's closest friends.

Born in Washington, D.C., Johnson grew up in Indiana. As a teenager, he worked as a telegraph operator and sometimes received messages from another operator named Thomas Edison. Johnson joined

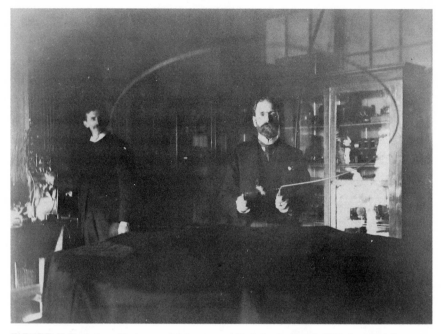

FIGURE 11.2. Robert Underwood Johnson and Tesla in the South Fifth Avenue laboratory. From NTM.

the staff of the popular magazine *Scribner's Monthly* in 1873 and, on occasion, visited Edison at Menlo Park in order to cover his inventions. In 1881, when *Scribner's* became *The Century Magazine*, Johnson was named associate editor, and he served as chief editor from 1909 to 1913. To boost *The Century*'s circulation, Johnson convinced Ulysses S. Grant to write a series of articles about Civil War campaigns. With the help of Mark Twain, he then convinced the general to write his memoirs, which subsequently became a bestseller. Johnson married Katharine McMahon of Washington, D.C., in 1876, attracted to her red hair, Irish ancestry, and fiery personality.[14]

Tesla immediately hit it off with the Johnsons, and in December 1893 he invited them to join him at the premiere of Antonín Dvořák's *New World Symphony*. "Upon receipt of your first note," wrote the Wizard to Robert, "I immediately secured the best seats I could for Saturday. Nothing better than the 15th row! Very sorry, we shall have to use telescopes. But I think the better for Mrs. Johnson's vivid imagination. Dinner at Delmonico's." Katharine reciprocated by sending Tesla flowers on 6 January, the day Orthodox Christians celebrate

Christmas. "I have to thank Mrs. Johnson for the magnificent flowers," Tesla
penned Robert. "I have never as yet
received flowers, and they produced
upon me a curious effect." In return,
Tesla sent Katharine a Crookes radiometer that he regarded "from the scientific viewpoint [as] the most beautiful
invention made."[15]

In the years that followed, the Johnsons regularly invited Tesla to dinners
and parties at their townhouse at 327
Lexington Avenue. There, Tesla met a
host of social and intellectual luminar-
ies including the sculptor August Saint-
Gaudens, the naturalist John Muir, the

FIGURE 11.3. Katharine Johnson.
From NTM.

children's author Mary Mapes Dodge, the pianist Ignace Padrewski,
and the writer Rudyard Kipling. At one soiree, an English lady turned
to the inventor and asked, "And you, Mr. Tesla, what do you do?"
"Oh, I dabble a little in electricity," responded Tesla. "Indeed," came
the reply. "Keep at it and don't be discouraged. You may end by doing
something someday."[16]

Knowing that Johnson was a poet, Tesla recited Serbian poetry for
him, including Jovan Jovanović Zmaj's poem "Luka Filipov." In this
ballad, Zmaj recounts the heroic deeds of Luka and his death in an
1874 battle between the Serbs and the Turks. Enthralled, Johnson had
Tesla prepare English translations of this and other poems by Zmaj for
The Century and he included "Luka Filipov" in his anthology, *Songs of
Liberty*. Thereafter, Tesla always referred to Robert as Luka and Katharine as Madame Filipov.[17]

Since Tesla could not entertain the Filipovs in his bachelor quar-
ters at the Gerlach, he had them visit his laboratory on South Fifth
instead. "We were frequently invited to witness his experiments," recalled Robert, in which "[l]ightning-like flashes of electrical fire of the
length of fifteen feet were an every-day occurrence, and his tubes of
electric light were used to make photographs of many of his friends
as a souvenir of their visits."[18] But rather than let the photographs
just be souvenirs for friends, wondered Johnson, why not have special

FIGURE 11.4. "First photograph ever taken by phosphorescent light. The face is that of Mr. Tesla, and the source of the light is one of his phosphorescent bulbs. Time of exposure, eight minutes. Date of photograph January 1894."

From TCM, "Tesla's Oscillator and Other Inventions," *The Century Magazine* 49:916–33 (April 1895), Fig. 3.

pictures taken using one of Tesla's inventions and publish them as a "first" in *The Century*? In particular, Johnson was intrigued by Tesla's efforts to use "phosphorescent light"—what we would now call fluorescent light—for photographic purposes (Figure 11.4).

Having just published a biographical sketch of Tesla written by Martin, Johnson suggested to Martin that he might write another story for *The Century* about Tesla that would include portraits taken under this new light source.[19] Martin readily agreed but recommended that they take precautions lest news of the photographs leaked out. "I will lock [them] up or put [them] in a safe deposit vault, if you wish until the hour of publication," Martin told Johnson. "But I want to get one of the *first* as a historical souvenir." Martin realized that the most likely source of a leak would be Tesla himself and that he would have to coach his protégé; as he wrote to Tesla, "I think that we ought to have a little talk about giving to the daily newspapers a hint that [you have] succeeded in taking photos by phosphorescence. It will leak out some hour and then someone . . . with customary arrogance [will place it] in the papers. . . . [We need] to get our priorities established. I think R. U. Johnson feels the same way."[20]

Heeding Martin's advice to keep quiet about the pictures, Tesla got to work preparing the necessary electrical apparatus. As he wrote to Johnson in February 1894,

> I have been hard at work to-day replaning and adjusting. I think we can make some trials to-morrow.
>
> I have prepared a tube for you and expect that it will show up well. . . . We may try to get a photo of the magnificent profile of Mr. Clemens. I have not yet communicated with the photographers because I have to try something in the morning. I shall immediately let them know if everything is all right. The best time to come would be 4 PM.[21]

The photographers were from Tonnele & Company who had done previous work for *The Century*.[22] Along with Mark Twain, Johnson invited the actor Joseph Jefferson and the novelist Francis Marion Crawford to pose in the photographs.[23] Tesla had each guest participate by holding a large loop of wire in their hands. When the loop was placed over the resonating coil in the center of the laboratory (see Figure 10.4), enough current was wirelessly transmitted from the coil to the loop so as to illuminate two or three lamps placed between the visitor's hands (Figure 11.5). "Strange as it may seem," observed Martin, "these currents, of a voltage one or two hundred times as high as that employed in electrocution, do not inconvenience the experimenter in the slightest. The extremely high tension of the currents which Mr. Clemens is seen receiving prevents them from doing any harm to him."[24]

When the prints came back from Tonnele, Tesla was thrilled with all of them, but he especially liked one of Jefferson, finding it "simply immense. I mean the one showing him alone in the darkness. I think it is a [work] of art." To celebrate this successful project, Katharine Johnson proposed a dinner at Delmonico's before she and Robert left for summer vacation at the Hamptons on Long Island. While Tesla was becoming cautious about accepting every invitation that came his way,

FIGURE 11.5. Mark Twain in Tesla's laboratory, 1894. Tesla is in the background to the left. From TCM, "Tesla's Oscillator and Other Inventions," *The Century Magazine* 49:916–33 (April 1895), Fig. 13.

he could not resist an evening with his favorite couple: "Even dining at Delmonico's is too much of a high life for me and I fear that if I depart very often from my simple habits I shall come to grief. I had formed the firm resolve not to accept any invitations, however tempting; but in this moment I remember that the pleasure of your company will soon be denied to me (as I am unable to follow you to East Hampton where you intend *camping out* this summer)—an irresistible desire takes hold of me to become a participant of that dinner."[25] As they had planned, Johnson and Martin waited for the right moment to release the photographs, and they did not appear in *The Century* until April 1895.[26]

NEWSPAPER NOTORIETY AND HONORARY DEGREES

Through 1894, Tesla began to enjoy more newspaper coverage, most likely due to the efforts of Martin and Johnson, who had connections at various papers. James Gordon Bennett's *New York Herald* had been covering Tesla's exploits for several years, but the Wizard was now a subject of feature articles in Joseph Pulitzer's *New York World* (as recounted in the introduction), the *New York Times*, and the *Savannah Morning News*. Indeed, since Tesla was so popular, some reporters resorted to writing bogus stories without bothering to interview the inventor. "For example," reported Martin, "one vivid young lady of the press, in her anxiety to be instructive, went so far as to depict herself undergoing a brilliant electrical ordeal that is possible only with the body entirely naked." Such an episode, Martin assured his readers, had never happened and was highly unlikely given how reticent Tesla was around women (see Chapter 12).[27]

Illustrated with his portrait, these feature stories recounted Tesla's early years, described his physical appearance, and commented on his style of invention. In talking to reporters, Tesla was much like a modern-day professional athlete who often tries to strike a balance between boasting about his or her ability (which is, after all, the purpose of the interview) and displaying some modesty about his accomplishments. "It is an embarrassment to me," Tesla told one reporter, "that my work has attracted as much public attention, not only because I believe that an earnest man who loves science more than all should let his work speak for him . . . but because I am afraid that some of the scientists whose friendship I value very much suspect me of encouraging newspaper notoriety."[28]

While his professional colleagues were undoubtedly suspicious that Tesla was courting the press, they were nonetheless impressed with him as a pathfinder, as someone who was challenging them to rethink the nature of electricity and possible new applications. When the Franklin Institute in Philadelphia presented Tesla with the Elliott Cresson Gold Medal in 1893 for his work on new forms of electric lighting, it noted that he had "developed a new and very important field of research in that direction in which little had been done before, and one which opened the way to very valuable results, the most important of which is . . . the rational generation of artificial or 'cold light' as it is often called. Although he has not yet solved the problem from a commercial standpoint, he has . . . opened a probable way for the solution of this most important and difficult problem."[29]

The Cresson Medal was followed by honorary degrees. Several universities were anxious to recognize Tesla since he had been "covered with honors while in England and France" and it would be an embarrassment not to recognize "a man who lives under our very eyes." One invitation came from the University of Nebraska, but Tesla was probably not interested in taking time away from his work to make the long train trip from New York to Lincoln. "I have urged him to accept," complained Martin to Johnson, and "I want you and Mrs. Johnson to bring your influence on him also. Her spell is now a potent one, I fancy, with him, so far as any woman's can be, next [to] his sisters."[30]

Although neither Martin nor Katharine could persuade Tesla to accept the honorary degree from Nebraska, Robert did write on Tesla's behalf to Columbia University:

> I think it may truly be said that there are few men . . . whose work promised more for the amelioration of the hard conditions of the poorer classes. Having seen a great deal of Mr. Tesla during the last six months . . . I have never heard a subject of scientific [importance] mentioned in his presence upon which he did not seem to be thoroughly well informed. As you undoubtedly know, he is on terms of intimate friendship with Crookes, Helmholtz, Lord Kelvin & others. Hertz was his friend. . . .
>
> As to his general culture, I may say that he knows the language and is widely read in the best literature of Italy, Germany and France as well as much of the Slavic countries to say nothing of Greek and Latin. He is particularly fond of poetry and is always quoting Leopardi or Dante or Goethe or the Hungarians and Russians. I know of few men of such diversity of general culture or such accuracy of knowledge.

Such a portrayal would have made Tesla's father proud, and Johnson closed by describing Tesla's character as "one of distinguished sweetness, sincerity, modesty, refinement, generosity, and force." Convinced by Johnson's recommendation, Columbia presented Tesla with an honorary doctorate in June 1894, and this was followed by another from Yale University.[31]

PUTTING THE PATENTS UP FOR SALE:
THE NIKOLA TESLA COMPANY

With growing coverage in both newspapers and the technical press as well as medals and degrees to show off, it was time to take the next step in a promotion strategy and form a company that would market and license his patents. To create this company, Tesla turned to Edward Dean Adams. As we saw in Chapter 9, Adams was the driving force behind the promotion of hydroelectric power at Niagara. At a critical moment in 1893 when his company had to decide between using AC or DC for Niagara, Adams had followed Tesla's advice.

After visiting the lab and seeing several demonstrations, Adams agreed to promote Tesla's latest inventions, and together they launched the Nikola Tesla Company in February 1895. Adams fancied himself as being eminently capable of helping Tesla; as his biographer gushed, the financier had assisted several "struggling geniuses who afterward saw [with] clearer eyes and performed with greater zeal and skill because they took counsel with Edward Dean Adams."[32]

Since this new company was going to promote not only Tesla's recent high-frequency patents but also those assigned earlier to Peck and Brown, Adams and Tesla included Alfred Brown as a director in the company. In addition, they invited another Niagara promoter, William Rankine, as well as Charles F. Coaney of Summit, New Jersey, to serve as directors. Adams hoped that his son Ernest, who had just published a story about Tesla, would join the company after he finished studying engineering at Yale.[33]

The Nikola Tesla Company planned to "manufacture and sell machinery, generators, motors, electrical apparatus, etc.," and the directors planned to issue stock to capitalize it at $500,000.[34] If fully subscribed by investors, this level of capitalization would have certainly provided Tesla with the funds he would need to develop his high-frequency inventions fully. However, it would not have been sufficient to undertake manufacturing on a commercial scale. Hence, despite the claim that it was going to manufacture electrical apparatus, it would appear that the Nikola Tesla Company was part of a promotion-oriented strategy. Once Tesla's lighting system and oscillator had been perfected, then either the patents or the entire company could be sold; this is what Tesla had done with the European rights to his motor patents in 1892.[35] Although Adams eventually invested about $100,000 in Tesla's work, he most likely saw himself not as an investor but as a promoter—someone who made his fortune by combining Tesla's technology and other people's money into an attractive enterprise.[36] This is what Adams's career on Wall Street was all about—he was a master at reorganizing railroads and other companies so that people would invest in them.

Together, Tesla and Adams waited for other investors to join the Nikola Tesla Company, but no one took up their offer. Why were there no takers for Tesla's inventions circa 1895?

In large measure, Tesla was stymied by business conditions. In the five years following the Panic of 1893, the American economy was in a recession. During the mid-1890s, neither the existing electrical manufacturers nor utility companies were especially profitable.[37] If the companies using Edison's DC incandescent lighting system or Westinghouse's AC power equipment were not earning money, why should investors take a chance on Tesla's next-generation technology of high-frequency AC?

While part of the problem was the recession, another part of the problem was Tesla himself. Having launched the Nikola Tesla Company for the purpose of selling or licensing his inventions, the next step was to show that these inventions could readily be converted into commercially feasible products. In this phase—commonly called development—the inventor has to know when to move from generating lots of alternative designs to focusing on perfecting the most promising version. In other words, the inventor needs to shift from divergent to convergent thinking.[38] For both geniuses and mere mortals, divergent thinking is much more fun than convergent thinking; it's a lot more enjoyable to dream up new alternatives than to face the difficulties associated with making a device reliable, efficient, and cost-effective.

Tesla, I suspect, had a genuine problem making the shift from divergent to convergent thinking. "A notable faculty of Tesla's mind is that of rushing intuition," observed one reporter. "You begin to state a question or proposition to him and before you have half formulated it he has suggested six ways of dealing with it and ten of getting around it."[39] In the mid-1890s, Tesla seems to have put off doing the essential work of development. In his lectures, he was never satisfied with demonstrating just a few of the most promising versions of his lamps; he felt compelled to show a dozen variations. Moreover, every few months Tesla would let reporters visit his laboratory so they could write up his latest discovery. Tesla may very well have thought that sheer variety conveyed the power of his genius, but it actually sent the wrong message to investors. If they were going to risk capital on an inventor and his patents, investors needed to feel confident that the inventor was willing to get down to the nitty-gritty of creating a marketable product.

Tesla's business partners also had an impact on his failing to shift from divergent to convergent thinking at this point. In the development of the AC motor, Tesla had relied heavily on Peck for guidance about how to patent, promote, and ultimately sell his invention. Unfortunately, Peck died unexpectedly in 1890, just as Tesla was starting his work on high-frequency AC. While Tesla's other early business partner, Brown, was involved with the Nikola Tesla Company, he does not appear to have had any significant input into the development of Tesla's later inventions. Both Adams and Rankine were certainly astute businessmen, but they were extremely busy and their expertise was in finance, not patent strategy or engineering. Hence there was no one who could help Tesla focus his work on a few select designs and then vigorously promote those to investors and entrepreneurs.

FROM WIRELESS LIGHTING TO RESONANT POWER

In response to the lack of interest by investors to his wireless lighting system and his oscillator, Tesla began to rethink what he might do with high-frequency AC. And rather than concentrate on developing one particular aspect, he decided to expand the scope of his efforts: he would go from lighting a few rooms to ambitiously powering the earth.

Although visitors were astounded by his incredible demonstrations with the phosphorescent lamps and oscillating transformer, Tesla now started thinking that this apparatus did not represent the way forward. Here the illusions—the special effects—were distracting people from fully understanding the ideal that was evolving in Tesla's mind. "They could not understand these manifestations of energy and thought that it was a genuine transmission of power," Tesla explained. "I told them that these phenomena were wonderful, but that a system of transmission, based on the same principle, was absolutely worthless. It was transmission by electromagnetic waves. The solution lay in a different direction."[40]

Practical problems prompted Tesla to reject the idea of transmitting power using electromagnetic waves through the ether or atmosphere. In the course of building resonating apparatus in his laboratory, he was frustrated by the challenge of tuning the transmitter and receiver to exactly the same frequency. "It would be wearisome," observed Martin, "to dwell on the difficulty often experienced in establishing the relation of 'resonance,' and the instantaneity with which it can be disturbed. . . . This harmonizing is deftly accomplished by varying either of the two elements which chiefly govern the rapidity of vibration, viz., the so-called 'capacity' and the 'self-induction' [of the transmitting and receiving circuits]. . . . In very exact adjustments, minute changes will completely upset the balance, and the very last straw of fine wire, for example in the induction-coil which gives the self-induction will break the spell." Although Tesla "could pick up wire, coil it up, and tell what the vibration would be without any test," the difficulty of tuning the two circuits to respond to the same frequency suggested to Tesla that it would not be easy to convert his laboratory demonstration equipment into a reliable commercial system.[41]

But from a theoretical standpoint, Tesla came to doubt the value of electromagnetic waves passing through the ether or atmosphere. Here he looked more closely at the experiments in which his transmitter was connected to an antenna and to the ground (see Chapter 10). In his mind, two things happened as vibratory energy flowed through his transmitter: electromagnetic waves radiated out from the antenna and a current passed into the ground. As Maxwell had argued, the new electromagnetic waves and light waves were the same, and hence, like light, these waves propagated in straight lines. But this meant that the waves traveled into space in all directions and away from the receiver. "That energy which goes out in the form of rays," remarked Tesla, "is . . . unrecoverable, hopelessly lost. You can operate a little [receiving] instrument by catching a billionth part of it, but except this, all goes out into space never to return."[42] Because so much of the energy was wasted by the electromagnetic waves, it made little sense to Tesla to investigate them. (What Tesla did not know at the time was that electromagnetic waves propagate through the earth's atmosphere by bouncing off of a layer of charged particles in the ionosphere known as the Kennelly-Heaviside layer. This layer was predicted in 1902 and detected in 1924.)

Instead of worrying about electromagnetic waves, Tesla decided to focus on the ground current produced by the transmitter. After all, Tesla's flash of insight during the thunderstorm in Serbia in 1892 had been to try and harness the electrical forces in the earth. Why not, wondered Tesla, have the transmitter send waves of current through the Earth to a receiver and then use electromagnetic waves in the atmosphere for the return circuit? (See Figure 11.6.) By using the ground current in this way, Tesla believed more energy could be sent from the transmitter to the receiver.

In making this decision, we see Tesla thinking like a maverick since the other early wireless pioneers—Hertz, Lodge, and Marconi—focused on transmitting electromagnetic waves through the air. When, on occasion, these pioneers grounded their equipment, they gave little thought to what might be done with the ground current.[43] Just as Tesla had invented his revolutionary AC motor by reversing standard practice, now with wireless power he looked to get ahead by inverting the roles played by electromagnetic waves and the ground current in his high-frequency apparatus. Just as in an AC motor he had decided to have the magnetic field change in the rotor rather than in the stator, now he decided that the ground current should transmit energy and the electromagnetic waves should simply serve as the return needed to complete the circuit.

Of course, in many modern applications of radio—such as FM or communicating with aircraft—the transmitter and receiver circuits do

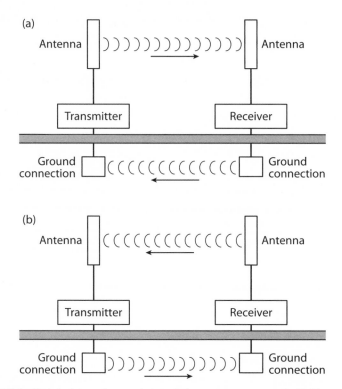

FIGURE 11.b. Tesla's vision of transmitting wireless power contrasted with his contemporaries in the 1890s.

Diagram (a) depicts how most early investigators (such as Marconi) thought that wireless communication would work. They assumed that the transmitter would generate electromagnetic waves that would be sent through the air to the receiver. The circuit would be completed because both the receiver and transmitter were grounded, and a return current would travel from the receiver back to the transmitter. Diagram (b) shows Tesla's vision in which the transmitter sent an oscillating current through the earth's crust to the receiver. The circuit was then completed by having electromagnetic waves flow through the atmosphere from the receiver back to the transmitter. As we will see in Chapter 12, Tesla later decided that the circuit from the receiver to the transmitter was completed by assuming that an electric current could be conducted through the upper atmosphere.

not need to be grounded. In insisting on a complete circuit through the Earth and returning through the atmosphere, Tesla reveals here that his thinking was based more on nineteenth-century practices in power and telegraphic engineering (which emphasized complete circuits) and not on the electromagnetic theory then being developed by the Maxwellians (see Chapter 6) that is widely accepted today. Thinking like a maverick has advantages and disadvantages.

Having decided to maximize the ground current and minimize the electromagnetic waves radiating from his apparatus, Tesla now adjusted the basic elements in his circuits, the capacitance of the capacitors and the inductance produced by the coils. In order to get a large ground current and minimum energy in electromagnetic waves, Tesla found it best to use a very large inductance and a very small capacitance in his transmitting circuit.[44]

In his first experiment with ground currents, Tesla used a tall, cone-shaped coil powered by a high-frequency current from an alternator and a bank of condensers. While one terminal of the coil was grounded, the other was left free in space. When the power was turned on, marvelous "purple streamers of electricity [were] thus elicited from the earth and [poured] out to the ambient air" (Figure 11.7).[45]

But what caused this outpouring of electrical streamers? For Tesla, they were evidence that he was tapping the earth's electrical energy. If this was the case, then rather than simply sending a current from one point to another on the earth's surface, might it be possible to transmit power by using resonance? By pumping electrical oscillations into the ground at the earth's resonant frequency, Tesla thought he might be able to broadcast power around the entire planet. According to Martin, Tesla explained this grand vision by using a vivid metaphor: "With this coil [Tesla] does actually what one would be doing with a pump forcing air into an elastic football. At each alternate stroke the ball would expand and contract. But it is evident that such a ball, if filled with air, would, when suddenly expanded or contracted, vibrate at its own rate. Now if the strokes of the pump be so timed that they are in harmony with the individual vibrations of the ball, an intense vibration or surging will be obtained."[46] Here Tesla realized the idea of a "sensitive trigger" that had come to him during the thunderstorm: a small force, properly applied, could be used to harness the tremendous forces of the earth. Tesla believed that he would not need to pump huge amounts of electrical energy into the earth; only a small amount was needed, at the right frequency, to serve as the trigger, and resonance would do the rest. With the whole Earth pulsing like his metaphorical football, Tesla was confident that he could annihilate distance and send power and messages around the world.

FIGURE 11.7. "Tesla coil for ascertaining and discharging the electricity of the Earth. The Streamers at [the] top of the coil are of purple hue, and in form resemble filaments of seaweed. The effect of mass being caused by prolonged exposure of flash-light negative."

From TCM, "Tesla's Oscillator and Other Inventions," *The Century Magazine* 49:916–33 (April 1895), Fig. 15.

For Tesla, it was a sublime moment to see the purple streamers coming off his coil, combining intellectual satisfaction with the opportunity to change human existence. "[O]ne can imagine the pleasure which the investigator [i.e., Tesla] feels when thus rewarded by unique phenomena," wrote Martin.

After searching with patient toil for two or three years after a result calculated in advance, he is compensated by being able to witness a most
magnificent display of fiery streams and lightning discharges breaking
out from the tip of the wire with the roar of a gas-well. Aside from their
deep scientific import and their wondrous fascination as a spectacle,
such effects point to many new realizations making for the higher welfare of the human race. The transmission of power and intelligence is
but one thing: the modification of climatic conditions may be another.
Perchance we shall "call up" Mars in this way some day, the electrical
charge of both planets being utilized in signals.[47]

ON THE EDGE AND INTO THE DEPRESSION

As Tesla's public persona grew during the 1890s, so did his social circle. Along with T. Commerford Martin and the Johnsons, he could now count on Mark Twain and Joseph Jefferson as friends, but another new acquaintance was the architect Stanford White. The son of a Shakespeare scholar, White studied architecture under Henry H. Richardson. In 1879 he joined Charles Follen McKim and William Rutherford Mead to form what became one of the most famous firms in American architectural history. Drawing on the Beaux Arts movement, McKim, Mead, and White designed many of the most important public buildings of their era. White's own masterpieces included the Washington Square Arch, the second Madison Square Garden, the Boston Public Library, and the restoration of the Rotunda at the University of Virginia. White designed houses for the very rich that included not only their "cottages" at Newport, Rhode Island, but also Fifth Avenue mansions of the Vanderbilts and Astors.[1]

Another of White's projects was the main power plant at Niagara, and so it is likely that Edward Dean Adams introduced White to Tesla. The architect and the inventor became good friends, and in 1894 White invited Tesla to join his club, The Players, on Gramercy Park. Founded by the actor Edwin Booth in 1888, The Players combined people in the arts—actors, writers, sculptors, architects, and painters—with bankers, lawyers, and businessmen. By having artists socialize with members of the business world, Booth hoped to raise the standing of the arts in New York society. When White moved to a townhouse on Gramercy Park in 1892, The Players became his favorite club, and so it is not surprising that White asked Tesla to join: "Will you not let me put you up for membership? It is an inexpensive club and the character of men I think you would like, and I know it would give me the greatest pleasure to meet you there now and then." Tesla immediately accepted and White nominated him, telling the club that "Nikola Tesla is one of the great geniuses and most remarkable men who have ever had anything to do with electricity."[2]

The Players became a regular haunt for Tesla, and he would drop by to see Mark Twain, who was a member, as well as Joseph Jefferson, who succeeded Booth as the club's president. Tesla also continued to socialize with White; in February 1895, for instance, White invited him "for a little supper for the artist, Ned Abbey, in my room in the Tower." A few weeks later, Tesla returned the invitation by doing a demonstration in his South Fifth Avenue laboratory for White, his wife, and his son, Lawrence.[3]

In visiting Tesla's lab, White would have seen that the Wizard was engaged in four main lines of research. One was his oscillator (i.e., the combination steam engine and electric generator), which Tesla regarded "as a practically perfected machine, but which of course, suggests many new lines of thought every day." A second was his new wireless lighting system while a third "was the transmission of intelligence any distance without wires. A fourth, which is an ever present problem for every thinking electrician touches on the nature of electricity."[4]

But in pursuing four lines of research at once, Tesla was wearing himself out, and during his visit in March 1895 White would have seen an overworked inventor. Interviewing Tesla about the same time, a reporter described him in the following manner: "I was a trifle shocked the first time I saw Nikola Tesla as he suddenly appeared before me . . . and sank into a chair seemingly in a state of utter dejection. Tall, straight, gaunt, and sinewy of frame like a true Slav, with clear blue eyes and small, mobile mouth fringed with a boyish mustache, he looked younger than his thirty-seven years. But what arrested my attention chiefly at the moment was the pallid, drawn, and haggard appearance of the face. While scanning it closely I plainly read a tale of overwork and of tremendous mental strain that must soon reach the limits of human endurance."

Tesla was aware that he should rest, but as he explained to the reporter, he could not stop working: "I would like to talk with you, my dear sir, he said, but I feel far from well to-day. I am completely worn out, in fact, and yet I cannot stop my work. These experiments of mine are so important, so beautiful, so fascinating, that I can hardly tear myself away from them to eat, and when I try to sleep I think about them constantly. I expect that I shall go on until I break down altogether."[5]

It was in this manic state that Tesla received a terrible emotional blow. At 2:30 A.M. on 13 March 1895, a fire broke out in the building containing his laboratory. The fire gutted 33–35 South Fifth and Tesla lost everything. "In a single night," reported the *New York Herald*, "the fruits of ten years of toil and research were swept away. The web of a thousand wires which at his bidding thrilled with life had been twisted by fire into a tangled skein. Machines, to the perfection of which he gave all that was best of a master mind are now shapeless things, and vessels which contained the results of patient experiment are heaps of pot sherds." Along with his apparatus, Tesla lost all his notes and papers, as he had just brought them to the laboratory to start organizing them. Tesla estimated that he had invested $80,000 to $100,000 in the laboratory, and, regrettably, he had no fire insurance.[6]

In response to the disaster, Tesla became severely depressed: "Utterly disheartened and broken in spirit, Nicola Tesla, one of the world's greatest electricians, returned to his rooms in the Gerlach yesterday morning and took to his bed. He has not risen since. He lies there, half sleeping, half waking. He is completely prostrated."[7] Knowing his delicate mental state, Tesla's friends feared for his well-being. When she had not heard from him for several days, Katharine Johnson wrote, "Today with the deepening realization of this disaster and consequently with increasing anxiety for you, my dear friend, I am even poorer in tears, and they cannot be sent in letters. Why will you not come to us now—perhaps we might help you, we have so much to give in sympathy."[8]

The newspapers portrayed Tesla's loss as having both personal and public significance. As Charles A. Dana said in the *New York Sun*, "The destruction of NIKOLA TESLA'S workshop, with its wonderful contents, is something more than a private calamity. It is a misfortune to the whole world. It is not in any degree an exaggeration to say that the men living at this time who are more important to the human race than this young gentleman, can be counted on the fingers of one hand; perhaps on the thumb of one hand."[9]

In the days after the fire, Tesla returned to the wreckage and set his men to work to salvage anything they could. To the reporters, he put on a brave face, but his heart was not in the work. A few nights later Tesla dropped into The Players, where he found the usual gathering of actors, musicians, and artists. "With quick and kind sympathy," reported the *New York Times*, the group "immediately organized an impromptu 'benefit concert' for his sole gratification, with an aggregation of talent that, had the public only known about it, would have given a substantial endowment for his new laboratory."[10]

Deep in depression, Tesla felt numb yet knew he had to find his own way out: "I have been overwhelmed with generosity and sympathy this week, and feel this kindness deeply, even if I can make no response," he told the *Electrical Review*. "But I must carve my way through or over the mountain suddenly planted in front of me."[11]

Tesla's way through the mountain was electrotherapy.[12] During his earliest work with high-frequency AC, Tesla had noted how such currents affected the body, and during his spectacular demonstrations, he may have observed how shocks altered his mood. Moreover, there was a tradition in popular medicine in mid-nineteenth-century America of using electric shocks from Ruhmkorff coils to treat a variety of ailments; Elihu Thomson's father, for instance, took shocks in the 1860s as a medical treatment.[13]

Over the next few months Tesla gave himself regular shocks, probably using one of his oscillating coils, in order to keep "from sinking into a state of melancholia." "I was so blue and discouraged in those days," he later told a reporter, "that I don't believe I could have borne up but for the regular electric treatment which I administered to myself. You see, electricity puts into the tired body just what it most needs—life force, nerve force. It's a great doctor, I can tell you, perhaps the greatest of all doctors."[14]

As he pulled out of his depression, Tesla put himself on a regular schedule, hoping to prevent a relapse. "He is a man of very regular habits," noted the *New York Sun*, "wherein he differs from Edison, who works fifty or seventy-five hours at a stretch, sometimes longer, when he has on hand something that interests him. Tesla is up every morning at 6½ o'clock. He has a lot of gymnastic exercises that he goes through with regularity. He has a light breakfast and then he loses little time getting to his work. He takes an hour for his luncheon in the middle of the day and the afternoon is devoted to hard work. He usually works until 8 o'clock in the evening, but often it is until midnight."[15]

INVESTIGATING X-RAYS

With his depression finally waning, Tesla rented a new laboratory on two floors of a building at 46 East Houston Street in July 1895 (Figure 12.1). There he employed "a clerk who attends to visitors, keeps away cranks, keeps a scrapbook, and sees that everybody who has real business with the inventor is provided with the latest copy of some scientific paper until Mr. Tesla is disengaged. He also has a dozen or more mechanics who are as loyal to him as Edison's men are to him; but the nature of his work and the magnitude of the problems he sets for himself to solve do not permit their rendering him the same sort of assistance that the Wizard's [i.e., Edison's] men furnish to their employer."[16]

Since he lost all his apparatus in the fire, Tesla took his efforts in new directions. While he continued to promote wireless lighting and the oscillator to potential investors, he now placed more effort on developing his ideas for the wireless transmission of power as well as two new areas, X-rays and radio control.

Realizing that investors and manufacturers were not interested in licensing or buying the patents for his system of wireless lighting (see Chapter 11), Tesla shifted from promoting an entire system to emphasizing components. Through 1895 and 1896, he redesigned his oscillating transformer (what others were now calling a Tesla coil) into a compact device that could take power from existing electrical circuits and step it up to a high voltage and high frequency. Tesla then used

FIGURE 12.1. Tesla's laboratory at East Houston Street. Note that the large spiral coil is visible in the back of the room. From NTM.

this improved oscillating transformer to power a new vacuum tube lamp, which he claimed gave out more light and was more efficient than Edison's incandescent lamps or the gas-filled tubes then being touted by D. McFarlane Moore. To demonstrate the power of his new lamp, Tesla posed for a portrait that now required only a two-second exposure from this new light source (Figure 12.2). Taken by Tonnele & Company, the portrait shows Tesla reading Maxwell's *Scientific Papers* while seated in a chair given to him by Edward Dean Adams; in the background was a large spiral coil that Tesla was using in his wireless power experiments.[17]

As he was working on these two components, however, Tesla also became excited about the recent discovery of X-rays. His involvement

FIGURE 12.2. "Tesla in His Laboratory—Portrait Obtained by an Exposure of Two Seconds to the Light of a Single Vacuum Tube without Electrodes, Having a Volume of about 90 Cubic Inches, Giving Approximately a Light of 250 Candle-Power—Photographed by Tonnele & Co."

From "Tesla's Important Advances," *Electrical Review,* 20 May 1896, 263 in TC 11:68, NTM.

with X-rays began with two missed opportunities. At the end of 1894, Tesla decided to investigate whether his lamps affected photographic plates in the same way as light coming from the sun or other sources of illumination. To do so, he sought the assistance of Dickenson Alley, a photographer employed by Tonnele & Company. Over a period of several months they tried a great variety of phosphorescent lamps, Crookes tubes, and vacuum bulbs with different kinds of electrodes. Since this was not a major project, Tesla and Alley worked on it periodically, and Alley stored spare glass photographic plates in a corner of the laboratory. However, they noticed that the unexposed plates had "unaccountable marks and defects" indicating that they had somehow been spoiled. Tesla wondered, in passing, if the plates might have been affected by cathode rays, which were a stream of charged particles that passed between the electrodes in some of his vacuum tubes when a voltage was applied across the electrodes. Tesla had recently read reports about how a Hungarian student of Heinrich Hertz, Philipp Lenard, was getting interesting results using tubes with an aluminum window that allowed the rays to pass out of the tube. However, before he could follow up on this hunch, the laboratory fire occurred and depression kept Tesla from working.[18]

The second missed opportunity came a few months later. In 1895, Tesla was discussing these photographic experiments with Edward Ringwood Hewitt, who was the son of the mayor of New York, Abraham Hewitt, and the brother of Peter Cooper Hewitt, who would invent the mercury vapor lamp in 1902. Through his brother's research, Edward was familiar with Crookes tubes, and in the course of conversation Tesla and Edward decided to try taking some photographs using these tubes as the light source. Perhaps knowing that Mark Twain had posed for a similar photograph (see Figure 11.5), Hewitt arranged for Twain to come to the lab. Because the light coming from the Crookes tube was weak, Twain had to sit still for a fifteen-minute exposure with his head supported by a headrest; to keep Twain amused during the sitting, Mrs. Hewitt read to him. A few days later, Hewitt checked to see how the portrait had turned out, and Tesla reported that the experiment had failed as the glass photographic plate had somehow been spoiled.[19]

Hewitt let the matter drop until he heard a few months later about the discovery of X-rays. X-rays are a form of electromagnetic radiation produced when a stream of fast electrons hits a metal target in an evacuated glass tube. Located beyond ultraviolet light on the electromagnetic spectrum, they were discovered by the German physicist Wilhelm Conrad Roentgen. Like Lenard, Roentgen was investigating cathode rays, and in November 1895 he was surprised to find that a barium platinocyanide screen fluoresced in the presence of a Crookes tube enclosed in a lightproof cardboard shield. Roentgen deduced that the fluorescence was caused by invisible radiation, which he termed X Strahlen, or X-rays, to indicate their unknown nature. In exploring this new phenomenon, he found that various materials were transparent to the rays but that photographic plates were sensitive to them. Roentgen combined these two observations to make a shadow-gram of the bones in his wife's hand. Roentgen announced his discovery at a meeting of the Physical-Medical Society of Wurzburg in December, and news of Roentgen's X-rays spread quickly. By 6 January 1896, the *New York Sun* reported that Roentgen had discovered "the light that never was," which could photograph hidden things such as bones inside a body.[20]

Upon reading about Roentgen's discovery, Hewitt rushed to Tesla's laboratory and begged to see the photographic plate they had taken a few months earlier. As Hewitt recounted,

> [Tesla] brought it out of the dark room and held it up to the light. There I saw the picture of the circle of the lens, with the adjusting screw at the side—also round dots, which represented the metal wood screws in the front of the wooden camera.
>
> Tesla gave one look. Then he slammed the plate on the floor, breaking it into a thousand pieces, exclaiming, "Damned fool! I never saw it."[21]

What Tesla and Hewitt had missed was that the Geissler tube had produced not only visible light but also invisible radiation—X-rays—that had spoiled the plate before the cap had even been taken off the lens and the exposure begun. Mark Twain had sat still for nothing and Tesla had missed making a major scientific discovery. "Too late," Tesla lamented, "I realized that my guiding spirit had again prompted me and that I had failed to comprehend his mysterious signs."[22]

Miffed that he had failed to notice X-rays first, Tesla sought to make up for lost time. As he told the *New York Times* a few weeks later, he began his experiments "within half an hour after the news of Prof. Roentgen's discovery was cabled to this country." Just as he had done when he had learned of Hertz's experiments with electromagnetic waves, Tesla now repeated Roentgen's experiments. Quickly noticing that other researchers were limited in their efforts by only using low-powered Ruhmkorff coils or electrostatic generators to power their tubes, Tesla began powering his tube using his new compact oscillating transformer. By taking advantage of the higher voltages and frequencies generated by this device, Tesla was able to produce more powerful X-rays than many of his contemporaries could. "I am producing shadows at a distance of 40 *feet*. I repeat 40 feet and even more," reported Tesla in March 1896.[23] During the next few months, Tesla kept his glassblower busy as he experimented with dozens of different tubes and corresponded with Hewitt about ways of testing them.[24]

With his powerful apparatus, Tesla concentrated on making the best possible images—which he called shadowgraphs—of parts of the human body. One of his first shadowgraphs was the right shoulder of a man, showing the ribs, shoulder bones, and bones of the upper arm. Another image was a picture of a foot with a shoe on, and "every fold of the leather, trousers, stocking, etc., is visible, while the flesh and bones stand out sharply" (Figure 12.3). Undoubtedly aware that

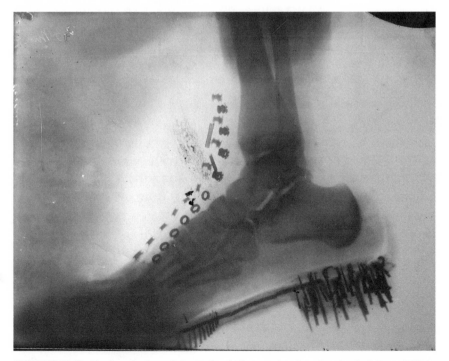

FIGURE 12.3. Shadowgraph made by Tesla of a human foot in a shoe in 1896. From NTM.

Edison had seen only "curvilinear murkiness" when he X-rayed the brain for William Randolph Hearst's *New York Journal*, Tesla secured an outline of the skull by exposing his own head for twenty to forty minutes. During the exposure, Tesla reported, "there is a tendency to sleep and the time seems to pass away quickly. There is a general soothing effect, and I have felt a sensation of warmth in the upper part of the head."[25]

Like other early investigators, Tesla initially regarded X-rays as benign. However, both he and his assistants soon experienced eyestrain, headaches, and burns on the skin of their hands. At first Tesla attributed these injuries to the ozone produced when running the tubes at high voltages, but he came to realize that the rays themselves were causing damage. Tesla was particularly upset when "a dear and zealous assistant" suffered severe burns on his abdomen after being exposed for five minutes to an X-ray tube positioned eleven inches from his body. "Fortunately," Tesla reported, "frequent warm baths, free application of vaseline, cleaning and general bodily care soon repaired the ravages of the destructive agent, and I breathed again freely." Nevertheless, in his later papers on X-rays, he recommended using a grounded aluminum shield around the X-ray tube, that people avoid getting too close to the tube, and that exposure times be limited.[26]

Through 1896 Tesla produced a steady stream of reports on his X-ray research not for his friend Martin's *Electrical Engineer* but for the rival *Electrical Review*.[27] However, Tesla's interest in X-rays soon waned. Initially he took up this topic because, by drawing on the skill he had acquired in manipulating Crookes tubes, he would either make a further scientific discovery or quickly develop a new product. To be sure, he insisted that his compact oscillating transformer was ideal for powering X-ray tubes, but he did not come up with a commercially viable X-ray tube. Instead, the firms that were able to develop X-ray products—either power sources or tubes—were those like General Electric that had expertise in manufacturing incandescent lamps or small scientific-instrument manufacturers that had the distribution network to reach scientists and medical doctors.[28] Hence Tesla probably dropped X-ray research when it became clear that he could not compete with the companies that were moving into this new field.

I also suspect that Tesla stopped experimenting with X-rays because they did not advance his evolving interests in wireless transmission.

Tesla probably checked out X-rays to see if they could be useful for transmitting power in the ways that he had started to think about before the fire (see Chapter 11), and he moved on to new topics when these experiments did not contribute anything new.

DEVELOPING RADIO-CONTROLLED BOATS

Though X-rays did not contribute to his thinking about wireless power, Tesla was already working on a new project that did: the development of radio-controlled automatons. While we might call these devices robots (a term the Czech writer Karel Čapek introduced in 1920), Tesla coined his own word for this new field: telautomatics.[29]

Tesla's interest in automata dates back to his childhood. As a boy, he suffered from nightmares that he overcame by developing his willpower. Struck by the fact that the frightening visions were often the result of some external stimuli that he could identify, Tesla concluded that all thoughts and emotions were the result of outside factors and that the human organism was no more than a "self-propelling machine, the motions of which are governed by impressions received through the eye." His efforts to understand and control his intense visions, as he explained in his autobiography, "led me finally to recognise that I was but an automaton devoid of free will in thought and action and merely responsible to the forces of the environment."[30] But if he were merely an automaton, wondered Tesla, why not build one as well?

"The idea of constructing an automaton, to bear out my theory, presented itself to me early," Tesla recalled, but he did not take it up in earnest until he had begun to perfect his wireless inventions and realized that his receivers could serve as surrogates for the eye or other sensory organs. As he explained in 1898, "Endeavoring to construct a mechanical model resembling in its essential, material features the human body, I was led to combine a controlling device, or organ sensitive to certain waves, with a body provided with propelling and directing mechanism, and the rest naturally followed."[31]

Tesla began working on a remotely controlled device in 1892, which, like his wireless lamps, was controlled by electrical induction. These early contrivances were destroyed in the fire at his laboratory,

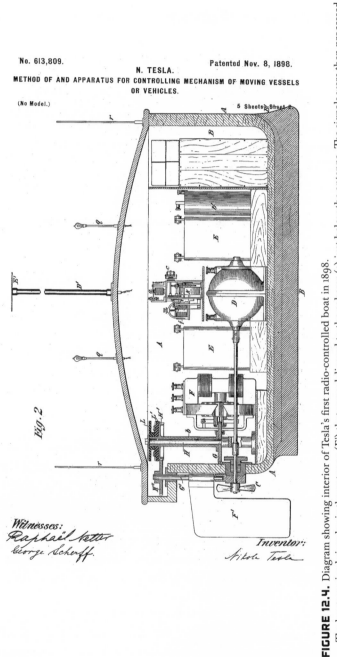

No. 613,809.

N. TESLA.

Patented Nov. 8, 1898.

METHOD OF AND APPARATUS FOR CONTROLLING MECHANISM OF MOVING VESSELS
OR VEHICLES.

(No Model.)

5 Sheets—Sheet 2.

Fig. 2

Witnesses:
Raphael Netter
George Scherff.

Inventor:
Nikola Tesla

FIGURE 12.4. Diagram showing interior of Tesla's first radio-controlled boat in 1898.

The boat received signals via the antenna (E') that were delivered to the coherer (c) just below the antenna. The signals were then processed by the disk mechanism in the stern, marked by (L). Depending on the number of times the signal was interrupted, the disk (L) advanced a certain number of "clicks" which in turn regulated the current delivered to the motor (F) that controlled the rudder (F'). The boat was propelled by another motor (D) and the current to the motors was provided by storage batteries (E). The signal lamps are marked by q and the space marked B was where an explosive charge could be carried.

From NT, "Method of and Apparatus for Controlling Mechanism of Moving Vessels or Vehicles," US Patent 613,809 (filed 1 July 1898, granted 8 Nov. 1898), Figure 2.

and so after moving to Houston Street he fashioned some new proto-types with which he was able to make "more striking demonstrations, in many instances actually transmitting the whole motive energy to the devices instead of simply controlling the same from [a] distance." Unfortunately, there are no descriptions of these early prototypes.[32]

In 1897 Tesla began building a test model with the help of one of his assistants, Raphael Netter. While he anticipated that his control mechanisms could be used in any kind of vehicle or flying machine, he chose to build the model in the form of a boat in response to the naval armaments race then under way.[33] In 1889 the dominant naval power, Britain, had decided that it should build a new fleet of battleships superior to the combined fleets of its rivals, France and Russia. As these three powers raced to build new ships, the United States, Germany, Spain, and Japan followed suit in order to protect themselves.[34] Powered by new triple-expansion steam engines and protected by hardened steel armor, this new generation of battleships carried main batteries of twelve-inch guns in turrets supported by additional batteries of smaller guns. Fast, well-protected, and bristling with weapons, these new warships looked as if they were invincible.[35]

To attack these new battleships, Tesla designed an unmanned torpedo boat that could carry an explosive charge and be directed by electromagnetic signals (Figures 12.4 and 12.5). One visitor to the Houston Street laboratory offered the following description of the boat and how Tesla demonstrated it:

> Elevated on stocks on a table in the centre of the laboratory stood a model of a screw-propelled craft, about four feet long and some-what disproportionately wide and deep. Mr. Tesla explained that it was merely a working model which he had made in order to exhibit to President McKinley, and that no attempt had been made to follow the usual sharp lines of a torpedo boat. The deck was slightly arched and surmounted by three slender standards, the centre one being considerably higher than the other two, which carried small incandescent bulbs, a third bulb being fixed at the bow.
>
> The keel consisted of a massive copper plate, the propeller and rudder being in their usual positions. Mr. Tesla explained that the boat contained the propelling machinery, consisting of an electric motor actuated by a storage battery in the hold, another motor to actuate the

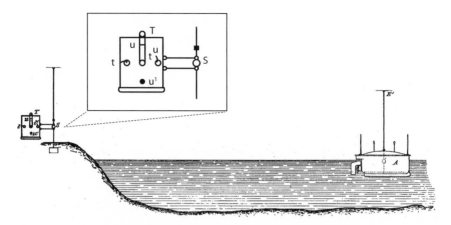

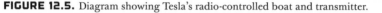

FIGURE 12.5. Diagram showing Tesla's radio-controlled boat and transmitter.

S is a generator producing continuous electromagnetic waves and is connected to an antenna. To its left is the control box with a lever (T). This lever could be rotated in one direction, making contact with contacts u, t', u', and t.

From NT, "Method of and Apparatus for Controlling Mechanism of Moving Vessels or Vehicles," US Patent 613,809 (filed 1 July 1898, granted 8 Nov. 1898), Figure 9.

rudder and the delicate mechanism which performs the function of re-ceiving through the central standard the electric impulses sent through the atmosphere from the distant operating station, which set in motion the propelling and steering motors, and through them light or extin-guish the electric bulbs and fire the exploding charge in the chamber of the bow in response to signals sent by the operator.

"Now watch," said the inventor, and going to a table on the other side of the room on which lay a little switchbox, about five inches square, he gave the lever a sharp turn. Instantly the little bronze pro-peller began to revolve at a furious rate. "Now I will send the boat to starboard," he said, and another quick movement of the lever sent the helm sharp over, and another movement turned it as rapidly back again. At another signal the screw stopped and reversed.[36]

To control this boat, Tesla had the transmitter generate a continu-ous electromagnetic wave at one frequency, which the boat detected using a coherer that had been invented in 1890 by the French physi-cist Édouard Branly and subsequently improved by Oliver Lodge. At the transmitter, Tesla could rotate the lever, touching from one of four contacts; in doing so, he would interrupt the signal being sent to the boat. Inside the boat, these interruptions caused a special disk to rotate and engage different contacts on the surface of the disk, thus activating the rudder and the motor to the propeller (Figure 12.4). For instance, rotating the transmitter's lever to a first contact might cause the rudder to turn right, moving it to the next contact would stop the rudder turning and start the propeller, and shifting the lever to the third contact would turn the propeller to the left. Because the contacts controlling the rudder and propeller were placed in a specific sequence on the disk, Tesla could not pick one function and directly execute it; instead, he had to move the lever from contact to contact in order to get the boat to do what he wanted. Hence, in the modern technical sense, Tesla's first boat was not "remote-controlled" but "radio-controlled," as the former term describes situations in which different signals are sent to execute different functions.[37]

Despite this distinction, Tesla's radio-controlled boat was a remarkable achievement. Working around the limitation of sending signals at only one frequency, Tesla came up with an ingenious electromechanical solution. The mechanism found in this invention is among the most sophisticated devices Tesla created in his career. And up to 1897, no one had conceived of using electromagnetic waves to operate an unmanned vehicle; Tesla introduced the idea into popular culture and engineering practice.[38]

Tesla's inspiration for this solution came from telegraphy. During the late nineteenth century, telegraphy was used not only for sending telegrams between cities but also for requesting messenger boys, the police, or the fire department using what was called a district telegraph. Seeking a reliable mechanism for his boat, Tesla drew on the circuitry typically used in district call boxes:

> How such an apparently complicated mechanism can be operated and controlled at a distance of miles is no mystery. It is as simple as the messenger call [box] to be found in almost any office. This is a little metal box with a lever on the outside. By moving the crank to a certain point it gives vibrating sounds and springs back into position, and its momentary buzzing calls a messenger. But move this same crank a third further around the dial and it buzzes still longer, and pretty soon a policeman appears, summoned by its mysterious call. Again, move the crank this time to the farthest limit of the circle and scarcely has its more prolonged hum of recoil sounded when the city fire apparatus dashes up to your place at its call.

Now, my device for controlling the motion of a distant submarine
boat is exactly similar. Only I need no connecting wires between my
switchboard and the distant submarine boat, for I make use of the now
well-known principle of wireless telegraphy.[39]

Confident that he had a great invention on his hands, Tesla began
drafting a patent application for his radio-controlled boat in 1897. At
the same time, he delighted in demonstrating the boat in his laboratory "to visitors who never ceased to wonder at the performances."
According to one source, these visitors included J. P. Morgan, William
K. Vanderbilt, John Hays Hammond Sr., and Charles Cheever. These
private demonstrations prompted one visitor, the successful mining
engineer John Hays Hammond, to invest $10,000 in the project.[40]

While Tesla was working on his radio-controlled boat, the United
States declared war on Spain in April 1898, and consequently there
was a great deal of excitement about new weapons for attacking
battleships. Anxious to trade on the wartime hysteria, another wireless entrepreneur, W. J. Clarke of the United States Electrical Supply
Company, arranged for a demonstration of his company's equipment
at the Electrical Exhibition taking place at Madison Square Garden in
May. His demonstration featured a radio-controlled underwater mine
that could blow up a ship. "By touching an instrument placed in the
southern gallery [of the exhibition]," reported the *New York Times*, "a
miniature Spanish cruiser anchored in the fountain lake on the lower
floor, 90 feet away, was blown into the air, together with a considerable quantity of water, which fell on those who were not quick enough
in getting out of the way." This dramatic demonstration proved to be
popular, and Clarke was soon repeating it four times a day.[41]

As the war with Spain continued, Tesla finished and filed his patent application for the boat in July 1898. Even though he had begun
to develop more advanced circuitry using "the joint action of several
circuits" (discussed later in the chapter), Tesla's attorney, Parker Page,
advised him to emphasize how he employed a single frequency since
they had not yet drafted patents protecting the more sophisticated
circuits. Because the examiners in the Patent Office could not believe
what Tesla claimed in his application, the examiner-in-chief came to
New York to see for himself how the boat operated. Satisfied, he let
Tesla's patent for the boat issue in November 1898.[42]

ABOLISHING WAR, ENDING A FRIENDSHIP

With a patent in hand, Tesla now cranked up the publicity, and stories soon appeared in the technical press and popular newspapers about his radio-controlled boat (see Figure 13.11). Rather than talk just about how his boat could destroy battleships, though, Tesla boldly insisted that his boat would bring about an end to war. As he told the *New York Herald*,

> War will cease to be possible when all the world knows to-morrow that the most feeble of nations can supply itself immediately with a weapon which will render its coast secure and its ports impregnable to the assaults of the united armadas of the world. Battle ships will cease to be built, and the mightiest armorclads and the most tremendous artillery afloat will be of no more use than so much scrap iron. . . .
>
> Imagine, if you can . . . what an irresistible instrument of destruction we have in a torpedo boat thus [remotely] controlled, which we can operate day or night on the surface or below it, and from any distance that may be desired. A ship thus assailed would have no possibility of escape. . . .
>
> But I have no desire that my fame should rest on the invention of a merely destructive device, no matter how terrible. I prefer to be remembered as the inventor who succeeded in abolishing war.[43]

Tesla was confident that all nations—both feeble and mighty—would be interested in his invention. As he informed Page in December 1898, "I have received from a number of countries propositions for the rights on my invention for controlling the movements and operation of bodies from a distance, and I am generally requested to name the price. In view of this it is very desirable for me that the patents for the chief European countries should be applied for without delay." Over the next year, Page filed patents on the radio-controlled boat in thirteen countries.[44]

Hearing that Tesla was securing foreign patents for the boat, Mark Twain penned him a quick letter from Europe:

> Have you Austrian & English patents on that destructive terror which you have been inventing?—& if so, won't you set a price upon

them & commission me to sell them? I know cabinet ministers of both countries—& of Germany, too; likewise William II. . . .

Here in the hotel the other night when some interested men were discussing means to persuade the nations to join with the Czar & disarm, I advised them to seek something more sure than disarmament by perishable paper contract—invite the great inventors to contrive something against which fleets and armies would be helpless & thus make war thenceforth impossible. I did not suspect that you were already attending to that, & getting ready to introduce into the earth permanent peace & disarmament in a practical & mandatory way.[45]

While the *Electrical Review* predicted that Tesla's radio-controlled boat would become one of the "most potent factors in the advance of civilisation of mankind," Tesla was surprised when this invention was sharply criticized by his professional colleagues who failed to see its novelty or practicality. As Princeton's professor of physics Cyrus F. Brackett complained, "There is nothing new about this. The theory is perfect, but the application is absurd. . . . Do you suppose that in the din of battle it would be possible to put into execution those minute and carefully adjusted mechanical experiments, all of which are presupposed by his theory, which require the quiet of an uninterrupted laboratory to work successfully?" More unkind words came from Amos Dolbear at Tufts College: "The announcement is most amazing, and coming as it does from Tesla, scientists are all the more chary about accepting it. During the last six years he has made so many startling announcements and has performed so few of his promises that he is getting to be like the man who called 'Wolf! wolf!' until no one listened to him. Mr. Tesla has failed so often before that there is no call to believe these things until he really does them. Meantime, we are all waiting with much patience and without solicitude. We will believe them when they are done."[46]

But the harshest criticism came from his friend T. Commerford Martin at the *Electrical Engineer*. Worried that his journal was losing market share and undoubtedly upset that Tesla was now sending material to the rival *Electrical Review*, Martin openly attacked Tesla in a November 1898 editorial.[47] Not only had Tesla not finished up inventions such as the steam-powered oscillator (which Martin now thought was destined for the scrap heap), Martin scornfully complained that there was nothing new in Tesla's radio-controlled boat and that he had simply appropriated the idea from Clarke's demonstration: "Last spring the ability to explode floating torpedoes under ships from a distance without any wires was brilliantly demonstrated at Madison Square Garden several times a day for a month. Taking that idea, Mr. Tesla has applied the same principle to the electro-mechanical steering of torpedoes." To add injury to insult, Martin rushed into print, without Tesla's permission, a paper he had recently given at the American Electro-Therapeutic Association.[48]

Tesla was furious that Martin published his paper but even more so that he impugned his honesty. In response, Tesla called Martin's attention not to his accomplishments but to his honor and academic degrees:

> Your editorial comment would not concern me in the least, were it not my duty to take note of it. On more than one occasion you have offended me, but in my qualities both as a Christian and philosopher I have always forgiven you and only pitied you for your errors. This time, though, your offence is graver than previous ones, for you have dared to cast a shadow on my honor.
>
> No doubt you must have in your possession, from the illustrious men [i.e., Brackett and Dolbear] whom you quote, tangible proofs in support of your statement reflecting on my honesty. Being a bearer of great honors from a number of American universities, it is my duty, in view of the slur cast upon them, to exact from you that in your next issue you produce these, together with this letter which in justice to myself, I am forwarding to other electrical journals. In the absence of such proofs, which would put me in the position to seek redress elsewhere, I require that, together with the preceding, you publish instead a complete and humble apology for your insulting remark which reflects on me as well as on those who honor me.[49]

In the next issue of *Electrical Engineer*, Martin published Tesla's letter as well as the "proofs" that Tesla had demanded. However, Martin opened with a highly revealing comment:

> One of the foremost electrical inventors of this country, whose name is known around the world, has been kind enough to say that The

Electrical Engineer made Mr. Tesla. This is an attribution that we naturally put aside, for it is a man's own work that makes or unmakes him, but we do plead guilty to the fact that for these ten years past we have done whatever mortals could do to bring Mr. Tesla forward and secure for him the recognition that was duly his. Not only in the columns of this and other journals, but in magazines and books we have striven with all the ability we possessed to explain Mr. Tesla's ideas. The record is before all men. If there is a line or a word in it that seeks to do Mr. Tesla "serious injury," who says we have ever in word or deed or thought tried to do Mr. Tesla any sort of injury, lies.

Within the last year or two Mr. Tesla has, it seems to us, gone far beyond the possible in the ideas he has put forth, and he has to-day behind him a long trail of beautiful but unfinished inventions. By mild criticism and milder banter, not being able to lend Mr. Tesla the cordial support of earlier years of real achievement, we have only very lately endeavored to express our doubts and to urge him to the completion of some one of the many desirable or novel things promised. We believe this to be true friendship.[50]

From Tesla's perspective, however, constructive criticism was not something that he wanted to hear from his friends, and as a result of this episode, Tesla and Martin went their separate ways.

While he was feuding with Martin in the columns of the *Electrical Engineer*, Tesla had begun work on a second and larger (six-foot-long) radio-controlled boat. In particular, he had become worried about security—that his boat should respond only to signals from his transmitter. Tesla came to this conclusion by thinking about the boat, the automaton, as being like a person:

[T]he automaton should respond only to an individual call, as a person responds to a name. Such considerations led me to conclude that the sensitive device of the machine should correspond to the ear rather than the eye of a human being, for in this case its actions could be controlled irrespective of intervening obstacles, regardless of its position relative to the distant controlling apparatus, and, last, but not least, it would remain deaf and unresponsive, like a faithful servant, to all calls but that of its master.

And to get the servant to be faithful to his master, Tesla turned his attention back to the problem of tuning: "I attained the result aimed at by means of an electric circuit placed within the boat, and adjusted, or 'tuned,' exactly to electrical vibrations of the proper kind transmitted to it from a distant 'electrical oscillator.' This circuit, in responding, however feebly, to the transmitted vibrations, affected magnets and other contrivances, through the medium of which were controlled the movements of the propeller and rudder, and also the operations of numerous other appliances."[51]

But how was he going to make sure that his boat responded only to his signals? As a solution, Tesla borrowed an idea he had been using in his wireless lighting demonstrations. Much to Tesla's annoyance, visitors to Houston Street often pointed out that several lamps lit up even though he wanted only one lamp to respond to his oscillating transformer. To overcome this problem, Tesla had his oscillator generate several different frequencies and then tuned the lamps so that they had to receive a combination of two frequencies before they would light up.[52]

Tesla now applied this technique to his second boat, and in the laboratory he devised several means for transmitting two signals to the boat:

> I did this in two ways: either by placing, in the inside of the boat, two circuits which I tuned and combined so as to cause the operation of the controlling mechanism when both of these circuits were energized by their respective vibrations which were passed through the cables around the room [i.e., the primary of his oscillating transformer] . . . or, in another way, by using two coils and connecting one end of each to the ground and the other to a metal plate or a bundle of wires [which would have functioned as two separate antennas] and then exciting them, either by means of the large coil in the form of a spiral [visible in Figures 12.1 and 12.2], . . . or by two circuits which were improvised or adopted for the experiment.[53]

With this technique, Tesla realized that he was not limited to using just two frequencies but that he could generate dozens of frequencies and use various combinations to individually control multiple vessels.

Hence he envisioned that one or several operators might simultane-ously direct fifty or one hundred vessels through differently tuned transmitters and receivers.[54]

Although Tesla initially promised "to exhibit a model of a torpedo boat at the [upcoming] Paris Exposition and direct all its movements from my office in New York" (Figure 12.6), he instead demonstrated the second boat to members of the Chicago Commercial Club in May 1899. When the distinguished guests arrived to hear his lecture, they were startled to see an artificial lake in the middle of the auditorium, and in the lake was Tesla's boat. Ever the master showman, Tesla invited the crowd to shout out questions and his automaton would answer them by flashing its lights, shifting the rudder, or setting off exploding cartridges. "This was considered magic at the time," Tesla recalled, "but was extremely simple, for it was myself who gave the replies by means of the device." In the lecture that followed, Tesla

WONDERS NIKOLA TESLA SAYS HE CAN PERFORM.
Shows How He Proposes by Electricity, Without Wires, to Control the Movements of a Model at the Paris Exposition from His Office in New York, and by Similar Methods Blow Up Ironclads and Send Lifeboats to Shipwrecked Vessels.

FIGURE 12.6. Newspaper sketch showing how Tesla planned to demonstrate his radio-controlled boat at the Paris Exposition.

While Tesla (left) sent signals from his laboratory in New York, the audience (right) would watch the boat maneuver in a tank in Paris.

From "Tesla Declares He Will Abolish War," *New York Herald,* 8 Nov. 1898, in TC 13:138–40, on 139.

described how he had conceived of this automaton and emphasized its potential to abolish war. Reflecting on the fact that the boat seemed almost to be alive, Tesla philosophized at some length about the nature of human thought, life, and death.[55]

WHY DID TESLA NEVER MARRY?

From 1896 to 1898, as Tesla threw himself into his work with X-rays and the radio-controlled boat, he still had bouts of melancholy. During a trip in July 1896 to Niagara, Tesla told a reporter, "I came to Niagara Falls to inspect the great power plant and because I thought the change would bring me needed rest. I have been for some time in poor health, almost worn out, and I am now trying to get away from my work for a brief spell."[56]

A few weeks later, another reporter came upon Tesla slumped in a café at a late hour, looking haggard and tired (Figure 12.7). "I am afraid," Tesla began, "that you won't find me a pleasant companion to-night. The fact is, I was almost killed today." Despite the precautions he took during his experiments, he had just received a shock of 3.5 million volts from one of his machines. "The spark jumped three feet through the air," said Tesla, "and struck me here on the right shoulder. I tell you it made me feel dizzy. If my assistant had not turned off the current instantly it might have been the end of me. As it was, I have to show for it a queer mark on my right breast where the current struck in and a burned heel in one of my socks where it left my body. Of course the volume of current was exceedingly small, otherwise it must have been fatal."

The reporter went on to ask Tesla if he was frequently depressed. "Perhaps not often," Tesla replied. "Every man of artistic temperament has relapsed from the great enthusiasms that buoy him up and sweep him forward. In the main my life is very happy, happier than any life I can conceive of." Tesla understood that some melancholy was the price of experiencing the exhilaration of invention, and he told his interviewer that "I do not think there is any thrill that can go through the human heart like that felt by the inventor as he sees some creation of the brain unfolding to success. . . . Such emotions make a man forget food, sleep, friends, love, everything."

FIGURE 12.7. Undated pencil sketch of Tesla at a café. From NTM.

Finding Tesla in a receptive mood, the reporter followed up with a risky personal question. Knowing that Tesla was a bachelor, he asked the Wizard about marriage. Was marriage suitable for persons of artistic temperament? Thinking for a moment, Tesla answered, "For an artist, yes; for a musician, yes; for a writer, yes; but for an inventor, no. The first three must gain inspiration from a woman's influence and be led by their love to finer achievement, but an inventor has *so intense a nature with so much in it of [a] wild, passionate quality*, that in giving himself to a woman he might love, he would give everything, and so take everything from his chosen field. I do not think you can name many great inventions that have been made by married men [emphasis added]." After giving his answer, Tesla hesitated and concluded the interview, remarking, "It's a pity too, for sometimes we feel so lonely."[57]

Over the next few weeks, both the tabloids and engineering papers puzzled over Tesla's explanation, as they thought it an "abnormal emotional condition" that he did not wish to marry. Since then, Tesla's biographers have continued to wonder about the Wizard's celibacy. "Tesla tried to convince the world that he had succeeded in eliminating love and romance from his life," noted biographer John O'Neill, "but he did not succeed. That failure . . . is the story of the secret chapter of Tesla's life."[58]

Though we may never know exactly why Tesla never married, the existing sources suggest several possible explanations. The first is, quite simply, that Tesla was more attracted to men than women.

With regard to women, it is clear that Tesla had a complex attitude toward them. At times, he put them on a pedestal, and in his later years he wrote popular articles suggesting that women might well be the superior sex. At other times he was clearly shy, even fearful, around women, especially when he was a young man. As he told a Serbian reporter in 1927, "I have never touched a woman. As a student, and while vacationing at my parents' home in Lika, I fell in love with one girl. She was tall, beautiful, and had extraordinary understandable eyes." Similarly, Martin intimated to Katharine Johnson in 1894 that he feared that Tesla "will go in the delusion that woman is generically a Delilah who would shear him of his locks. If you can manage it, I believe it would be a good scheme to have that Doctor get hold of him. . . . My prescription is a weekly lecture from Mrs. RUJ."[59]

Mrs. RUJ seems to have had some positive impact on the Wizard, as he learned to enjoy interacting with society women at the Johnsons' home and while out on the town. Among the women with whom Tesla socialized with were Mrs. John Jacob Astor, Mrs. Clarence McKay, the heiress Flora Dodge, Teddy Roosevelt's sister Corinne Robinson, and J. P. Morgan's daughter Anne. Over time, Tesla became sufficiently comfortable in talking to such women that he would approach them for money to support his inventions (see Chapter 15). Nevertheless, Tesla did not develop a deep relationship with any of these women. According to one female friend, the playwright Marguerite Merington, Tesla never went out with any women except for her, and one wonders if she might be boasting. As John O'Neill told Leland Anderson in the 1950s, "You can therefore, disregard any story you may hear from femmes that Tesla was interested in them. They were all double duds to him. He always treated them with the greatest respect. He did, however, have a mother-fixation, a situation that is quite understandable."[60]

In contrast to his relationships with women, Tesla was clearly attracted to men. As we saw earlier, Tesla found Anthony Szigeti to be physically attractive, they became close friends, and Szigeti followed Tesla from Budapest to Paris and on to New York. Sometime in 1891, Szigeti left and Tesla was deeply hurt. A few years later, Tesla befriended a young college graduate, Emile Smith, who was interested in engineering and seeking a position at the Westinghouse Company. Sadly, Smith died from typhoid fever only a few months after going to Pittsburgh, and one of Tesla's former colleagues wrote him about Smith, saying, "As he was a personal friend of yours I thought perhaps you would be interested in hearing of his death."[61]

One wonders if Tesla's attraction to men was merely platonic or if it was physical as well. The only evidence that speaks to this issue comes from a 1956 conversation that Leland Anderson had with Richard C. Sogge, a longtime member of the American Institute of Electrical Engineers. Sogge was glad that the institute was celebrating the centennial of Tesla's birth that year and told Anderson, "You know, it's a very good thing that the Institute is honoring Tesla in this way—it will go a long way toward diminishing his reputation for voyeurism which was embarrassing to the older members. The stories of Tesla's sexual episodes were at one time the talk of the Institute, and we didn't know how to deal with it if the matter should somehow become publicized. You must be aware, of course, that he never went out with women. . . . Anyway, the older members of the Institute are dying off so those stories will also die out eventually."[62] Sogge's remarks may help explain why Tesla's involvement with the AIEE declined in the 1890s. After being elected vice president in 1892–93, Tesla did not become president. He was only recognized by the institute for his contributions in 1917 when he was named a fellow and received the Edison Medal.[63]

In terms of looking for clues about homosexuality in historical documents, one has to keep in mind differences in the way language was used in the nineteenth century and today. To be sure, men in Victorian America often developed close emotional friendships, and sometimes used romantic and sexual language that in twenty-first-century

America would be reserved for heterosexual relationships; for instance, Stanford White included explicit sexual language and anatomical drawings in his correspondence with the sculptor Augustus Saint-Gaudens.[64] Moreover, since sexual degeneracy, like poverty, was viewed as proof that the poor were inferior, middle-class individuals were careful not to reveal anything that could be construed as unusual about their sexual conduct.[65] Consequently, it is not necessarily easy to know how to read the surviving documents for clues about Tesla's sexual orientation. To be sure, there is nothing in the scant material relating to Szigeti that indicates whether he and Tesla were lovers.

In contrast, the story is more complicated with another of Tesla's male friends, Richmond Pearson Hobson (1870–1937) (Figure 12.8). Born in Alabama, Hobson attended the U.S. Naval Academy at Annapolis, where he was an outstanding student. However, finding him headstrong and righteous, his fellow midshipmen refused to speak to him during his last two years at the academy. After graduation, Hobson was selected to become a naval constructor and spent several years studying naval architecture in Europe.

When the Spanish-American War broke out, Hobson was assigned to duty with Admiral William T. Sampson on the *New York*, and they sailed to Cuba to engage the Spanish fleet in Santiago harbor. Hoping to bottle up the Spanish warships in the harbor, Sampson decided in June 1898 to scuttle the collier *Merrimac* at the mouth of the harbor, and Hobson volunteered to lead this suicide mission. Hobson and his crew did sink the *Merrimac*, but not quite in the right location, and they were subsequently captured by the Spanish. However, the sunken ship did force the Spanish fleet to have to maneuver slowly out of the harbor, and a few weeks later Sampson's ships were able to destroy each Spanish warship as it tried to escape. The Spanish released Hobson in July, and he returned to America where he was heralded by the press as a war hero.[66]

Ever on the lookout for fresh material for *The Century Magazine*, Robert

FIGURE 12.8. Richmond P. Hobson.

From http://www.history.navy
.mil/photos/images/h00001/h00127
.jpg.

Underwood Johnson contacted Hobson shortly after his return to convince him to write a book about the *Merrimac* mission.[67] In August 1898 Johnson wrote the young lieutenant, inviting him to lunch and to meet Tesla: "Do you know Tesla? If not, would you not enjoy going to his laboratory tomorrow with me? He is a charming fellow and of course you have much in common. He is one of my best friends."[68]

Johnson must have known that Tesla would be drawn to this handsome war hero. According to his wife, Grizelda, Hobson was

> [a]lways physically fit muscularly. Powerful sloping shoulders and arms to match, deep chested, flat abdomen, heavy thighs powerfully muscled, powerful well formed calves. Physical power showed through his clothes. One look at him would have caused a fighter to hesitate to pick a fight with him. On the other hand, the great intellect which showed in his face and head, combined with a gentle friendliness of expression reflecting his real character were anything but invitations to fight. In a bathing suit he could have posed as "Tarzan." His entire make-up was suggestive of controlled power, mental, physical and spiritual.[69]

Over the next few months, Tesla had the opportunity to meet Hobson at dinners and parties with the Johnsons. Tesla took a liking to this naval officer, and he wrote Johnson in jest, "Remember, Luka, Hobson does not belong to the Johnsons exclusively. I shall avenge myself on Mme. Filipov by introducing him to Mme. Kussner and somebody will be forgotten."[70]

Still a commissioned officer, Hobson was assigned to duty in Hong Kong and Manila in late 1898 and so Tesla did not see much of his new friend. Once Hobson returned to the United States in September 1900, he was assigned to the Brooklyn Navy Yard and then the Navy Department in Washington. Frequently dining with mutual friends in New York, Hobson and Tesla grew close, as illustrated by an undated note from Hobson:

> My very dear Tesla:
>
> Thank you very much for your thoughtful and lovely note. I have engaged you for dinner tomorrow with the Van Beurens. . . . They are great friends of mine & my brother. Dinner is down for 7.30. . . .

Now, my dear fellow, if you are doing nothing for the next ¾ of an hour come over for a short tete a tete—I feel I have not seen half enough of you on this visit and I have so much to talk with you about—

But if you have to get up early, of course do not think about it—

Devotedly yours,
Richmond

And to celebrate the first day of the twentieth century (then regarded to have begun in 1901, not 1900), Tesla wrote to Hobson:

My dear Hobson,

This my first and heartiest greeting today is for you.

On the new page of human history just opened, you have already written your name in imperishable characters.

May the year have still greater opportunities and achievements in store for you.

Hoping to grasp your head soon I remain with renewed goodwishes,
Yours very sincerely
N Tesla

Tesla and Hobson continued to socialize with the Johnsons, and Tesla sent their daughter Agnes a New Year's card playfully signed "Nikola Hobson."[71]

Hobson did not hesitate to use his connections in the navy to help promote his friend's radio-controlled boat. In May 1902, he recommended that the boat be included as part of the navy's exhibit at the Pan American Exposition in Buffalo, and he urged Tesla to write to the navy as "I think this is a good opportunity for bringing your patents to the attention of the Navy without the usual difficulties of formalities. I think these patents have great value for our Navy and country and therefore my dear Tesla, do not fail in this matter of the first step toward their introduction." Hobson's efforts to promote Tesla's invention were, unfortunately, to no avail; apparently the invention was reviewed by several high-ranking officers but rejected because two of the officers were engaged in a bitter personal feud.[72] Despite this disappointment, Hobson and Tesla remained close, illustrating how Tesla was attracted to men and worked to develop an intimate friendship.

AN INVENTOR'S INTENSE AND WILD NATURE

So one answer for why Tesla never married was that he was more attracted to men than women. If he was going to have an emotional relationship, then it was most likely going to be with a handsome fellow like Hobson. However, there is a second answer that we also need to consider: marriage did not suit Tesla's approach to invention.

To do so, let's go back to that summer evening in the café with the reporter and take Tesla at his word, that an inventor—at least like himself—"has so intense a nature with so much in it of [a] wild, passionate quality." Invention for Tesla was an intricate dance between rigorous thinking and a vivid imagination, and what made it intense and wild for him was moving back and forth between thinking and dreaming.

Tesla explained how thinking and dreaming were part of the creative process for him in another interview he gave a few weeks later in the summer of 1896. There he offered an explanation of his creative process in the course of discussing his ideas about the "transmission of sight by wire," or what we might call cable television. Tesla's idea involved using a long series of tubes with mirrors inside that would reflect the program from the source to the viewer. This interview, however, is much more interesting for what it tells us about Tesla's cognitive style, and he began with a discussion of how an ideal—the fundamental principle for an invention—evolved in his mind:

> I am now able to indicate, in a general way, how far I have gone toward the solution of the problem. After many fruitless efforts, I have conceived of an idea. I have for a long time scrutinized it and found it agreed with all the established facts I knew, hence, as far as I was capable to decide it is possible. Next I have examined the difficulties which I had to overcome in carrying the idea into practice and have found that they were not insuperable; hence my scheme is practicable. Then I have searched for means of carrying it out and close analysis of these has led me to the conviction that my idea will probably be carried out. . . .
>
> I say that I conceived an idea. In reality I have conceived many, but for the benefit of my fellow men and myself they will never be known. They were either fallacious or impracticable, products of a heated inventor's imagination. But this particular idea is of a different kind.

It withstood my critical examination for weeks, months, and years. Now when through so long a time no flaw in an idea can be found, when through all the stages of excitation and subsequent relaxation it maintains its firm hold upon reason; when as the knowledge of the subject increases and the desire to accomplish grows more intense by approach to realization, it returns after each period of exhaustion with increased force then this idea is a truth. That is, it is a truth so far as the individual observer is concerned, for there is still the superior scrutiny of many [i.e., other experts] which may disclose errors which he [i.e., the inventor] was incapable of perceiving.

It is important to note here that Tesla says what constitutes an "idea" (his word) or an "ideal" (my word) is only true for him and that others may not understand the inventor's idea and may criticize it. For an inventor like Tesla, the idea or ideal served as an organizing principle that he could use to shape and guide his investigations.

In the interview, Tesla warned that inventors and scientists should not rush to announce their ideas; there might still be problems in how the idea should be realized and mistakes could be made. Instead, he suggested that the idea must be processed through the imagination, and he launched into an excursion revealing how his imagination worked:

Have you ever abandoned yourself to the rapture of contemplation of a world you yourself create? You want a palace, and there it stands built by architects finer than Michael Angelo [*sic*]—aye, even finer than my friends McKim, Mead & White. You fill it with marvelous paintings, and statuary and all kinds of objects of art. You summon fairies if you are fond of them. Now, perhaps, you want to sit on a throne, and there is your throne, greater than that of Great Britain! And all your subjects are around you—countless subjects. No fellows to run after you with a pistol, as fellows do after illustrious personages, like William and Nicholas or Li Hung Chang. And if they would, what do you care? You stop their bullets in midair.

Now you walk out in the streets of a wonderful city. Perchance it is one of my cities. Then you may see that all the streets and halls are lighted by my beautiful phosphorescent tubes, that all the elevated railroads are propelled by my motors, that all the traction companies' trolleys are supplied by my oscillators, or else that my friends of the Cataract Construction Company are transmitting all the power by my system from a far-off Niagara. And now, perhaps, you meet a tramp in the street and give him something. Five cents, you think. No, sir; you give him not less than $5,000,000.

Strangely, though, instead of collapsing at your generosity, he looks at you in an insolent way and turns the money in his hand and says contemptuously:—"Take it back, you mean skinflint." And then you throw down your royal insignia and you begin to grapple with him. You are endowed with giant strength, and he is no fellow to fool with either. At any rate, the issue is uncertain. He may be stronger, and then, well— then you wake up, saved, but badly used up. If you defeat him, then you recompense him royally by giving him your insignia and your throne, and you continue your adventurous voyage peacefully and contented.

Suddenly you throw yourself in the roar of a battle, you cut and slash, and a whole army of noble knights flees before you. And now something rattles in the bushes, and you, who know no fear, you run away. Then you may witness a terribly impressive scene of years gone by. You witness the death of your father or your mother, and you go through all the agonies again. You realize the immense gulf that sepa-rates you from them. Then overwhelming desire takes hold of you to be with them again. You know it is impossible to get them back; but never mind, you will invent something, you will discover some force which will reunite those separated molecules and make them form those lovely shapes so dear to you.

And now suddenly there comes a revulsion, and you are throwing a stick at a cat in a backyard. You miss it, too, aggravating circumstance. But years afterward you can tell the exact spot on the wall, you can tell every mark of the stick, and you see exactly how the cat's fur was brushed one or the other way. So your imagination leads you on, from sorrow to joy, from work to play, and all this world is ever present, ever ready for your pleasure and enlightenment, and at your wish and command.[73]

To develop an idea—such as the transmission of sight by wire—Tesla thus argued that one must complement rigorous analysis with exer-cising the imagination. To invent, one must be able to imagine the new device and a world within which it fits. Only by doing so can one perfect the invention and hone the ideal. At the same time, the world of imagination can give rise to desires, wishes, and visions—the illusions—that can be used to convince others to take up an invention.

For Tesla, then, invention required the inventor to have an intense and wild nature—intense in the sense that it required rigorous thinking to hone an ideal, wild in that one had to be able to freely engage and explore it in one's imagination. Both activities required solitude, and, consequently, marriage would not necessarily suit an inventor like Tesla.

Tesla may have come to the conclusion that an inventor needs solitude by drawing on his Orthodox religious background. To be able to discern the logos in the natural and man-made worlds, one has to learn not to be distracted by the temptations of life (see Chapter 1). To be open to the logos, one had to be willing to refine all of one's faculties—mental, physical, and spiritual—so that one was as perfect an instrument as possible for experiencing the Divine order. Perhaps for Tesla this preparation meant eschewing long-term commitments like marriage. Unlike Western Christianity where one overcomes distractions and prepares for enlightenment through asceticism and a rejection of the body, the Orthodox tradition does not presume that such a rigorous dichotomy of mind and spirit is necessarily required; rather, mindfulness can be achieved by living in the world and enjoying the material comforts as God's gifts.[74] Hence Tesla's spiritual preparation did not mean a withdrawal from the good life of New York but a careful management of it so that it did not get in the way of his efforts to hone his rational and imaginative faculties to discern ideals.

BECOMING UNSTUCK: SOLVING THE PUZZLE OF THE RETURN CIRCUIT

Tesla relied on a combination of rigorous thinking and imagination to perfect his ideas for wireless power transmission. As we saw in Chapter 11, by early 1895 Tesla had arrived at a basic scheme for transmitting power around the world without wires. Since electromagnetic waves traveled in straight lines and only a small amount of power carried by them was likely to reach the receiver, Tesla had decided to minimize the waves generated by his apparatus and maximize the ground current that passed between his transmitter and receiver (see Figure

11.6). Moreover, Tesla had hypothesized that if he could generate a ground current at the resonant frequency of the earth, then the power produced by his transmitter might easily travel to receivers located around the world.

Yet as promising as the idea of using ground current seemed, Tesla still had "to ascertain the laws of propagation of currents through the earth and the atmosphere."[75] Working with the transmitter in his Houston Street lab, Tesla set out to determine how electrical oscillations were transmitted through the Earth by once again carrying a small receiver around Manhattan. These local tests, he reported, "enable[d] me to reduce the determination of the effects produced at a distance to simple formulae or rules of electrodynamics. Having found these laws to be rigorously true in certain respects, further trials of this kind became unnecessary, and the dominating idea became to perfect a powerful transmitter."[76]

But while Tesla was pleased to find these formulae for ground transmission, he was still puzzled by what happened in the atmosphere. Yes, he was confident that ground currents could carry power from a transmitter to a receiver, but what completed the circuit from the receiver back to the transmitter? If one rejected electromagnetic waves as the means by which the circuit was completed in the atmosphere, then what made the system work?

Here, circa 1896–97, Tesla was stuck; he didn't have an answer. As he said in the August 1896 interview discussed earlier, "Finally, after a long study, mostly experimental, of all the means and conditions, I have arrived at a few precise facts, enough elements involved in a practical demonstration, and—here I am sticking, sticking since three years."[77] There were things he could do with wireless transmission, but he could not solve the puzzle of the return circuit.

Patents filed in 1896 show that Tesla concentrated not on developing a system employing ground currents but on improving his oscillator so that it could be used for wireless lighting and powering X-ray tubes. He also experimented with a host of circuit interrupters in order to adjust the frequency by which he could charge and discharge the capacitors in his system.[78]

Since several of these projects benefited by boosting the voltage of his high-frequency currents, Tesla continued to improve on his spiral coils made up of numerous turns of fine wire.[79] Using these spiral coils, Tesla filed a patent in March 1897 for a new system for transmitting power using one wire between the transmitter and the receiver (Figure 12.9). This system featured a transmitter and receiver, both of which were essentially transformers. The transmitter employed a generator that produced high-frequency AC. Just as he had done in his oscillator circuits (see Figure 10.3), Tesla fed this high-frequency current into the primary of the transformer, which consisted of a few turns of heavy cable. The secondary of the transformer on the transmitting side was the spiral coil. By using a few turns of thick cable for the primary and many turns of fine wire for the secondary in the transmitter, Tesla was able to step up the voltage to a very high level. While the terminal on the outside of the spiral coil was grounded, the terminal at the center of the coil was connected to a transmission line that carried the power to the receiver. At the receiver, Tesla created a similar transformer, only this time the spiral coil served as the primary and the heavy cable as the secondary. This caused the voltage to be stepped down so that it could be used in ordinary incandescent lamps and motors.[80]

With the system configured in this manner, Tesla again took up the puzzle of the return circuit. How could he eliminate the wire connecting the transmitter and the receiver to create a true wireless power system? To solve this puzzle, Tesla went back to thinking about why Crookes and Geissler tubes produce light when connected to an electrical source. While at atmospheric pressure, most gases oppose the passage of electricity and function as an insulator; however, to make his tubes light up, Crookes had evacuated most of the gas from the glass tubes. At low pressures, the gas glows when it is traversed by a high-voltage current. Reasoning by analogy, Tesla replaced the wire between the transmitter and receiver with what was effectively a giant Crookes tube. In his Houston Street laboratory, he erected a fifty-foot glass pipe between his transmitter and receiver (Figure 12.10). Using a vacuum pump, Tesla lowered the pressure to 120–150 mm of mercury (the pressure of the atmosphere at an altitude of five miles) and discovered that he could create a return circuit from the receiver back to the transmitter.[81] While power moved from the transmitter to the receiver through the ground, Tesla hypothesized that a return circuit

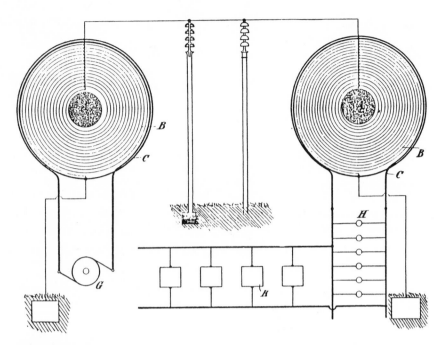

FIGURE 12.9. NT, "Electrical Transformer," US Patent 593,138 (filed 20 March 1897, granted 2 Nov. 1897).

    The transmitter was on the left and the receiver is on the right.

Key:
G   AC generator
C   primary winding of transmitter's transformer
B   secondary winding of transmitter's transformer
B'   secondary winding of receiver's transformer
C'   primary winding of receiver's transformer
H   incandescent lamps
K   electric motors.

was created in the evacuated pipe since the rarefied air allowed the passage of a current from the receiver back to the transmitter. Hence for Tesla, the secret to wireless transmission lay not with electromagnetic waves (i.e., radiation) passing through the atmosphere but that an oscillating current could be conducted through a gas at low pressure. "[T]he transmission of electrical energy," declared Tesla in October 1898, "is one of true conduction, and is not to be confounded with the phenomena of induction or of electrical radiation which have heretofore been observed and experimented with."[82] In emphasizing

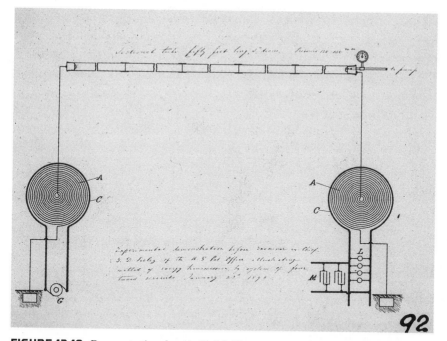

FIGURE 12.10. Demonstration done in Tesla's Houston Street laboratory to show the feasibility of conducting high-frequency currents through a low-pressure gas, 1898. On the left is Tesla's transmitter and on the right is his receiver. Across the top is the fifty-foot glass pipe with reduced pressure. From NTM.

that electrical oscillations moved through the atmosphere via conduction, Tesla was again distancing himself from most other inventors and scientists who felt that Hertzian waves were a form of radiation moving through the ether.

What really excited Tesla about this experiment showing how oscillating currents could move through gases at low pressures was that the process was so efficient; if the voltage and frequency were high enough and the atmospheric pressure low enough, a great deal of power could be transmitted. For Tesla, "the discovery of these new properties of the atmosphere not only opened up the possibility of transmitting, without wires, energy in large amounts, but, what was still more significant, it afforded the certitude that energy could be transmitted in this manner economically. In this new system it matters little—in fact, almost nothing—whether the transmission is effected at a distance of a few miles or of a few thousand miles."[83] As we shall see, this belief that distance was irrelevant figured prominently in how Tesla interpreted the results of the tests he subsequently conducted and the pronouncements he made about his system.

If he could set up a return circuit in a nearly evacuated tube, Tesla now reasoned that he could then do the same at high altitudes where the air was thinner.[84] All that was now needed was to connect the spiral coils in the transmitter and receiver to balloons with a large metallic surface area (Figure 12.11). Floating high in the air, these balloons would allow a current to pass from the receiver back to the transmitter. To depict Tesla's new wireless system, *Pearson's Magazine* ran an illustration showing balloons floating over the skyline of a city (Figure 12.12).

To avoid having balloons on mile-long tethers, Tesla believed that he could take two steps: first, raise the power of his system to millions of volts, and second, locate his transmitter and receivers on mountaintops. With regard to the first, Tesla began experimenting with increasing the power of his transmitter in the Houston Street laboratory. Using a primary consisting of two turns of heavy cable around the perimeter of the main workroom and his favorite spiral coil, Tesla was able to push the voltage up to 2.5 million volts and generate sixteen-foot sparks (Figure 12.13). He further tested the power of the system by taking a receiver out of the lab and traveling by boat up the Hudson River to West Point to see if he could pick up oscillations from his lab. Tesla found that he was able to detect oscillations thirty miles from the lab. During this test, Tesla concentrated on seeing if he could detect the continuous waves generated by his transmitter and he did not use the signal to send a message by Morse code or voice.[85]

Although revealing, these experiments did not tell him where best to locate his transmitter; at what voltages and what altitudes would his system work? What would it take to create "a transmitter of adequate power" that might "bridge the greatest terrestrial distances"?[86] To answer these questions, Tesla realized that he would have to move beyond the confines of his New York laboratory and build a pilot plant.

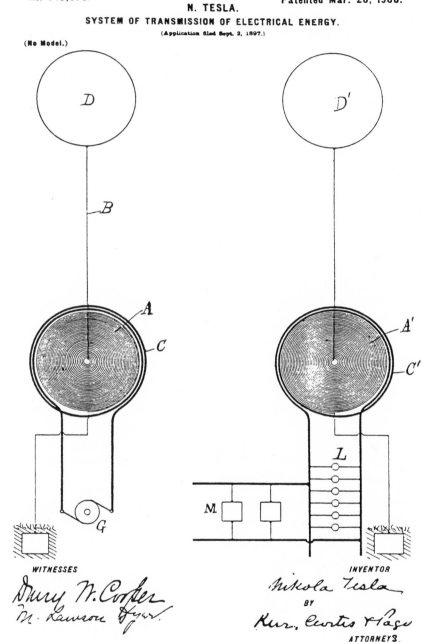

No. 645,576.   Patented Mar. 20, 1900.
N. TESLA.
SYSTEM OF TRANSMISSION OF ELECTRICAL ENERGY.
(Application filed Sept. 2, 1897.)

(No Model.)

FIGURE 12.11. NT, "System of Transmission of Electrical Energy," US Patent 645,675 (filed 2 September 1897; granted 20 March 1900).

D and D' are balloons attached to the transmitter and receiver.

FIGURE 12.12. "Tesla's proposed arrangement of balloon stations for transmitting electricity without wires."

From Chauncey Montgomery McGovern, "The New Wizard of the West," *Pearson's Magazine*, May 1899, 470–76, on 470, in TC 14:105–11.

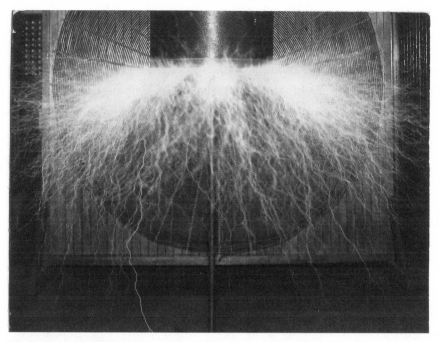

FIGURE 12.13. "Tesla's system of electrical power transmission through natural media.— View of model transformer, or "oscillator," photographed in action. — Actual width of space traversed by the luminous streamers issuing from the single circular terminal terminating the extra coil over sixteen feet. — Area covered by the streamers approximately 200 square feet. — Estimated electrical pressure two and one-half million volts."
    Published in *Electrical Review,* supplement, 26 Oct. 1898, in TC 13:127. From NTM.

COURTING JOHN JACOB ASTOR IV

To move beyond the limits of his New York lab and scale up his system, though, meant money. At first Tesla probably assumed that the funding would come through the Nikola Tesla Company that he had set up with Adams in 1895. Through this company, Adams and Tesla had hoped to entice businessmen who would buy or license the patents for Tesla's wireless lighting system and then put the system into production. Tesla would then be able to plow the profits from any deal into developing his new inventions. But as we have seen, through the mid-1890s, Adams and Tesla had no takers for this new venture (Chapter 11).

Consequently, Tesla sought to raise the necessary funds in other ways. Along with demonstrating his new system in August 1898 for Prince Albert of Belgium (whom he met previously in Paris), Tesla secured a loan of $10,000 from Crawford, a partner in the dry-goods firm of Simpson and Crawford.[87] But Tesla set his sights on catching a big fish: Colonel John Jacob Astor IV (1864–1912), who had served with Teddy Roosevelt and the Rough Riders in the Spanish-American War.

The heir to a $100 million fortune, Colonel Astor was the great-grandson of John Jacob Astor, who had become rich first in the fur trade and then in real estate in New York City. As one of the wealthiest families in America, the Astors ruled New York society; indeed, the social elite in late nineteenth-century America came to be called the Four Hundred because that was supposedly the number of guests who could fit into the ballroom in the New York home of Mrs. Astor, the colonel's mother. Educated at Harvard, Astor followed family tradition and invested in Manhattan real estate. Envious of the success that his cousin, William Waldorf Astor, was having with a new hotel, the Waldorf, Astor built his own luxury hotel next door in 1897 and named it the Astoria. The complex soon became known as the Waldorf-Astoria, and at the time, it was the largest hotel in the world.[88]

But along with building a grand hotel, Astor was fascinated by science and technology. Working in a laboratory at Ferncliff, the family estate, Astor tinkered with several inventions, including a bicycle brake, a "vibratory disintegrator" used to produce gasoline from peat moss, and a pneumatic machine for improving dirt roads. In 1894 he published a science-fiction novel, *A Journey in Other Worlds*, which described life in the year 2000 and travel to Saturn and Jupiter. In this novel, Astor speculated on new technologies such as a worldwide telephone network, solar power, and even a plan to modify the weather by adjusting the Earth's axial tilt.[89] Clearly a technological enthusiast, Astor must have seemed like a most promising patron to Tesla.

Astor was familiar with Tesla's work since he was a director of the Cataract Construction Company, the firm that had built the Niagara power plant. Astor presented Tesla with a copy of his novel in February 1895, and Tesla thanked him for "an interesting and pleasant memento of our acquaintance."[90] In addition, Tesla dined regularly at Delmonico's in order to be seen by the rich and powerful of New York, and he may have met the colonel over dinner. (*A Journey in*

Other Worlds opens with a meeting at Delmonico's of the Terrestrial Axis Straightening Company.) In the fall of 1898, Tesla moved to the Waldorf-Astoria, and that, too, might have given him access to Astor.

Tesla had invited Astor to buy shares in the Nikola Tesla Company in December 1895, but Astor had shown no interest.[91] Now three years later, Tesla was determined to woo Astor and employed a rhetorical strategy similar to the one he had used ten years earlier with Peck and the egg of Columbus (see Chapter 4). Like Columbus, Tesla was about to discover new worlds through his inventions, but just as Columbus had relied on Queen Isabella, Tesla needed a strong patron as well. Tesla hoped that such a role would appeal to the colonel. "My dear Astor," wrote Tesla in January 1899,

> It has always been my firm belief that you take a genuine, friendly interest in myself personally as well as in my labors. . . . Now I ask you frankly, when I have friend like J.J.A., a prince among wealthy men, a patriot ready to risk his life for his country, a man who means every word he says—who puts such a value on my labors and who offers repeatedly to back me up—have I not a foundation for believing that he would stand by me when, after several years of hard work I have finally brought to commercial perfection some important inventions which, even at the most conservative estimate, must be valued at several million dollars[?]

Although Westinghouse had given him $500,000 for his AC polyphase system and Adams had invested $100,000 to develop "14 U.S. and as many foreign patents" related to his oscillator, Tesla explained, there was still a "powerful clique" who opposed him (although it is not clear who exactly made up this clique). "And it is chiefly for this reason that I want a few friends, like yourself," he continued, "to give me at this moment their valuable financial and moral support."

Having established that he was seeking Astor to be his patron, Tesla introduced his marvelous inventions and how they stood to revolutionize the world. First, he extolled the virtues of his lighting system:

> I now produce a light superior by far to that of the incandescent lamp with one third of the expenditure of energy, as my lamps will last forever, the cost of maintenance will be minute. The cost of copper which

in the old system is a most important item, is in mine reduced to a mere trifle, for I can run on a wire sufficient for one incandescent lamp more than 1000 of my own lamps, giving fully 5000 times as much light. Let me ask you, Colonel, how much is this alone worth when you consider that there are hundreds of millions of dollars invested to-day in electric light in the various chief countries in which I have patented my inventions in this field?

Tesla's strategy was still to develop the patents for his lighting system to the point where they could be sold at a profit to companies that would in turn put them into production. "Sooner or later," he told Astor, "my system will be purchased by the Whitney syndicate [which was developing electric street railways], G.E., or Westinghouse, for otherwise they will be driven out of the market."

Next, Tesla took up all his other inventions:

> Then consider my oscillators and my system of transmitting power without wires, my method of directing the movement of bodies at a distance by wireless telegraphy, the manufactures of fertilizers and nitric acid from the air, the production of ozone . . . and many other important lines of manufacture as, for instance cheap refrigeration and cheap manufacture of liquid air, etc.—and you will see that, putting a fair estimate on all, I cannot offer to sell any considerable amount of my property for less than $1000 a share. I am perfectly sure that I will be able to command that price as soon as some of my inventions are on the market.

To underline that his inventions were a sound investment, Tesla reminded Astor that he had negotiated contracts with "the Creusot Works in France, the Helios Company in Germany, Ganz & Company in Austria, and other firms" for the manufacture of his motors. Tesla boasted that not only had his research in the past "paid $1500 for every $100 invested, on average" but "I am fully confident that the property which I have now in my hands will pay much better than this."

Having whetted Astor's appetite, Tesla moved in for the sale, inviting the colonel to invest $100,000. "If you do not take that much interest you will put me at a great disadvantage," Tesla wrote, and he hoped that if Astor came in, so would the colonel's associates, Clarence

McKay and Darius Ogden Mills. In closing, Tesla reassured Astor that if "after six months you should have any reason to be dissatisfied, it will be my first duty to satisfy you."[92]

Within a few days of receiving this carefully crafted sales pitch, Astor signed an agreement with Tesla. Having "placed faith" in Astor, Tesla had bought up enough shares in the Nikola Tesla Company so as to have majority control, leaving Adams, Rankine, Brown, and Coaney with minority interests.[93] In return for five hundred shares, Astor promised to invest $100,000 and was made a director of the company. When all of the shares were transferred to Astor, the colonel gave Tesla an initial payment of $30,000 and then promptly left on a trip to Europe.[94]

SPURRED ON BY MARCONI

Although Astor was primarily interested in having Tesla perfect a lighting system using his oscillator and new lamps, Tesla paid little attention to his patron's wishes. Instead, Tesla used Astor's support to pursue his vision of wireless power.

Tesla was now especially anxious to do so because he was becoming concerned about what a young Italian, Guglielmo Marconi (1874–1937), was doing with his wireless system. Like Tesla, Marconi had been fascinated by Hertz's apparatus and began experimenting with it in the attic of his father's house outside Bologna in 1894. From the start, Marconi sought to develop a system that could send telegraph messages, and he focused on increasing the distance over which he could send them. In order to finance and promote his system, Marconi traveled to England in 1896 where he could take advantage of the business connections via his mother's family, the Jamesons, who were prominent in the whiskey and grain business. Marconi steadily improved his apparatus, and by the fall of 1898 he could send messages over distances of eighty to one hundred miles.[95] Unlike Tesla, who demonstrated his apparatus privately to friends and an occasional reporter, Marconi offered regular public demonstrations of his system.

Impressed by these demonstrations, newspapers in England and America began touting Marconi's wireless telegraph as a breakthrough.

This positive coverage of Marconi annoyed Tesla since from his perspective, Marconi had done nothing new. As far back as 1890 Tesla had been experimenting with wireless apparatus, and in his 1893 lecture he had outlined how one could send messages over a distance. Careful to avoid using Marconi's name, Tesla complained in the *Electrical Review* in January 1899 that "One can not help admiring the confidence and self-possession of experimenters, who put forth carelessly such views and who, with but a few days', not say hours', experience with a device, venture before scientific societies, apparently unmindful of the responsibility of such a step, and advance their imperfect results and opinions hastily formed. The sparks may be long and brilliant, the display interesting to witness, and the audience may be delighted, but one must doubt the value of such demonstrations." And Tesla could not resist poking fun at the modesty of Marconi's apparatus. Unlike Tesla's sophisticated and powerful system, his rival's invention was "a worthless trap of interrupting currents, which usually consumes nine-tenths of the energy and is . . . just suitable for the amusement of small boys, who are beginning their electrical experience with Leclanche batteries and $1.50 induction coils."[96]

Although Tesla avoided mentioning Marconi in the *Electrical Review*, reporters suspected that he was concerned about his young Italian rival. Indeed, the gossipy *Town Topics* poked fun at Tesla, taking the view that while Tesla was making promises, Marconi was getting results:

> Tesla, America's Own and Only Non-Inventing Inventor, the Scientist of the Delmonico Café and Waldorf-Astoria Palm Garden, has been at it again. This time the news of young Marconi's success in telegraphing through space fired Tesla to feats hitherto undreamed of, and he filled columns in the *Herald*—which paper, I much fear me, inclines to help Tesla make a guy of himself—with profound droolings about volts and resistances and circuits and ampères and things and things. Tesla says he can do everything that Marconi has done. Of course, he doesn't really do them but that may be because he is afraid someone else may find out how they are done. He knows all about the theory and the practical machinery of Marconi's messages through miles of space and could prove it too—if old Bill Jones were alive. Indeed the actual results of the methods of the two inventors show only this slight

difference: Marconi telegraphs through space and Tesla talks through space.[97]

In March 1899, Marconi successfully sent a message across the English Channel from Wimereux in France to the South Foreland Lighthouse in England. Not to be outdone, Tesla announced that he was prepared to send messages instantaneously around the world. As he boasted in the *New York Journal*:

> The people of New York can have their private wireless communication with friends and acquaintances in various parts of the world.
>
> It will be no great wonder to have a cable tower [with a balloon tethered to it] than it is now to have a telephone in your house.
>
> You will be able to send a 2,000 word dispatch from New York to London, Paris, Vienna, Constantinople, Bombay, Singapore, Tokio [*sic*] or Manila in less time than it takes now to ring up "central."[98]

Having now promised worldwide wireless telegraphy, Tesla knew that he had to deliver results. To do so, he decided to use Astor's money to build the pilot plant he needed to work out the operating details of his wireless system. To build that plant, Tesla headed west to Colorado.

CHAPTER THIRTEEN
STATIONARY WAVES
(1899–1900)

Everything you can imagine is real.

PABLO PICASSO

By the spring of 1899, all the elements needed to realize his ideal wireless power system were falling into place for Tesla: he had perfected the circuitry needed to create a powerful high-voltage, high-frequency transmitter, he had discovered how to tune his transmitter and receivers by adjusting the capacitance and inductance, and he had become convinced that the atmosphere could serve as the return circuit for his system.

Yet to make this system practical, several areas required further investigation. First he had "to ascertain the laws of propagation of currents through the earth and the atmosphere" to ensure that his system could send power or messages from one point to another. Next Tesla sought "to develop a transmitter of great power," meaning that he had to figure out how to build coils and capacitors capable of working at millions of volts. And finally, knowing that he would need to deliver power or messages to specific users, Tesla sought to improve his methods of tuning, or as he put it, "to perfect means for individualizing and isolating the energy transmitted."[1]

To address these challenges, Tesla decided to move beyond the confines of his laboratory in New York City and relocated to Colorado Springs where he worked from May 1899 to January 1900. In Colorado, Tesla was at his peak as a creative experimenter, but overconfidence in the ideal system that he had been evolving in his imagination got in the way of rigorously testing his ideas and collecting the hard evidence he would need later to defend his patents and woo investors. Enamored of his ideal system, Tesla pounced on the first clues of success—illusions—rather than confront the problems and challenges that inevitably come with moving an idea from the imagination to the material world.

RELOCATING TO COLORADO SPRINGS

Located sixty-three miles south of Denver and at an elevation of six thousand feet, Colorado Springs was founded in 1871 as a posh mountain resort (Figure 13.1). With its natural scenic beauty (Pike's Peak is just west of town), high altitude, dry climate, and fluoride-rich waters, Colorado Springs attracted a well-to-do clientele seeking relief

FIGURE 13.1. Colorado Springs in the early twentieth century. From the postcard collection of Jane Carlson.

from a variety of ailments including tuberculosis. In addition to the well-heeled tourists, the nearby gold-mining districts of Cripple Creek and Victor produced a number of millionaires who built fine homes in Colorado Springs.[2]

One newspaper report suggests that Tesla had made a brief visit to Colorado Springs in 1896 to conduct a few wireless experiments; the 1899 move came at the recommendation of Leonard E. Curtis, who had been a partner of Parker Page, Tesla's patent attorney.[3] Curtis had moved to Colorado Springs to regain his health and invited Tesla to come and conduct experiments there. In rural Colorado Springs, Tesla could erect a larger system than the one he had in his New York laboratory and safely conduct experiments with higher voltages. "My coils [in New York] are producing 4,000,000 volts," Tesla told Curtis, and "sparks jumping from walls to ceilings are a fire hazard." Moreover, by being in the mountains, Tesla could study how currents were conducted through both the earth's crust and the atmosphere at high altitude.

Attracted by the chance to work away from the press, Tesla accepted Curtis's invitation to relocate temporarily and outlined his needs: "This is a secret test. I must have electrical power, water, and my own laboratory. I will need a good carpenter who will follow instructions. I am being financed for this by Astor, and also Crawford and Simpson. My work will be done late at night when the power load will be least." Delighted to have Tesla visiting, Curtis arranged for Tesla to get free power from the local electrical utility, the El Paso Power Company.[4]

Tesla traveled to Colorado Springs by way of Chicago, where he stopped to lecture before the Commercial Club, whose members constituted the Windy City's business elite. While the high point of the lecture was a demonstration of his radio-controlled boat (see Chapter 12), Tesla told his audience of his plans to broadcast power, signal Mars, and use electricity to convert nitrogen in the atmosphere into fertilizer. Although the reporter from the *Chicago Tribune* was disappointed that Tesla delivered this lecture "in a low voice and a defective accent," the reporter from the *Times-Herald* found him mesmerizing when interviewed one-on-one: "Tesla's bright eyes glowed as he spoke of his ends in his work. Leaning forward, peering almost each moment into the eyes of his interviewer to make sure that his meaning had been understood, he proved a talker from whose train of reasoning there was no escape while a man was under his influence."

In talking with the *Times-Herald* reporter, Tesla elaborated on his thoughts about interplanetary communication:

> Signaling to Mars? I have apparatus which can accomplish it beyond any question. If I should wish to send a signal to that planet I could be perfectly certain that the electrical effects would be thrown exactly where I desire to have them. . . . Further than that, I have an instrument by which I can receive with precision any signal that might be made to this world from Mars. Of course that is not the same as saying that I could establish communication with beings on Mars, but if they should know that I was signaling them, and had intelligence approximately similar to ours, communication would not be impossible.

Tesla used the *Times-Herald* interview to again position himself vis-à-vis Marconi without mentioning his rival by name. While Marconi was pursuing mere applications for money, Tesla argued that he was seeking the underlying principles of this new branch of technology:

> What I am doing is to develop a new art. Is that not more important than the attempt to elaborate an old art in some of its phases? I want to go down to posterity as the founder of a new method of communication. I do not care for practical results in the immediate present. Where I have time I stop to develop the application of the principles that I have announced, but that is part of the work which it is usually safe to leave to others. They will do it because there is money in it. For myself I am content to find the new principles through the knowledge of which the applications become possible.[5]

Leaving Chicago by train, Tesla arrived in Colorado Springs on 18 May 1899. At his hotel, the Alta Vista, he was immediately accosted by a reporter who asked him about his plans. "I propose to send a message from Pike's Peak to Paris," Tesla boldly replied. "I see no reason why I should keep the thing a secret longer," he continued. "I have been preparing for a long while to come here and carry on these experiments which have been so much to me. I am here to work out a system of transmission at a distance. I propose to propogate [*sic*] electrical disturbances without wires."[6]

To conduct these investigations, Tesla immediately began construction of an experimental station on the eastern outskirts of town. Located on an empty pasture known as Knob Hill, the station was positioned between the State Deaf and Blind Institute and the Printers' Union Home (Figure 13.2). (Today this is the intersection of Kiowa

FIGURE 13.2. View of the experimental station from the Pike's Peak side showing the tower and retractable pole that Tesla added in September 1899. Tesla used the smaller tower and hanging ball to measure how capacitance varied with distance from the earth. The building in the distance is the Union Printers Home.

Plate I, CSN, p. 299. From NTM.

and Foote streets.) From Knob Hill, Tesla had splendid views of Pike's Peak to the west and the rolling plains to the east.

Built by a local carpenter, Joseph Dozier, the experimental station was a sixty-by-seventy-foot wooden barn, consisting of one main room and two small offices on the front. Over the main room Tesla had Dozier fashion a roof that could be opened and closed, as well as a balcony for viewing the countryside. Though Tesla initially planned to use balloons to carry his aerials aloft, he soon realized that existing balloons could not lift the weight of hundreds of feet of wire. Consequently, he devised a telescoping mast that could hoist a thirty-inch copper-covered ball to a height of 142 feet. To stabilize the mast, Tesla added a twenty-five-foot tower to the roof of the station.

Tesla was especially anxious to keep his work in the experimental station secret. Dozier had included a single window in the rear wall of the station, but local boys kept peeking through it so Tesla had it boarded up. To keep other curious people away, Tesla ordered a fence be erected around the station, with signs reading "KEEP OUT. GREAT DANGER." One of Tesla's assistants added a final warning at the door, which quoted Dante's Inferno: "Abandon hope, all ye who enter here."[7]

To equip his experimental station, Tesla turned to two assistants who came from his New York laboratory: Fritz Lowenstein (1874–1922) and an assistant named Willie. A native of the Carlsbad region (today part of the Czech Republic), Lowenstein had studied engineering in Europe before immigrating to America in 1899 and coming to work for Tesla. Tesla took an immediate liking to Lowenstein and was soon confiding in him, sharing key ideas about wireless power. Willie, in contrast, was a mechanic who worked for Tesla in his Houston Street laboratory. Willie proved to be a disappointment and was soon discharged; to replace him, Tesla hired a local teenager, Richard B. Gregg, whose father knew Curtis. In the fall of 1899, Lowenstein left Colorado Springs and Tesla had another mechanic, Kolman Czito, come out from his New York laboratory to assist with the experiments.[8]

Under Tesla's direction, Lowenstein and Gregg built an enormous magnifying transmitter. In the station's main room, they constructed a circular wooden wall about six feet high and 49.25 feet in diameter (see Figure 13.7). Around the top of this wall they wound two turns of thick cable in order to create the primary winding of the transmitter.

In the center of the room they built the secondary coil using a hundred turns of finer wire.[9] One end of this secondary coil could be connected to either a spherical terminal inside the laboratory or the copper ball atop the mast while the other end was grounded.

To provide AC to the transmitter, Tesla tapped into the streetcar line that stopped just at the edge of the Knob Hill prairie. He stepped up this 500-volt current by employing a 50-kilowatt Westinghouse transformer that he rewound so that it converted the incoming current to 20,000 or 40,000 volts.[10] The transformer was connected to a large bank of capacitors that were automatically interrupted (and hence discharged) by a motorized breakwheel (Figure 13.3). Rounding out the equipment were several large coils that could be moved around the space between the secondary and the primary.[11]

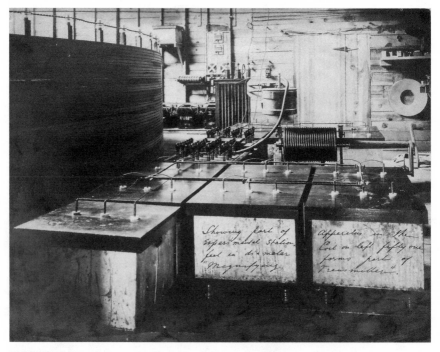

FIGURE 13.3. Interior of the experimental station showing the components that provided power to the primary coil of the magnifying transmitter.

In the front are the capacitors and behind them are the breakwheel and regulating coil. On the wall in back are the lightning arresters and the Westinghouse supply transformer is in the box on the wall to the left. From Plate III, CSN, p. 301. From NTM.

TAKING THE EARTH'S PULSE

As the experimental station took shape, Tesla came to appreciate not only the natural beauty but also the scientific potential of Colorado Springs. "I had not been there but a few days," he later wrote, "when I congratulated myself on the happy choice and I began the task, for which I had long trained myself, with a grateful sense and full of inspiring hope. . . . To this was added the exhilarating influence of a glorious climate and a singular sharpening of the senses. In those regions the organs undergo perceptible physical changes. The eyes assume an extraordinary limpidity, improving vision; the ears dry out and become more susceptible to sound. Objects can be clearly distinguished there at [great] distances . . . [and] claps of thunder [can be heard] seven and eight hundred kilometers away."[12]

In late June and early July, as Lowenstein and Gregg continued to fit out the station, Tesla began making observations in the clear, crisp environment. Since he intended to have his transmitter send currents through the Earth to a receiver some distance away, an immediate task was to study the earth's electrical potential and note how it varied. Because the numerous systems in New York—for telegraphy, telephony, lighting, and transportation—produced too much electrical interference, Tesla had not been able to make any reliable measurements of whether the Earth possessed a natural electrical potential or charge. If the Earth were uncharged, then Tesla would have to use his magnifying transmitter to introduce a tremendous amount of power in order to make the Earth vibrate electrically and transmit power over distances. To use the metaphor of a football (discussed in Chapter 11), an uncharged Earth would be the same as a football with little or no air inside it. However, if the Earth already possessed an electric potential, then Tesla would only need to add a small amount of electricity in order to transmit power; a charged Earth would be the equivalent of a fully inflated football.[13]

To study the earth's electrical potential, Tesla rigged up an instrument composed of a coherer with an ink recorder. The coherer consisted of a glass tube filled with loose iron filings between two terminals; whenever the tube detected a high voltage—such as from a spark or electromagnetic wave—the filings would line up to create a conducting path between the terminals. Because the filings tended to stay in position after detecting a signal, some experimenters added a tiny hammer that would jar the filings loose; in his design, Tesla added a clockwork that regularly rotated the coherer.[14]

To increase the sensitivity of his coherer, Tesla placed it in the secondary circuit of a transformer, while the primary of the transformer was connected to the ground and an elevated terminal of adjustable capacity. This meant that any variations of the electrical potential in the Earth would give rise to electric surgings in the primary winding, and these in turn would induce currents in the secondary winding and hence in the instrument. Using this arrangement, Tesla found that "[t]he earth was . . . literally, alive with electrical vibrations, and soon I was deeply absorbed in the interesting investigation."[15]

Pleased to discover that the Earth did indeed have an electrical potential, Tesla next needed to know how electric currents flowed through the earth's crust. To determine this, Tesla carefully monitored the vibrations with his receivers, and he soon noticed that his receivers were more strongly affected by lightning discharges taking place in far-off thunderstorms than they were by lightning from nearby storms; common sense would suggest that the more distant the lightning discharge, the weaker the signal that would be picked up by the receiver. "This puzzled me very much," Tesla recalled. "What was the cause?"[16]

Tesla soon came up with a hunch. As he walked back to the hotel with Lowenstein one evening, he suddenly realized that the variations could be caused if the lightning bolts set up stationary waves in the earth's crust. A stationary wave is created when two waves traveling in opposite directions add in phase to create a new single wave whose amplitude is stationary in time.[17] A simple example of a stationary wave is what happens when one shakes one end of a string while its other end is fastened to a wall. As one vibrates the loose end, a wave travels along the string to the wall; there the wave is reflected back along the string. If one adjusts the incoming vibrations so as to be the resonant frequency of the string, then the two waves will add together to create a single wave whose peaks and valleys appear to be standing still. Just as mechanical vibrations can travel along a string, so Oliver Lodge had demonstrated in 1887 that electromagnetic oscillations can travel along a wire or conductor, be reflected at some point, and set up a stationary wave.[18]

In the case of the lightning discharges, Tesla surmised that the lightning strokes set off an electromagnetic wave in the earth's crust that reflected back on itself to create a stationary wave. While preparing his Franklin Institute lecture in 1893 Tesla had first thought that electromagnetic stationary waves might be set up in the earth, but at that time he "dismissed it as absurd and impossible." Now, in Colorado, noted Tesla, "my instinct was aroused and somehow I felt that I was nearing a great revelation."[19]

Tesla confirmed his hunch by tracking a spectacular thunderstorm on 3 July 1899. That evening, a violent storm broke loose in the mountains to the west, passed over Colorado Springs, and then moved quickly east onto the plains. According to Tesla, the storm produced an "extraordinary display of lightning, no less than 10–12 thousand discharges being witnessed inside of two hours. The flashing was almost continuous and even later in the night when the storm had abated 15–20 discharges per minute were witnessed. Some of the discharges were of a wonderful brilliancy and showed often 10 or twice as many branches."

To measure these lightning discharges, Tesla connected his rotating coherer to the ground and an elevated plate. To magnify any electrical effects transmitted through the ground, he inserted a capacitor between the coherer and the ground. And in order to record each lightning discharge, the coherer caused a telegraph relay to sound. As he reported in his notes, "The relay was not adjusted very sensitively but it began to play, nevertheless, when the storm was still at a distance of about 80–100 miles, that is judging the distance by the velocity of sound. As the storm got nearer the adjustment had to be rendered less and less sensitive until the limit of the strength of the spring was reached, but even then it played at every discharge."[20]

As the storm passed overhead, Tesla quickly set up a second instrument. An electric doorbell was connected to the Earth and elevated terminal so that it rang in response to each lightning discharge. This arrangement was similar to a lightning detector used by the Russian physicist Alexander Popov in 1895.[21] To this instrument Tesla added a small spark gap that was bridged by a bright spark whenever lightning occurred. To get a sense of the strength of the current passing between the ground and the elevated plate, Tesla held his hands across the gap and felt the shock that came with each lightning stroke.

But "[a]s the storm receded," noted Tesla, "the most interesting and valuable observation was made." As the storm continued east over the plains, Tesla turned back to using his rotating coherer and relay. As he recorded in his notes:

> the instrument was again adjusted so as to be more sensitive and to respond readily to every discharge which was seen or heard. It did so for a while, when it stopped. It was thought that the lightning was now too far and it may have been about 50 miles away. All of a sudden the instrument began again to play, continuously increasing in strength, although the storm was moving away rapidly. After some time the indications again ceased but half an hour later the instrument began to record again. When it once more ceased the adjustment was rendered more delicate, in fact very considerably so, still the instrument failed to respond, but half an hour or so it again began to play and now the spring was tightened on the relay very much and still it indicated the discharges. By this time the storm had moved away far out of sight. By readjusting the instrument and setting it again so as to be very sensitive, after some time it again began to play periodically. The storm was now at a distance of greater than 200 miles at least. Later in the evening repeatedly the instrument played and ceased to play in intervals of nearly half an hour although most of the horizon was clear by that time.[22]

To explain why the signals started and stopped every half hour, Tesla concluded that he was observing stationary electromagnetic waves. He reasoned that the lightning strokes set off an electromagnetic wave in the earth's crust that then reflected back on itself to create the stationary wave. Tesla was not certain where the waves were reflected. "It would be difficult to believe that they were reflected from the opposite point of the Earth's surface, though it may be possible," he observed. "But I rather think that they are reflected from the point of the cloud where the conducting path began; in this case the point where the lightning struck the ground would be a nodal point."[23] Since this nodal point would change as the storm continued to move while Tesla's receiver stayed in one place, the receiver would respond periodically as a peak of the stationary wave passed through the ground underneath the receiver.[24]

As it turns out, it is possible to set up stationary electromagnetic waves, not necessarily in the earth's crust but between the ionosphere and the earth's surface in what is called the Schumann cavity. By using extremely low frequency waves (ELF), the U.S. Navy discovered that stationary waves penetrate deep into the ocean, making it possible to maintain radio contact with nuclear submarines. From the 1980s to 2004, the navy operated stations at Clam Lake, Wisconsin, and Republic, Michigan, that transmitted ELF signals to submarines. To transmit ELF signals, these stations required an underground antenna that extended twenty-eight miles.[25] The navy's ELF project suggests that Tesla probably did detect stationary waves produced by lightning storms. His observations were based on actual physical phenomena.

Tesla regarded the discovery of stationary electromagnetic waves to be "of immense importance," for he now knew not only that the Earth was electrically charged but how electromagnetic waves traveled through the earth. Prior to this discovery, Tesla thought that the Earth might behave "like a vast reservoir or ocean which, while it may be locally disturbed by a commotion of some kind, remains unresponsive and quiescent in a large part or as a whole." In this case, electromagnetic waves—such as those produced by lightning—would travel for a distance and then simply peter out, much like the waves created when a stone dropped in the ocean are strong around the point where the stone hits the water and then dissipate in concentric circles. However, the existence of stationary waves suggested to Tesla that the Earth did not behave as an ocean when it came to electromagnetic waves. "Impossible as it seem[s]," explained Tesla, "this planet, despite its vast extent, behave[s] like a conductor of limited dimensions." And if stationary waves could be set up by lightning, he concluded, then "it is now certain that they can be produced with an oscillator."[26] In subsequent experiments at Colorado Springs, Tesla strove to generate low-frequency waves to imitate those he had detected during lightning storms.

For Tesla, the discovery of stationary waves meant that his system would have far greater reach than Marconi's apparatus. Yes, Marconi had sent messages across the English Channel, but now Tesla felt he could transmit both messages and power around the world. "Not only was it practicable to send telegraphic messages to any distance without wires," Tesla later wrote, "but also to impress upon the entire globe the faint modulations of the human voice, [and] far more still, to transmit power, in unlimited amounts, to any terrestrial distance and almost without loss."[27]

Not only did lightning storms allow Tesla to discover stationary waves, but he continued to track storms over the next few months to determine how far his transmitter should reach. "This I did," he explained,

> by comparison with lightning discharges which occurred almost every day and which permitted me to determine the effect of my transmitter and to ascertain experimentally the energy which it was capable of transmitting, as compared with that energy which was transmitted from a certain great distance by a lightning discharge. These I could follow up to distances of many hundreds of miles, and I could at any time tell precisely how much of a fraction of a watt I would obtain with my transmitter in a circuit situated at any point of the globe. The energy ascertained by measurement agreed exactly with that determined by calculation.[28]

Here Tesla was reasoning by analogy. He would observe the movement of a lightning storm, determine how far away it was, and measure with his instruments how the strength of the stationary waves varied with the distance. Tesla then assumed that if a storm could transmit so much power over such-and-such a distance, there should be no problem in using his transmitter to send power over the same distance. "With these stupendous possibilities in sight," wrote Tesla, "I attacked vigorously the development of my magnifying transmitter, now, however, not so much with the original intention of producing one of great power, as with the object of learning how to construct the best one."[29] But before he took up this task, his instruments detected another interesting set of signals.

AN INTERPLANETARY MESSAGE?

Thrilled to have found that stationary electrical waves could be set up in the earth's crust, Tesla continued to improve his instruments for detecting feeble electrical disturbances as far away as 1,100 miles. In particular, Tesla connected the coherer to a second oscillator that injected a radio frequency (RF) voltage into the circuit so that the coherer was highly charged and ready to go off in response to the smallest voltage change. With this injected RF voltage, Tesla could connect a telephone receiver to the coherer circuit so that he would hear a beep each time the coherer detected electromagnetic oscillations.[30] In part, he did so thinking that such a sensitive instrument might be used for tracking the speed and direction of storms. Aware that Britain, Germany, and the United States were building up their navies (see Chapter 12), Tesla thought a storm-tracking device might be used by battleships to avoid bad weather.[31]

However, this highly sensitive receiver provoked another discovery. Working at night, Tesla was amazed to detect weak oscillations consisting of regular beeps: first one, then two, and finally three beeps. "My first observations [of these beeps] positively terrified me," Tesla recalled later, "as there was present in them something mysterious, not to say supernatural. . . . I felt as though I were present at the birth of a new knowledge or the revelation of a great truth."

Puzzled that the beeps had "such a clear suggestion of number and order," Tesla at first considered whether they were "electrical disturbances as are produced by the sun, Aurora Borealis and Earth currents, and I was as sure as I could be of any fact that these variations were due to none of these causes. The nature of my experiments precluded the possibility of the changes being produced by atmospheric disturbances." Rejecting these possible solar or terrestrial causes, Tesla appears not to have been able to determine the cause of these unusual signals while he was in Colorado. Over the next year or so (1899–1900) he continued to think about these unusual observations until "the thought flashed upon my mind that the disturbances I had observed might be due to an intelligent control. Although I could not decipher their meaning, it was impossible for me to think of them as having been entirely accidental. The feeling is constantly growing on me that I have been the first to hear the greeting of one planet to another."[32] At the end of 1900 he concluded that the beeps must indeed be from another planet, and he announced this conclusion in a letter to the American Red Cross in January 1901 (discussed in Chapter 14).

In his earliest interviews about the beeps, Tesla insisted only that the signals were extraterrestrial in nature, but reporters were quick to assume that the signals must have come from Mars.[33] In studying Mars in the late 1870s, the Italian astronomer Giovanni Virginio Schiaparelli had observed a network of long, straight paths or channels that he labeled "canali" on his maps of the red planet. Many people concluded that Schiaparelli's canals could not have been caused by natural forces and were an indication of intelligent life on Mars. The idea that Mars was inhabited was advanced further by the American amateur astronomer Percival Lowell, who built an observatory in Flagstaff, Arizona, specifically to view the Martian canals. In his 1895 book *Mars*, Lowell argued that the planet was suffering from severe drought and that the canals were an ingenious response by Martians to direct water from the polar ice caps to central parts of the planet.[34]

Lowell's ideas about intelligent life on Mars received wide circulation in newspapers and magazines, and Tesla was certainly aware of these theories. As we have already seen, Martin had speculated in his 1895 article that Tesla's oscillator might be used to "call up" the Martians (see Chapter 11) and Tesla had elaborated on these possibilities when lecturing in Chicago in May 1899.

Although Tesla asked George Scherff to send him an astronomy book during the summer of 1899, there is no mention of these extraterrestrial signals in the notes he kept in Colorado. Consequently, Tesla's biographers have puzzled over what he might have actually detected with his sensitive receiver. What could Tesla have heard with his receiver that he took to be a message from Mars?

One explanation advanced by biographer Marc Seifer was that Tesla picked up signals from tests Marconi was conducting with the British and French navies in July 1899.[35] The difficulty with this explanation is that Tesla's receiver was not tuned to the frequencies being used by Marconi. According to Kenneth L. Corum and James F. Corum, while Marconi was transmitting at an RF frequency, Tesla's receiver was set to detect very low frequency (VLF) waves in the 8 to 22 KHz range.[36] Tesla was working with waves in this range because he believed that waves with a low frequency would propagate more efficiently through the earth's crust. Moreover, powered only by batteries, Marconi's transmitter in 1899 probably did not have sufficient power to generate waves that could travel from England to Colorado.

Indeed, to transmit across the Atlantic in 1901, Marconi had to develop a system consisting of a 25-horsepower steam engine driving an AC generator that produced 2,000 volts that in turn was stepped up to 20,000 volts.[37]

Rejecting the explanation that Tesla had simply picked up Marconi's signals, the Corum brothers instead suggested that Tesla did indeed detect extraterrestrial radio signals. After establishing that Tesla's receiver was operating in the VLF range, the Corums investigated what sort of VLF signals might have been coming from space during the summer of 1899. As it turns out, one of Jupiter's moons, Io, emits a 10 KHz signal as it passes through a torus of charged plasma particles that surrounds the planet (Figure 13.4). First detected in 1955, radio signals from Io often come as a series of pulses. To test their explanation, the Corums rebuilt Tesla's receiver, used it during a Jovian radio storm in 1996, and recorded a series of beeps similar to what Tesla reported hearing in 1899.

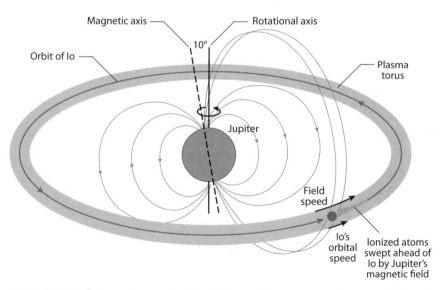

FIGURE 13.4. Diagram showing how Jupiter's moon Io passes through a torus of charged particles.

Electromagnetic waves are generated as Io pushes charged particles through Jupiter's magnetic field. Because Io's rotational axis is not parallel to the axis of Jupiter's magnetic field, Io sweeps through stronger and weaker parts of the magnetic field, thus causing pulses in the generated waves.

Adapted from http://physics.uoregon.edu/~jimbrau/BrauImNew/Chap11/FG11_20.jpg.

Finally, to explain why Tesla associated these signals with Mars, the Corums used astronomical software to determine where Jupiter and Mars would have been in the night sky over Colorado Springs during the summer of 1899. With this software, they found that on several nights in July 1899 Jupiter would have emitted a signal for part of the evening but would have stopped just as Mars was setting in the western sky. If Tesla had looked out the doors of the experimental station as he heard the beeps stop, he would have seen Mars disappear behind the mountains, making it easy for him to connect the red planet with the cessation of the signals. Just as Tesla detected stationary waves set up in the Schumann cavity, the Corums argue that Tesla was again observing real phenomena.[38]

RUNNING THE MAGNIFYING TRANSMITTER

In late July, as his assistants finished assembling the magnifying transmitter, Tesla shifted from listening to the Earth and sky to investigating how to best operate this big machine. In his laboratory on Houston Street, Tesla had been able to reach 3 million volts and produce 16-foot discharges; with this larger magnifying transmitter in Colorado, he hoped to generate 50 million volts and produce artificial lightning bolts between 50 and 100 feet long.[39]

For Tesla, a typical day at the experimental station began with a buggy ride from the Alta Vista Hotel out to Knob Hill. He would spend the morning in his office in one corner of the lab, making calculations, planning the next round of experiments, and engaging Lowenstein in long discussions. At midday, Tesla and Lowenstein enjoyed a lunch sent from the hotel. Since the El Paso Electric Company's primary business was providing electric lighting, it probably did not run its generators during the day, and hence Tesla had to wait until the power came on in the late afternoon to begin his experiments with the magnifying transmitter.[40] When Tesla did throw the switch, his teenage assistant, Gregg, recalled that "enormous sparks would pour between the balls up high. Often the sparks were 15 or 20 feet long, just like lightning. They made a big crash that echoed inside the lab and could be heard from some distance away."[41]

In operating the magnifying transmitter, Tesla sought to generate a powerful current that could be sent through the ground. To do so, he needed to push the voltage up as high as possible and determine the optimal frequency for transmitting through the earth. To raise the voltage and adjust the frequency, Tesla varied the values for each component that fed power to the primary loop of the magnifying transmitter (these components can be seen in Figure 13.3). At different times he raised the voltage coming in from the Westinghouse transformer, modified the speed of the breakwheel, which controlled the discharge of the capacitors, changed the size of capacitors feeding the primary coil, and sometimes used one or two windings in the primary (Figure 13.5). In addition, Tesla experimented with the components inside the circular wall: here he tried several different coils for the secondary as well as "extra coils" that were placed in the transmitter's circuit in different ways. During the first months he was in Colorado, both the secondary and the "extra" coil were located toward the center of the fifty-foot circle, but in the last few months, Tesla created a new secondary by winding twenty turns on the circular wall and instead located the extra coil in the center of the circle.[42]

Tesla also varied the capacitance of the secondary circuit by connecting the secondary and extra coil to different-sized copper-covered balls. To tune the secondary circuit to the desired frequency, Tesla adjusted the height of the spherical terminal and soon found that he had to elevate the ball through the station's roof in order to tune the transmitter sharply. Consequently, in September, he installed the telescoping flagpole that permitted him to raise the ball up to 142 feet. During the fall months, Tesla carefully measured how the capacitance of the secondary circuit varied the height of the ball.

For the modern electrical engineer, accustomed to standard electronic components either on a circuit board or in a computer simulation, it is easy to vary the components and play with different circuit configurations. However, for Tesla working on the magnifying transmitter in 1899, we need to remember that there were neither standard components nor any convenient instruments for measuring the value of components like coils or capacitors. Each coil, as we have mentioned, had to be carefully wound, measured, and then adjusted so that it had the necessary inductance. Moreover, given the voltages and frequencies that Tesla was after, many of these components were huge

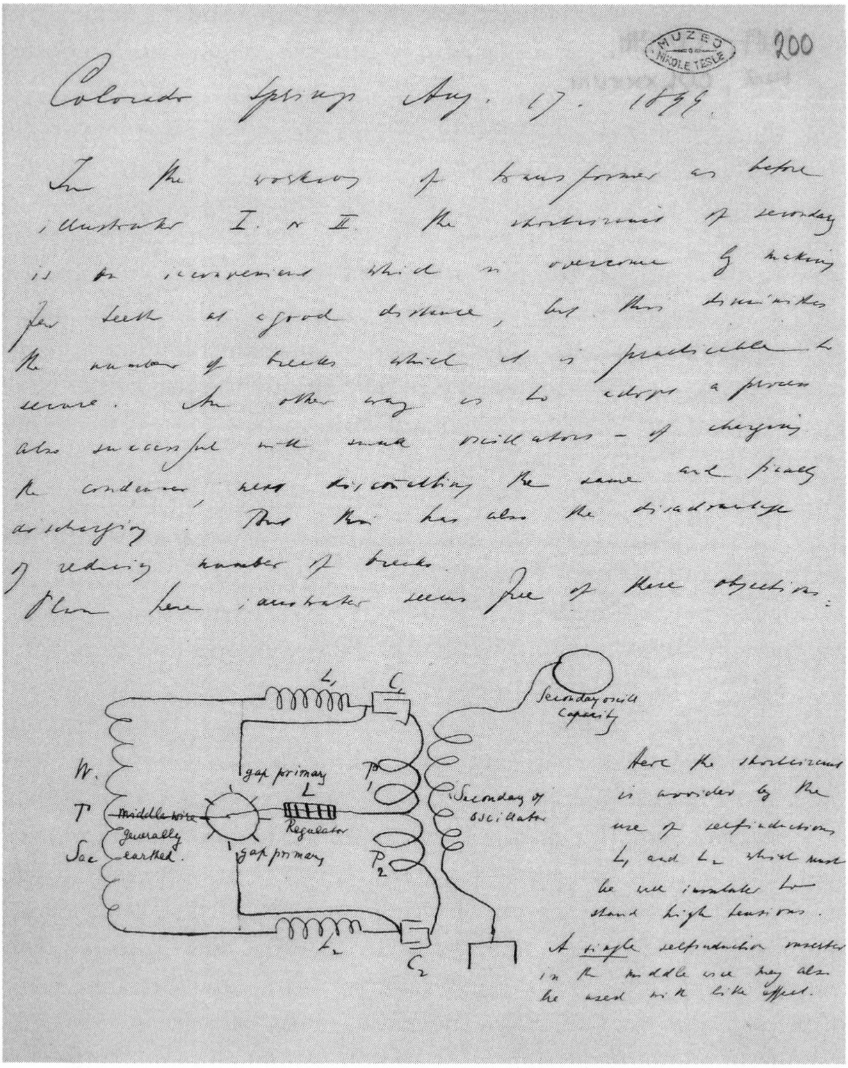
Colorado Springs Aug. 17. 1899.

In the working of transformer as before illustrated I. & II. the shortcircuit of secondary is an inconvenience which is overcome by making few teeth at a good distance, but this diminishes the number of breaks which it is practicable to secure. Another way is to adopt a process also successful with small oscillators — of charging the condenser, then disconnecting the same and finally discharging. But this has also the disadvantage of reducing number of breaks.

Plan here illustrated seems free of these objections:

Here the shortcircuit is avoided by the use of selfinductions L_1 and L_2 which must be well insulated to stand high tensions.

A single selfinduction inserted in the middle wire may also be used with like effect.

FIGURE 13.5. Notebook sketch of circuit typically used by Tesla at Colorado Springs.

Key:

W.T.	Secondary of Westinghouse transformer
L1, L2	inductance coils
C1, C2	capacitors
P1, P2	primary winding of magnifying transmitter

From NT, 17 August 1899, CSN, NTM.

(as compared to modern electronic devices) and could require days to modify. For instance, more than once, Tesla increased the size of the capacitors in the magnifying transmitter; to do so, his assistants had to build more tanks, fill them with a brine solution, and then add the requisite number of glass bottles. To determine how many dozen bottles—and whether they should be champagne bottles or those used to bottle water from nearby Manitou Springs—Tesla had to calculate the dielectric properties for the shape of the bottle. Consequently, changes to the magnifying transmitter could take days to make.[43]

To operate the magnifying transmitter, Tesla had to carefully tune all of the coils—the primary, secondary, and extra—so that each resonated with the preceding coil and hence stepped up the voltage in the manner he desired. At the same time, to adjust the wavelength of the oscillations, Tesla discovered a general rule of thumb: the length of windings in the secondary or extra coil should be one-quarter of the desired wavelength.[44]

To estimate the voltage at which the magnifying transmitter was operating, Tesla would quickly open and close the switch feeding power to the primary side of the transmitter; interrupting the current in this way caused huge sparks to fly off the sphere connected to the secondary or extra coil. According to Tesla scholar and engineer Aleksandar Marincic, when the sparks reached two to four meters in length, the transmitter was probably operating at two million volts, but before long Tesla was running the transmitter so that it produced streamers that jumped across sixteen-foot gaps. Because Tesla had calculated the dimensions of the building as closely as possible to save money on construction, the streamers came within six or seven inches from the sides of the building, and more than once these artificial lightning bolts set the station on fire.[45]

On one occasion Tesla was engulfed by high-voltage streamers while inside the circular wall that held the primary coil of the transmitter. As he recalled,

> For handling the heavy currents, I had a special switch. It was hard to pull, and I had a spring arranged so that I could just touch the handle and it would snap in. I sent one of my assistants down town and was experimenting alone. I threw up the switch and went behind the coil to examine something. While I was there the switch snapped in, when suddenly

the whole room was filled with streamers, and I had no way of getting out. I tried to break through the window but in vain as I had no tools, and there was nothing else to do than to throw myself on my stomach and pass under [the streamers]. The primary carried 500,000 volts, and I had to crawl through the narrow place [i.e., in the circular wall] with the streamers going. The nitrous acid was so strong I could hardly breathe. These streamers rapidly oxidize nitrogen because of their enormous surface, which makes up for what they lack in intensity. When I came to the narrow space they [i.e., the streamers] closed on my back. I got away and barely managed to open the switch when the building began to burn. I grabbed a fire extinguisher and succeeded in smothering the fire.

As frightening as this experience was, Tesla was nonetheless captivated by the forces with which he was wrestling. As he wrote to his friend Robert Underwood Johnson about his time at Colorado Springs, "I have had wonderful experiences here, among other things, tamed a wild cat and am nothing but a mass of bleeding scratches. But in the scratches, Luka, there lies a mind. MIND."[46]

While the streamers amply demonstrated the power of the transmitter, Tesla generally avoided producing them not only for safety reasons but because they wasted the energy of the magnifying transmitter and prevented him from efficiently transmitting power through the ground. As he explained, "The streamers, of course, cause frictional loss and thereby diminish the economy of the system and impair the quality of the results. They also cause a loss of pressure just as leaks in air or water pipes." According to Tesla coil builder Robert Hull, if we were to see the magnifying transmitter operating at night, lightning bolts would not be flying off the sphere elevated over the roof; instead, we would see a blue beam rising straight up over the station, the result of a corona of fine streamers surrounding the mast and sphere. As Tesla recalled, "[A]t night, this antenna, when I turned on to the full current, was illuminated all over and it was a marvelous sight."[47]

NAVAL OVERTURES

Even as Tesla was deeply engaged with his experiments in Colorado, he nonetheless kept an eye out for new sources of funding, and the military was one potential patron. Not only was Marconi at this time busy cultivating contacts with several navies around the world, but as we have seen, Tesla's friend Hobson had tried to interest American naval officers in Tesla's radio-controlled boat. Yet Tesla's biggest interaction with the U.S. Navy was via the Light-House Board.

Consisting of naval officers and civilian scientists, this board was responsible for maintaining the federal government's lighthouses and other aids to navigation. The board took a special interest in using new technology to improve navigation; for instance, in 1886, it used electric lights to illuminate the Statue of Liberty so that it could serve as a landmark for ships sailing into New York Harbor.

In May 1899, the Light-House Board asked Tesla if he could establish a system of wireless telegraphy between the Nantucket Lightship and the shore sixty miles away.[48] This lightship was the first landmark that transatlantic ships encountered as they approached the coast of the United States, and it would be helpful to merchants and steamship companies to have early information about which ships were about to arrive in New York or other ports.[49] Tesla initially agreed to provide experimental apparatus to test on the lightship, and he advised Scherff to push some work through the New York laboratory "as I am preparing myself for the plant at Nantucket (for the government), and want to have as much work done as possible before I return" from Colorado Springs.[50]

Tesla's relations with the Light-House Board soured within a few months, particularly when he discovered that his wireless apparatus was going to be compared with Marconi's system. By September 1899, the newspapers were filled with reports of Marconi's imminent arrival in New York to cover the America's Cup race and the board became worried about rumors that Marconi would be conferring with the Army Signal Corps and not the navy. Fearful that it would be subject to criticism if it delayed adopting wireless technology, the board asked Tesla to step up his efforts to install equipment on the Nantucket Lightship. In its request, the board emphasized that it wished to work with Tesla, preferring "home to foreign talent," but that "whatever is to be done should be done quickly."[51]

But Tesla did not respond quickly to this request, and after brooding for nearly two weeks, Tesla shot off an angry letter to the board. In his view, the board should not have to give him any special home advantage since he was the "first pioneer" who had "laid down certain novel principles on 'wireless telegraphy'" and was now busy perfecting devices based on these principles. As the pioneer in this field, Tesla was bothered that people saw him as competing with the upstart Marconi, who possessed "more enterprise than knowledge and experience." If his equipment was going to be compared with that of Marconi, then Tesla would need the money necessary to build apparatus that could be properly tested. Consequently, he now told the board: "When you first wrote me, not knowing anything of any competition, I persuaded myself that I might furnish you something intended only for temporary service and not free of defects, but now that a number of others are in the field, I would prefer to put out an outfit which will be able to stand crucial tests. But situated as I am [in Colorado Springs], I could not possibly undertake such work unless you could give me an order for, say, twelve sets at least, in which case I could devote myself to the task of furnishing you the apparatus as speedily as possible."[52] The board declined to place this large order, Tesla and the navy went their separate ways, and in 1902 the navy purchased its first wireless equipment from several French and German companies.[53]

TUNING FOR "SECRECY, IMMUNITY, AND SELECTIVITY"

In running the magnifying transmitter at Colorado Springs, Tesla devoted a great deal of effort to perfecting techniques for tuning that would provide "secrecy, immunity, and selectivity."[54] To develop a large-scale, capital-intensive communications system, Tesla reasoned that it would be necessary to make messages secure and private. As he explained a few years later, while in Colorado he thought about sending messages across the ocean, "and the greater the distance the more important and the more essential . . . it was to secure privacy and non-interference of messages; at small distance it does not matter so much whether the operation of a device is interfered with, but when costly apparatus, such as was necessary to employ in my contemplated undertaking was used to produce effects at great distances, the value of a plant and of the whole investment would be destroyed if the messages could not be kept private."[55]

Though Marconi could send messages from his transmitter to a receiver in 1899, he could not protect his messages from interference
from other transmitters operating in the vicinity. For instance, during
the 1901 America's Cup race off New York, Greenleaf Whittier Pickard
of the American Wireless Telephone & Telegraph Company sent a series of ten-second transmissions that jammed the signals being sent by
Marconi and Lee de Forest as they tried to cover the race.[56] Equally,
early users of radio were worried about their messages being picked
up by other receivers that might be listening, and this was a serious
problem for an army or navy using Marconi's system to send messages since it meant that the enemy could intercept messages. And
commercial users of a wireless telegraph system would want to be sure
that their messages would be just as secure as telegrams sent over the
existing wired network.

To create a secure system in Colorado, Tesla built upon the work
he had done in New York using two frequencies with his second radio-
controlled boat. As we saw previously, visitors to the Houston Street
lab often noticed that several lamps lit up at the same time when Tesla
was transmitting power and that he overcame this problem by tuning
the lamps so that they had to receive a combination of two frequencies
before they would light up.

Drawing on this solution, Tesla evolved a general scheme for how to
tune his wireless circuits. Tesla was an avid reader of the English social
scientist Herbert Spencer, and he was particularly taken with "Spencer's clear and suggestive exposition of the human nerve mechanism."
In his *Principles of Psychology*, Spencer had puzzled over the question of
how the brain with "a limited number of fibres and cells [can] become
the seat of a relatively unlimited number of perceptions." Spencer proposed that the brain overcame this problem by having nerve fibers that
were responsive not just to single impulses but to numerous and varied
impulses. Spencer compared the nerves in the brain to keys on a piano;
though it had only a limited number of keys (each sounding its own
note), the piano could produce a large number of complex chords when
multiple keys were depressed. Drawing on Spencer, Tesla observed that
"We are able to receive numberless distinct impressions because the
controlling nerve fibres lend themselves to innumerable combinations,
and we can distinguish an individual from all others by reason of a great
many characteristic features which in no other individual exist."[57]

Inspired by Spencer, Tesla began thinking about his tuned circuits as nerve strands. If nerve strands responded to multiple combinations of impulses, why not do the same with his transmitter and receiver? Instead of viewing the fact that the transmitter generated both a fundamental signal and harmonics as a problem, why not design a receiver to detect the complex signals given off by the transmitter? "In this reside both the essence and virtue of the invention," explained Tesla. "The transmitter is not primitively characterized by a single note, or peculiarity, as heretofore, but represents a very complex and therefore, unmistakable individuality, of which the receiver is the exact counterpart, and only as such can it respond. To go a step further, I make the operation of the receiving instrument dependent not only on a great number of distinctive elements in combination, but also on their order of succession and, if necessary, I go so far as to vary continuously the character of the individual elements."[58]

Now in Colorado, Tesla was ready to test this idea and see if he could send and receive signals employing a number of distinctive elements. For the transmitter, Tesla used the Westinghouse transformer and breakwheel to supply power not to a single bank of capacitors but instead to two separate banks, each of which was then connected to one turn of the primary loop that ran around the fifty-foot circle in the station (Figure 13.6). Inside the circle, Tesla then had two separate secondary coils. Since each secondary coil was wound with a different length of wire, each generated oscillations at a different frequency. Since the entire transmitter was grounded, both oscillations were sent through the Earth (Figure 13.7).[59]

On the receiving end, Tesla used a variety of arrangements involving one, two, or even more receivers. However, to test for security, he often used a receiver with two separate circuits, each of which was tuned to detect one of the two frequencies sent by his transmitter. In this arrangement, the telegraph sounder responded only when it received signals from both detection circuits. By sending and receiving two frequencies in this way, Tesla believed that his messages would be secure and private.[60]

As Tesla made progress with this new tuning technique, he began drafting a patent application and openly criticizing Marconi in newspaper interviews. In September 1899, Marconi traveled to New York

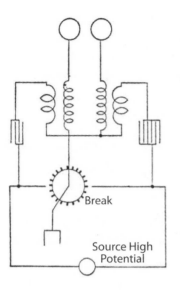

FIGURE 13.6. Diagram of tuned circuit used by Tesla in Colorado Springs.

The source of high potential was the Westinghouse transformer and the break[wheel] is the same as the one shown in Figure 13.7. Note that the breakwheel is grounded. Just above the breakwheel are two capacitors of different sizes. These are connected to two coils that are the two windings on the primary of the magnifying transmitter. The two other coils are the secondaries and they are shown in Figure 13.7.

From NT, 27 June 1899, *Colorado Springs Notes*, p. 55. Diagram redrawn by the editors.

Break

Source High Potential

to use his system to provide coverage of the America's Cup race. Tesla did not hesitate to point out that his rival's system could not overcome problems with interference but that his new system would; as he told reporters in Colorado Springs:

> I hope within a week or two to announce to the world a system of wireless telegraphy that will be nearly perfect. . . .
>
> I will not disparage the Marconi system, but every system in use at the present time has a serious defect in that its action is disturbed by similar instruments within a radius of a greater or less number of miles. Besides twelve words a minute seems to be a very fast time.
>
> With the system I have about completed I hope to attain by mechanical methods a speed of fifteen hundred or two thousand words a minute, and further, to have instruments absolutely free from extraneous disturbances. I have succeeded in eliminating outside influences and can make instruments work free from another almost within touch.
>
> Marconi's experiments are interesting, but they are not new. I am glad they are being successfully employed in the yacht races, but it is by no means improbable that before the next international race is pulled off,

FIGURE 13.7. The magnifying transmitter with several secondary coils energized by the primary coils on the circular wall. Since each secondary coil had a different number of turns, they produced two different wavelengths that were sent into the earth and transmitted to Tesla's receiver.

Plate V, CSN, p. 306. From NTM.

boats propelled by wireless power transmission will follow the races, and I hope the yachts themselves will have wireless telephones on board.[61]

TESTS AND WITNESSES

Although he conducted some experiments using the magnifying transmitter to send messages, Tesla focused most of his attention on the problem of transmitting power wirelessly, and he was gratified to discover that he could indeed send oscillating currents through the ground. On 23 July 1899, he noted that horses in the pasture near the experimental station were disturbed whenever the magnifying transmitter was run, presumably because their iron horseshoes picked up the ground current. The next day, with the secondary coil of the transmitter grounded, he noticed that the lightning arresters on a separate grounded circuit sparked whenever the secondary coil discharged. "There was no other possible way to explain the occurrence of these sparks," wrote Tesla excitedly, "than to assume that the vibration was propagated through the ground and following the ground wire at another place leaped into the line! This is certainly extraordinary for it shows more and more clearly that the earth behaves simply as an ordinary conductor and that it will be possible, with powerful apparatus, to produce the stationary waves which I have already observed in the displays of atmospheric electricity." Tesla noted that he detected the sparks sixty feet away from the secondary coil; assuming that the magnifying transmitter was already operating at 10,000 volts, he calculated that it should be possible to detect these sparks 114 miles away.[62]

Over the next few months, Tesla conducted additional tests to verify that his magnifying transmitter was sending currents into the ground and that they could be detected. In August he tried "arrangements for telegraphy," finding that "[t]he apparatus responded freely to [a] small pocket coil at a distance of several feet with *no capacity attached* and *no adjusted circuit*. Consequently will go at great distance." A few weeks later, he took a receiver outside and connected it to an underground water pipe; at 250 feet from the station, he drew one-inch sparks, and at 400 feet he got half-inch sparks. On 11 September 1900, Tesla carried a receiver a mile away from the station, to nearby Prospect Lake, where he was able to measure that the magnifying transmitter was operating with a wavelength of about 4,000 feet.[63]

In mid-December, as he was completing his experiments in Colorado, Tesla arranged for the photographer Dickenson Alley to come out from New York and photograph his magnifying transmitter in action. (These photographs are discussed more fully later.) As part of this series, Tesla had Alley make several images of lamps lit by the oscillator outside the experimental station. For one image, Tesla placed three lamps on the ground and connected them to a square of wire that was sixty-two feet on each side; this huge square picked up the ground

FIGURE 13.8. "Experiment to illustrate an inductive effect of an electrical oscillator of great power."
    This photograph shows three incandescent lamps located in a field a short distance from Tesla's experimental station. The lamps were connected to a large square loop of wire that detected ground currents sent out by the magnifying transmitter
    Plate XVII, CSN, p. 332. From NTM.

current from the oscillator and caused the bulbs to glow (Figure 13.8). Tesla reported that the bulbs were sixty feet from the secondary coil inside the station. In a second photograph taken outside, Tesla connected an incandescent lamp between a large coil and the ground and again the lamp glowed (Figure 13.9). For this second image, Tesla did not indicate how far away the coil was from the transmitter.[64]

These are the only experiments in Tesla's published notes from Colorado in which he recorded the distance that energy was transmitted. Instead of observing receivers operating in response to the transmitter at various distances, Tesla relied on calculations to evaluate the performance of the magnifying transmitter. "When the apparatus was in action," he later explained, it would produce "a current of many hundreds of amperes" that passed "away into the earth and creat[ed] a disturbance. I would measure the effects produced and calculate the

FIGURE 13.9. "Experiment to illustrate the transmission of electrical energy without wire." Here Tesla placed a coil outside the laboratory with the lower end connected to the ground and the upper end free. The lamp was lit by three turns of wire wound around the lower end of the coil. Plate XXVI, CSN, 340. From NTM.

distance at which a certain amount of energy, necessary for the operation of a receiving device, would be transmitted."[65]

In subsequent articles about his work in Colorado, Tesla emphasized that he had proven that power could be transmitted through the earth, but he did not specify in these reports how far power had been transmitted. While biographer John O'Neill claimed that Tesla lit two hundred incandescent lamps twenty-six miles from Colorado Springs, no evidence has been found to support this claim.[66]

One unusual aspect of the few distance tests that Tesla performed was that there were no witnesses. Despite his frequent interactions with journalists in New York, Tesla avoided talking to reporters while he was hard at work in Colorado; hence contemporary newspaper accounts offer few clues as to what Tesla was doing while out West.[67] When asked during a deposition if he had seen the receivers operating during transmission tests, his assistant Lowenstein reported, "As to receiving apparatus, Mr. Tesla never charged me with experimenting with it and always did that part himself whilst letting me operate the transmitter." According to Lowenstein,

> Mr. Tesla would at first go with the receiving apparatus in his room and experiment there, while I was handling the oscillator. Then he would take, after certain accomplishments in the room, the whole set in portable boxes and go outside of the building, leaving me instruction for continuously switching on and off in certain intervals. In these instances, I often ran to the door to see him for a moment, and I saw that he went as far as, say, a thousand or two thousand feet. I could not watch him closely, as I had to stick to my switch, and so it came that sometimes when just looking out to see him for a moment I couldn't see him, and then I don't know how far he went, but by the time he came back again, being to my recollection not more than the forenoon or the afternoon, you may easily build an idea how far Mr. Tesla could have gone at the time I was standing at the switch.[68]

As a trained engineer, Lowenstein would have been an ideal witness, but since Tesla insisted on having him operate the magnifying transmitter and not go with him to see the receivers, Lowenstein could not report on how the transmission experiments turned out (Figure 13.10).

Moreover, Tesla and Lowenstein had a major disagreement, resulting in Lowenstein's leaving Colorado in September 1899 before Tesla

FIGURE 13.10. Unidentified assistant [possibly Lowenstein?] at main power switch in the experimental station. From NTM.

undertook some of the major experiments with the magnifying transmitter. On the surface, it would appear that Lowenstein left because he found that the high altitude of Colorado Springs made him ill and because he wanted to go back to Europe to marry his fiancé. Tesla clearly regarded Lowenstein as more than a mere assistant; Tesla discussed ideas with him, ate lunch and dinner with him, and spent evenings with him back at the hotel.[69] Lowenstein may have been coming to be the sort of friend and confidante that Szigeti had been years earlier. However, Tesla found some letters that Lowenstein had written or received, prompting Tesla to conclude that Lowenstein was not the person that Tesla thought he was. One possibility is that Tesla may have been attracted to Lowenstein and was upset when he learned that Lowenstein wanted to get married. Tesla later forgave Lowenstein and employed him again for a few years beginning in 1902. Lowenstein subsequently started a company that built radio sets for the U.S. Navy during World War I. Nevertheless, whatever happened in Colorado Springs made Tesla reluctant to use Lowenstein as a witness in legal proceedings.[70]

The lack of witnessed distance tests in Colorado Springs is puzzling, especially given the fact that Marconi was regularly offering demonstrations of his wireless system to reporters and potential investors. Years earlier, Tesla had made good use of both private and public demonstrations for his inventions. With the AC motor, Peck and Brown had sent him off to see Professor Anthony in order to undertake rigorous tests and secure the endorsement of an expert. Similarly, Tesla had not hesitated in his public lectures in the early 1890s to show audiences his wireless lighting systems. Why, then, in Colorado Springs, did Tesla not undertake more extensive distance tests that would show that his system worked as well as Marconi's? Given that his Colorado Springs notes report signals detected a mile from the station, why didn't Tesla invite Lowenstein, reporters, or other witnesses to confirm that his receivers were detecting currents from the transmitter?

This lack of witnessed distance tests can be explained on two levels: the theoretical and the personal. From a theoretical standpoint, Tesla did not believe that such tests were necessary. Tesla had decided that stationary waves in the earth, unlike ordinary Hertzian or light waves, did not lose energy as they propagated; consequently, if they could be detected a short distance from the transmitter, these waves could be detected at *any* distance. Likewise, Tesla also thought that in the return circuit through the atmosphere the process of conduction was extremely efficient and that there would be minimal losses. If there were no losses as the waves traveled from the transmitter to the receiver and back again, then any test detecting the waves—no matter how short the distance—was sufficient for Tesla. Hence he concluded that "communication without wires to any point of the globe is practicable . . . [and] would need no demonstration."[71]

On a personal level, the lack of distance tests reflects Tesla's approach to invention. As a boy, Tesla had developed a powerful imagination, in which he was able to picture all sorts of wondrous things. As an inventor, he had continued to rely on his imagination to envision how new devices might work and to then seek confirmation in selected experiments. Had he not pictured his AC motor first in Budapest, only to have it confirmed when he built prototypes in Strasbourg and New York? Was it not enough for Tesla to get a tin can to spin in a rotating magnetic field in 1887 in order for him to be convinced that his idea was feasible? With wireless power, Tesla had spent years refining in his mind how it would be possible to transmit energy through the earth. Supremely confident of his ability to imagine new technology,

Tesla needed only a small amount of confirmatory evidence from the world around him to convince him that what he imagined was possible. Once he had detected stationary waves in the Earth and once he had detected sparks a mile away from his transmitter, Tesla had all the evidence *he* needed to be convinced that his system worked.

As we shall see, the consequence of not performing additional distance tests or demonstrating his system for witnesses was that Tesla was subsequently hard-pressed to convince others about the value of his system. Over the next decade, when testifying to defend his patents, Tesla was unable to provide hard data showing whether his system worked. He had only one witness, Lowenstein, who could talk about Colorado Springs, but Lowenstein never saw what happened with the receivers. Equally, Tesla had little to show potential investors; they had to take Tesla's word that the system was capable of transmitting messages and power across the seas and around the world.

PHOTOGRAPHING POWER

Rather than having people witness the performance of the magnifying transmitter, Tesla instead chose to rely on photography to document his work at Colorado Springs; photographs were the illusion of choice. In the late fall, Tesla telegraphed Richard Watson Gilder, the editor-in-chief at *The Century Magazine* (and Robert Underwood Johnson's boss): "Friendly motives prompt me to inquire would it pay you to send [a] photographer here with [the] object of obtaining material for illustrations to appear at your pleas[ur]e in Century my rights subsequently reserved[?] If so would like very much getting Mr Alley of Tonnele because of familiarity skill and discretion[.] Would join you in expense as he may do some work for me[?] Kindly answer experimental station."[72]

Gilder and Johnson readily agreed to share the costs and sent the art photographer Dickenson Alley out from New York. Alley had helped test Tesla's lighting tubes in 1894 (see Chapter 12) and had taken the photographs that appeared in the *Electrical Review in* the spring of 1899. Over the last two weeks of December 1899, Tesla and Alley created a remarkable series of sixty-eight photographs.[73]

To create these images, Alley used 11 × 14-inch glass plates that were exposed not only by the giant streamers produced by the magnifying transmitter but supplemented with illumination from an arc light or flash powder. Since power from the local electrical company was still only available at night, Alley and Tesla worked all night to set up the equipment for these pictures, which often meant they did so in freezing conditions. With the wind whistling through the thin walls of the experimental station, Alley appears in several pictures bundled up in his heavy winter coat and hat.[74]

Tesla and Alley began by documenting how the experimental station looked outside and inside (see Figures 13.2, 13.3, and 13.7).[75] They next photographed the magnifying transmitter operating at the normal level of excitation (see, e.g., Figure 13.7). To create various kinds of streamers, Tesla connected the free terminal of the extra coil to different disks, balls, a ring with protruding wires, as well as a fan of wires radiating out; at the same time, Alley varied the position of the camera so as to capture the streamers from various angles (Figure 13.11). In

FIGURE 13.11. "Discharge of 'extra coil' issuing from many wires fastened to the brass ring" [on top of coil].
Plate LIX, CSN, p. 366. From NTM.

order to generate streamers to photograph (remember that streamers represented wasted energy and so Tesla generally tried to avoid producing them), he had his assistant Czito quickly open and close the main power switch 50, 100, or 200 times. In the course of these experiments, Tesla was delighted to see that his transmitter could throw off streamers that spanned straight-line distances of 31–32 feet but whose curved path might be as long as 128 feet.[76] One streamer managed to strike Alley and his camera in one corner of the station, as far away from the extra coil as possible. Thrilled, Tesla confidently predicted that with more copper wire in the transmitter and a bigger building, he could have generated streamers with curved trajectories over 300 feet.

Tesla and Alley created several images in which they each sat in a chair next to the extra coil. Because it would have been far too dangerous for someone to actually sit through the electrical storm captured on camera, Alley used a bit of trick photography. "To give an idea of the magnitude of the discharge the experimenter is sitting slightly behind the 'extra coil,'" explained Tesla. "I did not like this idea but some people find such photographs interesting. Of course, the discharge was not playing when the experimenter was photographed, as might be imagined! The streamers were first impressed upon the plate in dark or feeble light, then the experimenter placed himself on the chair and an exposure to arc light was made and, finally, to bring out the features and other detail, a small flash powder was set off."[77] Hence, by using a double exposure on a single photographic plate, Alley created the memorable picture of Tesla calmly reading while his magnifying transmitter let loose a torrent of streamers (Figure 13.12).[78]

The inventor and photographer then experimented with how to capture the wireless transmission of power.[79] To do so, they concentrated on photographing standard sixteen-candlepower incandescent lamps that were connected to a receiving coil and the ground in order to show that the lamps were lit up by power sent through the ground from the magnifying transmitter (see Figures 13.8 and 13.9). For the first batch of photographs, Tesla set up the lamps and receiving coil somewhere outside the experimental station since he knew that "[t]aken in this manner they would be undoubtedly much more interesting to scientific men." For these images, Tesla carefully recorded many of the operational details but not the distance between the lamps and the transmitter. While these photographs typically show three or

FIGURE 13.12. Tesla seated in magnifying transmitter, with discharge passing from the secondary coil to another coil. This picture was a double exposure on a single glass plate; Tesla was first photographed and then the magnifying transmitter was turned on.

Plate XII, CSN, p. 322. From NTM.

five lamps lit, Tesla estimated that he could easily light as many as sixty bulbs.[80]

Indeed, suspecting that an increase in the capacitance in the receiving circuit would require a ground current with a lower voltage, Tesla had Alley photograph a second series of experiments inside the station while he varied the capacitance of the receiving circuit by connecting bigger metal balls as well as the telescoping flagpole. These tests confirmed his hypothesis about the capacitance of the receiving circuit, leading Tesla to boldly predict that, at full power, his magnifying transmitter could light over a thousand lamps.[81]

In the course of using the magnifying transmitter to light lamps outside the laboratory, Tesla managed to short-circuit the generator at the power company's station and cut off power to the rest of Colorado Springs. The short circuit was caused by the fact that Tesla was generating shorter waves than usual and these waves ruptured the insulation in the generator's windings. Impressed by the power of the short waves, he commented in his notes that they might well be used for sending signals over a thousand miles. To placate the local power company manager, Tesla took his crew down to the station and repaired the damaged generator. Nevertheless, for the remaining time he was in Colorado, the power company supplied electricity to Tesla using a backup generator.[82]

It was also in the course of taking these photographs that Tesla encountered fireballs or ball lightning produced by his magnifying transmitter. Along with the electrical streamers, Tesla occasionally saw luminous spheres, about 1.5 inches in diameter, which floated in the air for a few seconds. Fireballs such as these had a long history in mythology but had also been reported in the scientific literature; for instance, while duplicating some of Benjamin Franklin's experiments with lightning rods in 1753, Georg Richmann in St. Petersburg, Russia, was struck by a fireball that killed him, knocked out his assistant, and tore a nearby door off its hinges.[83]

As powerful as fireballs are, scientists still debate what causes them to appear.[84] While Tesla initially assumed that these were some sort of optical illusion, he soon developed a scientific explanation for their appearance. "[T]he phenomenon of the fireball," he wrote in his notes, "is produced by the sudden heating, to a high incandescence of a mass of air or other gas as the case may be, by the passage of a powerful discharge." According to Tesla, when a powerful electrical discharge (like a streamer) passes through the air, it heats and expands the air in some places but leaves a partial vacuum in other places. The atmosphere naturally rushes in to fill this vacuum. If, however, one of these partial evacuated spaces—which takes the form of a sphere—is struck by a second powerful streamer, then what little gas is in the partial vacuum is brought to a very high incandescence. This glowing ball exists for several seconds because it reaches a state of temporary equilibrium; as Tesla wrote, "It can not cool down rapidly by expansion, as when the vacuous space was being formed, nor can it give off much heat by convection. . . . All these causes cooperate in maintaining, for a comparatively long period of time, the gas confined in this space at an elevated temperature, in a state of high incandescence."[85] Tesla also noted that more fireballs seemed to be created when the air surrounding the magnifying transmitter was filled with vaporized carbon, such as when the high voltages in the transmitter caused the rubber insulation to break down. Fascinated, Tesla promised in his notes to continue to study fireballs using a more powerful magnifying transmitter and more sensitive photographic plates.[86]

Tesla was enthralled with the photographs that he and Alley took, finding them both aesthetically and technologically pleasing. He felt that these images effectively represented the amount of power that he could generate and manipulate with his magnifying transmitter. Over the next few years, Tesla presented copies of selected photographs to individuals whom he hoped would support his research. In reflecting on one of the experiments that Alley had photographed, Tesla noted a few days before leaving Colorado that "Nothing could convey a better idea of the tremendous activity of this apparatus and . . . shows that one of the problems followed up here, that is the establishment of communication with any point of the globe irrespective of distance, is very near its practical solution."[87]

CONFIRMATION, NOT DISCONFIRMATION

Taken together, Tesla's discovery of stationary waves, his detection of interplanetary messages, his puzzling indifference to distance tests, and his use of photography to document his work reveal a significant aspect of his style as a mature inventor. At Colorado Springs, Tesla appears to have sought only evidence to *confirm* his hypotheses and not look for anything that might *disconfirm* his theories.

Inventors, as well as scientists, are always looking for confirmatory evidence—that the device or experiment actually works. Put more precisely, both inventors and scientists are seeking proof that the ideas they have in their minds are confirmed by the action of objects and forces in the material world. Because so few ideas readily match up with the material world, it is understandable that inventors and scientists often develop a confirmatory bias; they want to see their ideas succeed and hence they are quick to seize upon evidence that supports their idea. Such hope is essential to technological and scientific work, for without it inventors and scientists would lack the optimism to keep trying new things.

Yet this hope must be complemented by a degree of tough-mindedness. Yes, an experiment seems to show that a particular hypothesis is true, but then the task for the scientist is to eliminate all other hypotheses. To do so, the scientist and inventor often tries to disconfirm his or her hypothesis—to show that it doesn't work.[88] In invention, disconfirmation is perhaps even more important than it is in science. First, because an inventor is promising patrons that his creation will work in certain ways and under certain conditions, he or she often has to "torture-test" an invention to establish when it works and when it fails. Second, the failure of an invention often provides an inventor with valuable clues about how to improve a device. By studying how a device fails, an inventor can identify ways needed to perfect an invention. As Edison remarked when his phonograph emitted recorded speech for the first time in 1877, "I was always afraid of things that worked the first time"—he feared that without something going wrong, he would not know how to go forward improving the device.[89]

But in contrast to Edison working on the phonograph, Tesla in Colorado Springs did not seek to disconfirm his ideas about transmitting power and messages through the Earth by using stationary waves. He did not, for instance, systematically eliminate all other explanations for why he heard the "one-two-three" beeps in his receiver. And Tesla did not measure directly how far he could actually send power through the earth. Instead, anxious to see his theories about wireless power confirmed, he eagerly seized on the evidence that supported his ideas. He did not seek to disconfirm his ideas while at Colorado Springs—to "torture-test" them so that he really knew where they did and did not work. Didn't Alley's photographs reveal that he was controlling unbelievable amounts of power? This is not to say that what Tesla discovered in Colorado Springs was "wrong" in some way; indeed, he was observing actual phenomena—like the stationary waves created by lightning storms or the pulses emanating from Jupiter's moon Io—during his stay there. What is significant—and ultimately tragic—about his time in Colorado was that Tesla let a small amount of confirmatory evidence suffice when he needed to be open to the alternatives. High hopes and a small amount of confirmatory evidence created illusions in his mind.

Satisfied in his own mind that power could be transmitted around the world without wires, Tesla triumphantly returned to New York in January 1900. With wireless power, Tesla was now at the same point he had been in 1887 with his AC motor. In both cases he had observed phenomena he believed were ideal for shaping into major inventions. In 1887, he had just succeeded in getting a shoe polish tin to spin in a rotating magnetic field whereas in 1900, Tesla was convinced that he could set up stationary waves in the Earth and transmit power and messages. It was now time to convert the phenomenon he had observed in Colorado into a major invention, and to do so he needed to create a network of people, ideas, money, and resources. It was time to implement the business strategy he had learned from Peck and Brown: secure strong patents, promote his work to build public interest, and then sell to the highest bidder. It was time to talk about the potential of wireless power and get people excited.

But would it all go well this time, just as it had in 1887? Would promotion lead to the money and ultimately to success for Tesla? To be sure, the situation confronting Tesla in 1900 was different than it had been in 1887. Since Peck had died ten years earlier, Tesla could no longer rely on him for advice, and he was not especially close to Edward Dean Adams, his patron since the mid-1890s. Who might serve as his business mentor and patron? Could he count on John Jacob Astor, who seemed interested in his work? Unlike his motor, did wireless power have a keystone component on which Tesla could focus his efforts and secure strong patent coverage? Indeed, wireless power was a system, and as such, Tesla would need not thousands of dollars (which had been sufficient for his work on the motor) but hundreds of thousands to build a demonstration plant. Moreover, what kind of publicity should he pursue? Could Tesla work with the professional electrical engineering community, or should he take advantage of his celebrity in the popular press? And above all, in 1900, he had to deal with strong rivals; with the motor, Tesla had easily outmaneuvered Ferraris and Dolivo-Dobrowolsky, but with wireless technology, he was competing with Marconi, Reginald Fessenden, and Lee de Forest. Could Tesla assemble a network of ideas and resources fast enough that would allow him to beat these rivals? These were the challenges that Tesla would confront over the next five years.

MAKING BOLD PLANS

While he was glad to settle back into the luxurious Waldorf-Astoria Hotel after his stay in Colorado, Tesla was even more excited about getting back to his laboratory on Houston Street and begin preparing patents. As a first step to protecting his Colorado discoveries, Tesla drafted a patent that summarized how his magnifying transmitter could create stationary waves in order to broadcast power and messages through the earth. To illustrate the potential of this system, Tesla proposed an early form of radio navigation and described how his system could set up two stationary waves with different wavelengths that ships at sea could detect and then use to calculate their position. Tesla was very pleased with this patent, recalled Scherff, and he regarded it as "one of the best he had ever written." Once this broad application was ready, Tesla then prepared three more applications detailing his method of tuning using two different signals (see Chapter 13).[1]

As he was drafting his patent applications, Tesla also did not hesitate to tell the world what he planned to do. He boldly repeated the goal that he had told the press when he had arrived in Colorado

Springs in May 1899, that he would send messages to Paris: "My experiments have been most successful, and I am now convinced that I shall be able to communicate by means of wireless telegraphy not only with Paris during the [upcoming] Exhibition [of 1900], but in a very short time with every city of the world."[2]

When Tesla announced this ambitious plan to transmit messages across the Atlantic, his rival Marconi had only been able to send messages 86 miles and was hoping to increase that distance to 150. Consequently, Marconi had "no belief in Nicola Tesla's promise to communicate across the Atlantic," reported one electrical paper. "He believes that science, while gradually progressing, will be unable to obtain such great results before the preliminary results have been surmounted. Personally, he does not expect, yet, to girdle the Atlantic."[3]

Ignoring Marconi, Tesla knew he would need to design and build a larger commercial plant to girdle the Atlantic and reach Paris. Borrowing a metaphor from Hobson, he explained, "The plant in Colorado was merely designed in the same sense as a naval constructor designs first a small model to ascertain all the quantities before he embarks on the construction of a big vessel."[4] Based on his experiments in the mountains, Tesla could now calculate the size of the components he needed in his system in order to transmit across the Atlantic.

Since this system would require large-scale AC-generating equipment of the sort manufactured by the Westinghouse Company, Tesla naturally turned to George Westinghouse for help. His success in Colorado, Tesla boasted to his old patron, "has been even greater than I anticipated, and among other things I have absolutely demonstrated the practicability of the establishment of telegraphic communication to any point of the globe." But to move forward, he would need a Westinghouse steam engine and dynamo. Knowing the high cost of this equipment and recognizing that there was at that moment a "panicky feeling all around" on Wall Street, Tesla was worried that it would be difficult to raise the necessary capital. Hence he asked Westinghouse "whether you would not meet me on some fair terms in furnishing me the machinery, retaining the ownership of the same and interesting yourself to a certain extent" in the new enterprise? "I have been so enthused over the result achieved and have worked with such passion," Tesla confessed, "that I have neglected to make such provisions for money. . . . Being thus compelled to borrow money I turn to you to ask whether your company will not advance me say $6000 on the guarantee of my [motor] royalty rights from England, or if preferable whether they would buy outright my claims on the royalty for a sum of $10,000."[5] Tesla explained that he preferred not to sell his rights since doing so might hurt his old partners Brown and Peck (presumably Peck's widow), who owned five-ninths of the English rights.

Though Westinghouse declined to take a stake in Tesla's new enterprise, he did loan Tesla the money; doing so made sense since the loan was tied to the future purchase of Westinghouse equipment.[6] At the same time, Westinghouse probably wanted to curry favor with Tesla since the company was embroiled in a series of lawsuits over the validity of the Tesla motor patents, and it was counting on Tesla to testify as an expert witness in these cases.[7]

Along with asking Westinghouse for a loan, Tesla continued to court John Jacob Astor. Though Astor had agreed to help finance Tesla's experiments, he appears to have made only one payment. During his time in Colorado, Tesla had anxiously waited for more money and regularly asked Scherff for news about "JJA." From Colorado, he had sent Astor a set of the photographs showing the magnifying oscillator in action.[8] Now back in New York, Tesla sent the colonel copies of his wireless patents and assured Astor that these patents gave him an absolute monopoly on transmitting power and messages. Astor, however, had lost interest in Tesla and ignored his entreaties. Overall, Astor invested only $30,000 in Tesla's wireless experiments.[9]

COLD PHILOSOPHICAL STONES OR HOT THROBBING FACTS?

Certain that other investors besides Westinghouse and Astor would flock to his magnificent wireless plans, Tesla spurned the scientific press and instead promoted them in newspapers and popular magazines. In particular, he poured his energy into writing the article he had proposed in the fall of 1899 when he had asked *The Century Magazine* to send Dickenson Alley to Colorado to photograph his work. Tesla presented a first version at the end of January at a dinner given by the Johnsons, but since it was too short, the article could not be published in the March or April issue of the magazine.[10]

Rather than focus the article on his recent accomplishments in Colorado, Tesla began expanding it, determined to show how his inventions constituted a grand intellectual scheme. For years he had been pondering how his inventions would change history, and he now decided that together they represented a comprehensive way of increasing the physical energy available to humanity. The article in *The Century* was his chance to show the world the significance of his work. "I knew," wrote Tesla with his usual confidence, "that the article would pass into history as I brought, for the first time, results before the world which were far beyond anything that [had ever been] attempted before, either by myself or others."[11]

But as Tesla's article grew voluminous, Johnson became worried that the Wizard was serving up a mess of cold philosophical stones instead of a dish of hot throbbing facts.[12] "I just can't see you misfire this time," wrote an anxious Johnson to Tesla. "Trust me in my knowledge of what the public is eager to have from you. Keep your philosophy for a philosophical treatise and give us something practical about the experiments themselves." In response, Tesla penned back, "Dear Robert, I heard you are not feeling well and hope that it is not my article that makes you sick."[13]

The article went back and forth from author to editor, growing from four to sixteen chapters. To satisfy Johnson, Tesla added final sections highlighting wireless telegraphy and his plans for transmitting power. When it finally appeared in the June issue of *The Century Magazine*, Tesla's treatise had reached thirty-six pages and was illustrated with photographs of electrical streamers pouring out of the magnifying transmitter as well as images of incandescent lamps sitting out in the Colorado countryside while being powered without wires.[14] (The *Century* illustrations included Figures 13.3, 13.7, 13.8, and 13.9.)

Titled "The Problem of Increasing Human Energy," the article was Tesla's interpretation of the role of energy and technology in the course of human history. In florid Victorian language, he began,

> Of all the endless variety of phenomena which nature presents to our senses, there is none that fills our mind with greater wonder than that inconceivably complex movement which . . . we designate as human

life. Its mysterious origin is veiled in the forever impenetrable mist of the past, its character is rendered incomprehensible by its infinite intricacy, and its destination is hidden in the unfathomable depths of the future. Whence does it come? What is it? Whither does it tend? are the great questions which the sages of all times have endeavored to answer.[15]

Tesla was inspired to tackle such questions, having read John William Draper's *History of the Intellectual Development of Europe*. Drawing on his study of physiology, Draper had sought to demonstrate "that civilization does not proceed in an arbitrary manner or by chance, but that it passes through a determinate succession of stages, and is a development according to law." Determined to find the laws guiding human progress, Tesla framed his approach in mechanical and mathematical terms. "Many a year," Tesla explained, "I have thought and pondered, lost myself in speculations and theories, considering man as a mass moved by a force, viewing his inexplicable movement in the light of a mechanical one, and applying the simple principles of mechanics to the analysis of the same."[16]

For his law guiding human development, Tesla proposed the equation $E = mV^2/2$, where E was total human energy, m equaled the mass of humanity, and V was the velocity of human change. While these terms were hypothetical, what counted for Tesla were the relationships embodied in this equation. In particular, the equation suggested to Tesla that there were three ways that human energy might be enlarged: by increasing the human mass (i.e., improving society), by eliminating any forces retarding humanity, and by increasing the velocity (i.e., the rate of progress). Tesla discussed each of these three approaches at some length in the article.

To increase human mass, Tesla argued that it was necessary to pay attention to public health, education, and the availability of pure water and wholesome food while suppressing gambling and smoking. He was especially concerned about the speed and haste by which most city-dwellers experienced life. To improve water purification, Tesla advocated using the ozone produced by his electrical oscillators to kill germs. To expand the food supply, Tesla recommended vegetarianism and described how electricity could be used to capture nitrogen from the atmosphere in order to create cheap fertilizer.[17]

Turning to the forces retarding the human mass, Tesla listed ignorance, deceit, and war. To eliminate war, Tesla described at length his
radio-controlled boat and how he could use electromagnetic oscillations to give boats and other devices a "borrowed mind." Representing the start of a new field, telautomatics, Tesla argued that the steady
development of radio-controlled devices would result in wars that
were fought by machines with few human casualties. The advent of
telautomatics, wrote Tesla, "introduces into warfare an element which
never existed before—a fighting-machine without men as a means of
attack and defense. The continuous development in this direction
must ultimately make war a mere contest of machines without men
and without loss of life—a condition which would have been impossible without this new departure, and which, in my opinion, must be
reached as preliminary to permanent peace."[18] Indeed, Tesla believed
that future wireless weapons would become so powerful and dangerous that humanity would be prompted to outlaw war.

In order to increase the velocity of humanity—to speed up
progress—Tesla wanted to harness greater amounts of energy. Arguing that nearly all of the energy available on Earth comes from the
Sun, Tesla was confident that humanity could tap vast amounts of the
solar energy and turn it into cheap electricity. Among the many ways
to produce electricity, Tesla discussed windmills, solar-powered boilers, geothermal energy, hydroelectric plants, and ideal heat engines.
As electric power became more plentiful, he believed that electricity
could revolutionize the production of iron and steel since cheap power
could be used to break water down into hydrogen and oxygen, the hydrogen could be used as fuel in blast furnaces, and the oxygen could
be sold as a byproduct. Tesla was equally excited about the potential use of aluminum since it could be easily smelted using electricity.
Knowing that all of this cheap power would need to be transmitted
efficiently, Tesla thus set the stage for introducing his plans for the
wireless transmission of energy. In the closing sections, he described
his discoveries in Colorado and how he anticipated that he could send
both power and messages around the world without losses. Absolutely
convinced of the importance of wireless power, Tesla concluded his
Century article with a poetic flourish from Goethe:

> I can conceive of no [other] technical advance which would tend to
> unite the various elements of humanity more effectively than this

one. . . . It would be the best means of increasing the force accelerating the human mass. . . .

I anticipate that [some people], unprepared for these results . . . will consider them still far from practical application. . . . [Yet] the scientific man does not aim at an immediate result. He does not expect that his advanced ideas will be readily taken up. His work is like that of the planter—for the future. His duty is to lay the foundation for those who are to come, and point the way. He lives and labors and hopes with the poet who says:

> Daily work—my hands' employment,
> To complete is pure enjoyment!
> Let, oh, let me never falter!
> No! there is no empty dreaming:
> Lo! these trees, but bare poles seeming,
> Yet will yield both food and shelter![19]

Tesla's article in *The Century Magazine* elicited great interest in the popular press, and excerpts appeared in newspapers and magazines across Europe and America during the summer of 1900.[20] In the scientific community, the article was, not surprisingly, greeted with skepticism; in a letter to *Popular Science Monthly*, "Physicist" growled that the public should be protected from such wild speculation passing for scientific fact."[21]

One trenchant response to *The Century Magazine* story came from Tesla's former friend, T. Commerford Martin. Still angry with Tesla about their 1898 feud over the radio-controlled boat and the publication of Tesla's medical electricity paper, Martin wrote to the editor of *Science* that "I have had so much satisfaction in the review and criticism recently published in *Science*, of Mr. Tesla's magazine article on 'Human Energy' that I cannot avoid making public acknowledgement of my appreciation of its justice and timeliness." Martin observed that the combination of decreasing publication costs and increasing competition among newspapers and magazines during the 1890s had resulted in the publication of a great deal of sensational and unreliable scientific reporting. "Among many other evils growing out of . . . 'Newspaper Science,'" warned Martin, "not the least is the manufacturing and maintaining of false reputations. The constant appearance of a name in connection with the development of a given art, science, discovery or invention makes an impression which is difficult to destroy, and this is true even among the most intelligent classes. To find who is really and truly eminent in any field of human activity one must go to the specialists in that field. The popular verdict is more than likely to be wrong because it is based on fictitious, newspaper-created renown."[22]

On the one hand, Martin's remarks may seem disingenuous, given how hard he had worked a decade earlier to build up Tesla's reputation through strategically placed stories and the publication of a book on Tesla. On the other hand, Martin signals to us that the world of electrical technology in 1900 was vastly different from that of 1890. To establish electricity as a legitimate commercial technology, it may have been acceptable in 1890 for practitioners like Tesla to make all manner of bold claims; there was a naïve and democratic faith that the public and investors would be able to figure out on their own which claims were trustworthy. In contrast, by 1900 the electrical industry was well established and worth tens of millions of dollars; hundreds of professional electrical engineers were in key positions in this industry. Determined to protect this new industry and this new profession, leaders of electrical engineering now viewed themselves as the gatekeepers—whom Martin referred to as the "specialists"—and along with Martin, they were vitally aware that sensational newspaper science could do more harm than good. The first decades of the twentieth century would be the age of the professional expert whose authority was grounded in scientific training, professional affiliations, and hard evidence.[23] As we shall see, over the next few years Tesla found it more and more difficult to mobilize the evidence he needed to maintain his credibility with professional scientists and engineers.

Tesla paid little heed to Martin's concerns about the danger of using newspapers to create his reputation. Through the summer of 1900, he followed up the publicity generated by the *Century* story with additional pronouncements. In August, newspapers reported that he was testing his oscillator on tuberculosis patients to see if the high frequencies could kill the bacteria causing the disease. A few weeks later, Tesla announced that he had patented a new electrical transmission cable that used ice to reduce power losses and that this new cable could permit power to be transmitted across the ocean from Niagara Falls to London.[24]

THE WIZARD AND THE GREAT MAN

In November 1900, Tesla got a lucky break. He was able to meet with the most powerful man on Wall Street, J. P. Morgan (1837–1913), and to convince Morgan to loan him $150,000 to support his wireless work.

The son of a prominent financier, Morgan was educated in Boston and at the University of Göttingen. He followed his father into banking, and in the 1870s Morgan drew on his father's connections in London financial circles to secure much-needed capital for American railroads. In the 1880s, as railroads engaged in cutthroat competition and bankrupted themselves, Morgan frequently stepped in to refinance and reorganize these lines. In doing so, he protected the investments made by his firm (as well as those of his clients) by retaining large of blocks of stock and demanding seats on the boards of the railroads he reorganized. By the end of the century, Morgan controlled most of the railroads operating in the eastern portion of the United States. From railroads, Morgan moved into industrial mergers; in 1892, he assisted the Boston banker Henry Lee Higginson in combining the Thomson-Houston Electric Company and Edison General Electric to form the General Electric Company. By the time Tesla met Morgan, he had become the dominant figure in American capitalism.[25]

It is not entirely clear how Tesla came into contact with Morgan, but it may have been the result of Morgan and his partners developing an early interest in wireless telegraphy. As an avid yachtsman, Morgan probably became aware of wireless technology during the 1899 America's Cup races. During this race, Morgan was the commodore of the New York Yacht Club and the chief sponsor of the defending yacht, *Columbia*. As we saw previously, Marconi came to New York in order to demonstrate his wireless system by transmitting messages from the races to reporters from the *New York Herald*. Just prior to the race, Edward C. Grenfell and Robert Gordon from Morgan's London branch approached the directors of Marconi's company, the Wireless Telegraph & Signal Company, offering to buy Marconi's American patents for 200,000 pounds.[26] As part of the deal, the Morgan men insisted that it include "Ocean rights" that they could exercise "if ever Wireless telegraphy could communicate from England to New York." The Marconi directors, however, were not satisfied with that price and kept changing the terms of the deal; the Morgan representatives eventually gave up in disgust.[27]

Unable to buy Marconi's American patent, Morgan may very well have regarded Tesla as an alternative way to investigate this new technology. Consequently, about a year later, Morgan met briefly with Tesla on 23 November and 7 December 1900 at his home. During these meetings, Tesla introduced Morgan to his wireless technology, boldly suggesting that it would replace both the telegraph and the telephone. He proposed that he and Morgan form one or two companies to develop wireless technology, insisting that Morgan take a controlling 51% share of the stock in these new ventures.[28]

But beyond broaching these ideas, Tesla appears to have kept these meetings with Morgan short, perhaps because he found it difficult to deal simultaneously with the power of Morgan's personality and Morgan's large, deformed nose, which had been scarred by rhinophyma and was covered with warts. With regard to Morgan's powerful presence, Annette Markoe Schieffelin recalled that when he entered a room "you felt something electric: he wasn't a terribly large man but he had a simply tremendous *effect*—he was the king. He was *it*." Likewise, the art dealer James Henry Duveen described his first encounter with Morgan in the following manner: "I was unprepared for the meeting. . . . I had heard of a disfigurement, but what I saw upset me so thoroughly that for a moment I could not utter a word. If I did not gasp I must have changed colour. Mr. Morgan noticed this, and his small, piercing eyes transfixed me with a malicious stare. I sensed that he noticed my feelings of pity, and for some time that seemed centuries, we stood opposite each other without saying a word. I could not utter a sound, and when at last I managed to open my mouth, I could produce only a raucous cough. He grunted."[29] Tesla would have been equally squeamish about Morgan's nose.

Unable to make his case fully in person, Tesla followed up each encounter with letters. In the first, Tesla built a case for developing his wireless technology by pointing out that it could compete effectively with the transatlantic telegraph cables. Knowing that Morgan relied on messages sent between his New York and London offices by undersea cable, Tesla reported that he was now able to manipulate electrical pressures of a hundred million volts and hundreds of thousands of horsepower of electrical energy so that it was no longer necessary to rely on those "long and expensive cables" to send messages. Implicitly citing what he had learned by observing the stationary waves set up by thunderstorms in Colorado, Tesla assured Morgan that "Long practical experience with apparatus of this kind and exact measurements embracing a range of nearly seven hundred miles, enable me to construct plants for telegraphic communication across the Atlantic and, if need be, across the Pacific Ocean, with the fullest measure of success." Referring to his work on tuning, he informed Morgan that he could selectively operate "a great number of instruments without mutual interference, and can guarantee the absolute privacy of all messages." Tesla estimated that construction of two temporary stations for transmitting across the Atlantic would cost $100,000 and require six to eight months of work. To span the Pacific, Tesla would need $250,000 and he could accomplish that in a year. But most of all, he wanted to assure Morgan that he was willing to trust him completely. "Although the development of these inventions has consumed years of effort," wrote Tesla, "knowing that I have to deal with a *great man*, I do not hesitate to leave the apportionment of my interest and compensation entirely to your generosity [emphasis added]."[30]

In a second letter, Tesla discussed the patent situation. He had already secured broad patents covering his inventions in America, Australia, and South Africa, and he was free to make new arrangements to exploit these patents. In response to questions raised by Morgan concerning Marconi and other rivals, Tesla assured the Great Man that even though the British Post Office favored Marconi's technology, this nonetheless demonstrated that there was a potential market for wireless technology. Further, Tesla provided quotes from leading scientists—Lord Kelvin, Sir William Crookes, and Adolph Slaby— praising his previous inventions and genius. Confident of his legal position, Tesla told Morgan that with "my patents in this still virgin field, should you take hold of them, you will command a position which . . . will be legally stronger than that held by the owners of Bell's telephone inventions or by those of my own discoveries in power transmission by alternating currents."

Just as he had done previously with both Peck and Astor, Tesla skillfully played up the appeal of supporting a pioneer by providing both the money and business acumen needed to convert his ideas into revolutionary new inventions. "Permit me to remind you," said the Wizard to the financier,

> that had there been only faint-hearted and close-fisted people in the world, nothing great would ever have been done. Rafael [*sic*] could not have created his marvels, Columbus could not have discovered America, the Atlantic cable could not have been laid. You, of all, should be the man to embark boldly on this enterprise, only seemingly hazardous, prompted by superior insight as well as desire to advance an art of inestimable value to mankind.
>
> Coming to the financial question – please remember: These inventions,—the results attainable only by their means—which now I alone am able to accomplish—in your strong hands, with your consummate knowledge and mastery of business—are worth an incalculable amount of money.
>
> Although I have expressed myself in my last letter, I will be more explicit relative to my share and compensation. The control is yours, the larger part is yours. As to my interest—you know the value of discoveries and artistic creations—your terms are mine.[31]

Anxious that he had not yet persuaded Morgan, Tesla drafted a third letter, ten pages long, but before he could send it he received word that the Great Man was willing to support him.[32] Thrilled, Tesla quickly penned a note to Morgan on 12 December:

> How can I begin to thank you in the name of my profession and my own, great generous man! My work will proclaim loudly your name to the world!
>
> You will soon see that not only am I capable of appreciating deeply the nobility of your action, but also of making your primarily philanthropic investment worth a hundred times the sum you have put at my disposal in such a magnanimous and princely way![33]

Although Morgan nominally agreed to help Tesla in December 1900, they did not finalize their agreement for two months, probably because Morgan was deeply involved in negotiating the creation of United States Steel with Andrew Carnegie.[34]

In the meantime, Tesla used the occasion of the opening of the twentieth century (then regarded as starting on 1 January 1901) to announce that while he had been in Colorado, he had received an interplanetary message (see Chapter 13). In response to an inquiry from the American Red Cross asking him what he considered the most important scientific development that would affect the new century, Tesla informed the group that the major challenge ahead would be figuring out how humans could establish contact with other worlds.[35] The Wizard was careful not to claim that the messages he had detected had come from Mars, but enterprising reporters quickly interpreted his statements to mean that Tesla believed he had received emanations from the red planet. Tesla was roundly criticized by leading scientists for making these claims, and one can only wonder what Morgan thought as these Martian reports turned up everywhere in newspapers and magazines.[36] Was Tesla simply confident that Morgan was going to support him no matter what? Or was it that he could not resist an opportunity to put forward such a compelling illusion?

While waiting for Morgan in the early weeks of 1901, Tesla sought to use the Great Man's endorsement to entice Astor to support his wireless lighting inventions. Since his return from Colorado, Tesla had been tinkering with his wireless lamps that now took the form of a rectangular spiral of glass tubing. "Hearty wishes for the new Century," wrote Tesla to Astor. "Mr. Morgan's generous backing, for which I shall be grateful all my life[,] secures me my triumphs in wireless telegraphy and telephony, but I am still unable to put my completed inventions on the market. I can hardly believe that you, my friend since years, should hesitate to join me in introducing them which I can offer you ten times better returns on your investment than anyone else." Despite their long friendship and promises of huge profits, Astor was suspicious, having read in the newspapers that Marconi and others might have stronger patents in the wireless field than Tesla. "Do not be misled by what the papers say, Colonel," replied Tesla. "I have the controlling rights. Why not come in with Mr. Morgan and myself[?]" Still cautious, Astor declined any further involvement with Tesla. To compensate for Astor's withdrawal, Tesla promptly launched a newspaper blitz, emphasizing how his lamp produced an "artificial sunshine" that purified the atmosphere, killed germs, and exerted "a "Soothing Effect Upon the Nerves."[37]

Unaware that Morgan was busy combining most of the steel industry into a single "billion dollar trust," Tesla grew worried, as the weeks
passed, that he had not heard from his new patron. Perhaps to force
Morgan's hand, Tesla began a new publicity campaign for his wireless
telegraph system in mid-February. Tesla told reporters that his system
was now complete and that he would be transmitting messages across
the oceans in eight months' time. Further stories indicated that Tesla
was thinking about locating his transmitting station on the New Jersey
coast and that he had dispatched an agent to look for a location for a
receiving station in Portugal. Although Tesla had not yet worked out
the business details, reporters noted that he "was most enthusiastic in
his veiled references to the financial assistance he has received."[38]

Never a fan of the New York tabloids, Morgan was unimpressed
with Tesla's newspaper coverage and perhaps even annoyed.[39] As the
Literary Digest noted, "The daily papers treat [Tesla's] various pronun-
ciamentos each after its kind, the yellow journals with weird pictures
and big headlines, the more serious ones with skeptical paragraphs."[40]
However, as the negotiations with Andrew Carnegie and the other steel
barons came to a close, Morgan assigned one of his partners, Charles
Steele, to work with Tesla. Steele asked Tesla to draft a letter of agreement in which Morgan would advance funds to Tesla in return for a
51% share of Tesla's wireless patents. No mention was made of forming any companies or sharing the stock between the Great Man and
the inventor. Tesla readily assented to "sign any document approved
by Mr. Morgan" but asked if the agreement could cover not only the
development of wireless telegraphy but also his wireless lighting system. After meeting with Tesla again on 26 February, Morgan agreed
to include lighting along with wireless telegraphy in the agreement.[41]

Consequently, on 1 March 1901 Tesla submitted a letter to Morgan
summarizing their arrangement. Having spent several years working
to perfect an electric lighting system as well as investigating wireless
telegraphy and telephony, Tesla was

> anxious now to construct the apparatus necessary for putting my dis-
> coveries and inventions to practical use and for continuing my investi-
> gations on the subjects named.
>
> For this I desire to procure the sum of [blank space] and I hereby
> agree that if you will advance such sum to me as hereinafter stated, I

will assign to you an interest of fifty-one percent in all of said patents and inventions, and also in any patents or inventions which I may hereafter secure.

On 4 March—the day after the formation of United States Steel was announced—Morgan accepted Tesla's letter and instructed Steele to insert $150,000 in the blank space in the letter.[42]

It is important to pause and consider what this agreement probably meant to each man when it was signed. For Morgan, it was, as Tesla recalled later, "a simple sale"; in return for $150,000, Morgan was assigned 51% of Tesla's patent rights. The amount involved was probably not significant for Morgan; in April 1901, he bought sight unseen the famous painting *The Duchess of Devonshire* by Thomas Gainsborough for $150,000. A few months later he purchased Raphael's altarpiece, *Colonna Madonna*, for $400,000, and after glancing at some architectural plans, he readily gave the Harvard Medical School $1 million for three buildings.[43]

While it is often assumed that Morgan invested in Tesla in order to gain control of the future wireless industry (and hence protect his investments in the existing wired telegraph and telephone networks) by taking preemptive control of wireless technology, there is nothing to support this supposition. Indeed, if this had been the case, Morgan would most likely have waited until this industry had further matured; for instance, he did not work with Higginson to form General Electric until after others had invested heavily and built the Edison and Thomson-Houston companies into thriving enterprises. Morgan made his money not by investing in high-tech startups but in consolidating and financing companies in growing, well-established industries.

Instead, Morgan's involvement was, as Tesla hinted in his 12 December 1900 letter, largely "philanthropic." Morgan periodically provided money to various individuals to advance artistic and scientific projects; for example, in 1902 he paid for Bashford Dean to move from the American Museum of Natural History (where he had studied armor-plated fish from the Devonian era) to the Metropolitan Museum of Art so that he could catalogue a collection of medieval armor.[44] In all likelihood, Morgan saw Tesla as an interesting artist or scholar and wireless telegraphy as a promising research venture, and as such, Morgan was not especially worried about whether Tesla's project succeeded commercially.

Although Tesla must have been delighted to secure $150,000 from Morgan, he did not get everything that he wanted in this agreement. For Tesla, the agreement was not supposed to be about selling his patent rights for $150,000 but about forming companies and establishing a partnership. In Tesla's mind, while he provided the technical wizardry, Morgan was supposed to deliver the financial genius needed to make his wireless technology into a marvelous new business. What Tesla was probably hoping was that Morgan would take the place of Peck, his old patron from the AC motor days—that Morgan would take the time to understand Tesla's dreams, nurture them, and help connect those dreams to practical concerns in the business world. At the time he signed the agreement, Tesla did not complain to Morgan about their arrangement, either because he was too excited about getting the money or because he did not want to offend the Great Man.[45]

Yet Tesla remained anxious about his relationship with Morgan. In accepting Morgan's terms, Tesla wrote to Steele in March 1901 that he hoped "that all manner of conveying a wrong impression on Mr. Morgan is removed by his kind acceptance of my proposal" and "that in a time not distant I may be able to prove myself worthy of the confidence he has placed in me."[46]

THE LABORATORY AT WARDENCLYFFE

Having secured funding from Morgan, Tesla eagerly set to work planning a new laboratory for demonstrating his wireless telegraphy. To design this new laboratory, Tesla turned to his friend Stanford White. White sketched an imposing one-story brick building that was ninety-six feet square with a tall chimney in the center (Figure 14.1). Though the building was largely functional, White provided arched windows and decorated the chimney with a cast iron wellhead inspired by one that he had seen in Italy. White estimated that construction of the laboratory would cost about $14,000.[47]

With plans in hand, Tesla now had to choose where to build his new wireless facility. He flirted with the idea of locating this laboratory

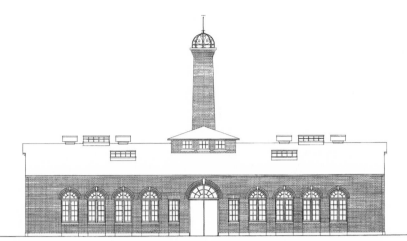

FIGURE 14.1. Tesla's laboratory at Wardenclyffe.
Reproduced with permission from Christopher J. Bach, Architect.

near Niagara Falls in order to transmit the excess power generated there, but he soon focused on picking a location on the Atlantic coast near New York City. As rumors spread that Tesla was going to break ground on a new lab, he was approached by James S. Warden, a lawyer and banker from Ohio who had relocated to Suffolk County on the north shore of Long Island. Believing that a real-estate boom would follow the recent extension of the Long Island Railroad's Northern Branch from Port Jefferson to Wading River, Warden had purchased 1,600 acres of farmland near the tiny village of Woodville Landing. Warden's property was located sixty-five miles from New York City and could be reached by fast train in one-and-a-half hours. Hoping that it might be attractive to New Yorkers as a summer retreat, Warden christened his property Wardenclyffe.

Anticipating that Tesla might employ between 2,000 and 2,500 workers, many of whom would need houses, Warden offered Tesla 200 acres for his laboratory across from the train station. Tesla accepted Warden's offer in August 1901 and began construction the following month. Tesla made regular trips out to the site, arriving by train at 11 AM and staying until 3:30 PM, in order to supervise construction. On these inspection trips, he was frequently accompanied by a Serbian manservant who carried a hamper filled with food prepared by the chefs at the Waldorf-Astoria. Once construction was completed, Tesla stayed in a rented cottage.[48]

Tesla divided the interior of his laboratory at Wardenclyffe into four large rooms: a machine shop, boiler room, engine and dynamo room, and electrical room. Located on the side of the building facing the railroad station, the machine shop was extensively equipped with a blacksmith's forge, lathes, drill presses, a milling machine, and a planer (Figure 14.2). Tesla installed these machine tools anticipating that he would need to fabricate many of the components of his new system. In the boiler and engine rooms were two boilers that provided steam to a 400-horsepower Westinghouse reciprocating steam engine that was directly connected to a custom-designed dynamo. In addition, the engine room was equipped with another dynamo for lighting, an air compressor, and water pumps. Surviving photographs indicate that there was a balcony level in the center of the building, above the

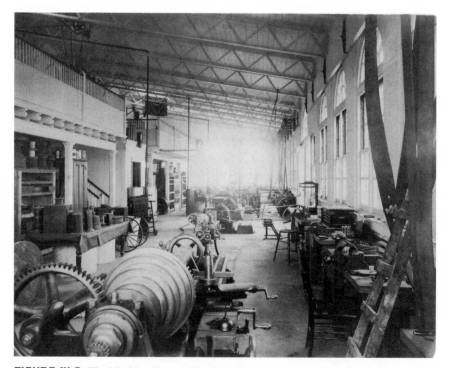

FIGURE 14.2. The Machine Shop at Wardenclyffe.
The balcony running across the center of the building can be seen on the upper left side. From NTM.

FIGURE 14.3. The Engine and Dynamo Room at Wardenclyffe showing ornate woodwork on the mercury interrupter and control panels.
The Westinghouse dynamo can be seen in the back. The staircase probably connected to the gallery or balcony that ran across the center of the building. From NTM.

machine shop. In addition, the photographs reveal that Tesla insisted not only on high-quality apparatus but also on details such as handsome woodwork (Figure 14.3).

By far the most prominent space inside the laboratory was the electrical room that ran across the back of the building, on the side closest to the tower (Figure 14.4). Here the equipment included four massive Westinghouse transformers that could operate at 60,000 volts, four large capacitors, a motorized mercury rectifier, and a specially built control unit, which, according to Tesla, could "give every imaginable regulation that I wanted in my measurements and control of energy." In the center of the room, Tesla displayed dozens of pieces of experimental apparatus.[49]

A key feature of Wardenclyffe was the large tower located 350 feet from the laboratory building (Figure 14.5). Tesla knew that the range

FIGURE 14.4. The Electrical Room at Wardenclyffe showing apparatus on display. From NTM.

of the plant was proportional to the size of the tower and hence dependent on the amount of money available: "If your capitalists are willing to go deep into their pockets, you can put up a tremendous antenna because . . . as I pointed out in 1893, . . . the effects will be proportionate to the capital invested in that part."[50]

In planning the tower, Tesla had to take into account two factors: the amount of electrical energy that could be stored in the elevated terminal and the length of the waves he hoped to transmit through the ground. In Colorado, Tesla had used wooden balls sheathed in metal as his elevated terminal. At Wardenclyffe, he now wanted to vastly increase the amount of energy stored in the elevated terminal so that he could transmit power around the world. "Drawing on an old truth which [was] recognized 200 years ago" in electrical science, Tesla knew that it is possible to store more electrical charge on a sphere than on other geometrical shapes; consequently, he planned that the tower at Wardenclyffe be crowned with a metal terminal with

FIGURE 14.5. The laboratory and tower at Wardenclyffe.
The tower was located 350 feet away from the laboratory since Tesla calculated that, at that distance, it was unlikely that an electrical streamer could leap from the tower to the lab. From NTM.

as large a radius of curvature as possible and then studded on the surface with numerous hemispheres.[51] (See Figure 14.6 for one version with hemispherical studs.) Toward this end, his initial sketches from May 1901 show a curved, adjustable cupola in the shape of a mushroom (Figure 14.7).

In terms of the wavelength, Tesla had learned in Colorado that the length of the conductor in the secondary of his magnifying transmitter should be one-quarter of the wavelength that one wanted to generate. Since he initially wanted to transmit power using low-frequency long waves on the order of 2,400 feet, Tesla calculated that his secondary—which would be housed in the Wardenclyffe tower—should be 600 feet and, in turn, that would be the height of the tower. At 600 feet, Tesla's proposed tower would be two-thirds the size of the Eiffel Tower, and White estimated that it could cost $450,000.[52]

Alarmed by how much his tower would cost, Tesla asked Morgan in September 1901 for more money.[53] Morgan, however, refused to lend

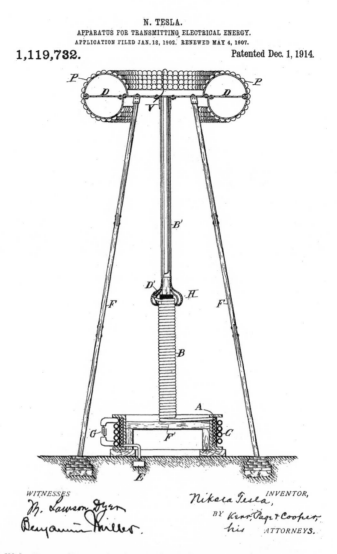

N. TESLA.
APPARATUS FOR TRANSMITTING ELECTRICAL ENERGY.
APPLICATION FILED JAN. 18, 1902. RENEWED MAY 4, 1907.

1,119,732.
Patented Dec. 1, 1914.

FIGURE 14.6. Patent diagram for the Wardenclyffe Tower showing one version of the elevated terminal as well as the circuitry Tesla planned to use.

Key:
D is donut-shaped terminal studded with hemispheres P
G is a source of electric current (probably a capacitor)
C is the primary coil of the magnifying transmitter
A is the secondary coil of the magnifying transmitter
B is another coil connected to secondary coil A
E is the ground connection.

Tesla did not build a terminal on top of his tower in the donut form shown here but rather used this form in his patent "to illustrate the principles." See NT, Radio Testimony, 145. From NT, "Apparatus for Transmitting Electrical Energy," US Patent 1,119,732 (filed 18 Jan. 1902, granted 1 Dec. 1914).

FIGURE 14.7. The tower at Wardenclyffe showing hemispherical terminal on top. From KSP, Smithsonian Institution.

any more money, and Tesla was forced to scale back his design. "One thing is certain," wrote Tesla to White, "we cannot build that tower as outlined. I cannot tell you how sorry I am, for my calculations show, that with such a structure I could reach across the Pacific."[54]

Although he considered using two or three smaller towers together, Tesla settled on a smaller design topped with a single large hemisphere fashioned out of steel girders (Figure 14.7). Tesla planned that this terminal be covered with copper plates and perhaps again studded with smaller hemispheres. Although the shape of the terminal had changed, it still possessed a large radius of curvature, which Tesla believed would enable him "to produce with this small plant [at Wardenclyffe] many times the effect that could be produced by an ordinary plant of a hundred times the size."[55]

Construction of the tower at Wardenclyffe began in November 1901, and photographs indicate that the structure was in place by September 1902. Tesla was not able to complete the tower as quickly as he would have liked because Morgan apparently delayed paying the final installment of $50,000 by two months.[56]

One of the challenges in building the tower was that no matter what the shape of the terminal on top, it was extremely heavy; the final hemispherical version was 68 feet (20.7 meters) in diameter and weighed 55 tons. In addition to the weight, the terminal also had a large surface area so that it functioned like a sail in response to the winds coming off Long Island Sound. To support the weight of the terminal and offset the forces of the wind meant building a large and strong tower. To achieve this strength, White resorted to an octagonal form that gradually tapered upward. Since the terminal had to be insulated from the ground, the tower could not be constructed using iron or steel girders and White instead used unfinished pine timber. Top to bottom, the Wardenclyffe tower rose 187 feet (57 meters) and could be seen from New Haven, Connecticut, across the sound.[57]

While the tower was certainly imposing—a reporter from the *New York Times* called it "very 'stagey' and picturesque"—even more impressive were the well and tunnels beneath it.[58] Because Tesla intended to transmit electric power through the ground, it was essential that his system "get a grip of the earth, otherwise it cannot shake the earth. It has to have a grip on the earth so that the whole of this globe can quiver, and to do that it [was] necessary to carry out a very expensive construction."[59] The ground connection in Colorado Springs had consisted of several metal plates buried in the ground outside; at Wardenclyffe, Tesla decided that he needed to make a stronger connection with the water table below the tower. To accomplish this, Tesla sank a ten-by-twelve foot well that went down 120 feet, well below the water table. To provide access to the well, there was "a wooden affair very much like the companionway on an ocean steamer" at the base of the tower and a circular staircase to the bottom.[60]

While one end of the magnifying transmitter's secondary coil was connected to the extra coil and then to the elevated terminal at the top of the tower, the other end of the transmitter's secondary coil was grounded (see Figure 14.6). To make this ground connection between the magnifying transmitter at the base of the tower and the bottom of the well, Tesla placed "a big shaft again through which the current was to pass, and this shaft was so figured in order to tell exactly where the nodal point is, so that I could calculate every point of distance. For instance I could calculate exactly the size of the earth or the diameter of the earth and measure it exactly within four feet with that machine."[61]

At the bottom of the well, Tesla completed the ground connection by linking the metal shaft that came down from the magnifying transmitter with an elaborate system of horizontal pipes. As he explained, "the real expensive work was to connect that central part [i.e., the metal shaft] with the earth, and there I had special machines rigged up which would push the iron pipe, one length after another, and I pushed these iron pipes, I think sixteen of them, three hundred feet, and then the current through these pipes takes hold of the earth."[62] Tesla used compressed air to drive these pipes into the earth, and so he placed a compressor in the laboratory's engine room and transmitted the compressed air out to the well via a special pipeline that ran alongside the electrical conduit.

One newspaper reported that at the bottom of the well, Tesla planned to keep the water warm, perhaps hoping that this would improve the connection with the earth.[63] In addition, he had workmen dig four stone-lined tunnels, each a hundred feet long, from the bottom of the well, each of which gradually sloped upward to the surface. At the end of each tunnel there was a brick igloo-like exit.[64] It is not clear what role these tunnels played in Tesla's overall plan.

HOW DID TESLA THINK WARDENCLYFFE
WOULD OPERATE?

As construction continued at Wardenclyffe into 1902, Tesla began to
see that the development of wireless power would involve several distinct steps. As he explained to Morgan, these steps would consist of:
"(1) the transmission of minute amounts of energy and the production
of feeble effects, barely perceptible by sensitive devices; (2) the transmission of notable amounts of energy dispensing with the necessity of
sensitive devices and enabling the positive operation of any kind of
apparatus requiring a small amount of power; and (3) the transmission of power in amounts of industrial significance. With the completion of my present undertaking the first step will be made."[65] So,
given this development plan, how did Tesla plan to use Wardenclyffe
to transmit power?

Although Tesla left no complete description of how the Wardenclyffe station operated, all indications are that he planned to run the
station in a manner similar to how he ran his system at Colorado
Springs, which was based on the principles in the fundamental patents he had filed in 1897 (see Figure 12.11).[66]

At Wardenclyffe, the boilers produced steam for the Westinghouse
combination steam engine and AC dynamo. Current from the dynamo
was then stepped up to 44,000 volts by the four transformers in the
electrical room that were arranged to deliver four-phase power. Tesla
used this high-voltage current to charge four large capacitors in the
room, and he regulated the frequency of the current by using several
variable inductance coils, resistance boxes, and a large mercury-arc
interrupter. With these controls, he could vary the frequency from
200,000 down to 1,000 cycles per second, and he could produce a
continuous train of undamped waves.[67]

This high-voltage, high-frequency current was conducted out to
the tower via cables in a special underground conduit. As shown in
Figure 14.6, Tesla installed his magnifying transmitter at the base of
the tower. Although the primary and secondary coils were smaller in
diameter than their counterparts in Colorado, the extra coil appears to
have been much longer. To connect the extra coil to the elevated terminal, Tesla used a large metal shaft. While he was primarily concerned

with using the magnifying transmitter to broadcast power, Tesla could send telephone or telegraph messages by connecting either a microphone or a telegraph key to the transmitter's primary coil "such that by speaking into it, or actuating it by hand or otherwise, variations in the intensity of the waves are produced."[68] Tesla estimated that the Wardenclyffe magnifying transmitter generated 200 kilowatts.[69]

Operating at this level of power, Tesla hoped that the magnifying transmitter would be able, via the well below the tower, to "get a grip of the earth" and set up a stationary current wave in the earth's crust. To conduct these current waves from the secondary of the magnifying transmitter into the ground, Tesla used another large metal shaft that extended from the base of the tower to the bottom of the well, and from there, the waves traveled into the Earth via the sixteen horizontal pipes.

Tesla believed that this current wave produced by his magnifying transmitter would travel through the Earth to a point opposite Wardenclyffe and reflect back on itself; if this wave was at the resonant frequency of the earth, then the reflected wave would be in phase with the original wave and thus establish a stationary wave. Based on measurements he made in Colorado, Tesla calculated that the lowest resonant frequency of the Earth was 6 Hz. Modern theory predicts this frequency to be 10.5 Hz (neglecting losses), and actual measurements give a resonant peak at approximately 8 Hz, suggesting that Tesla was on the right track in terms of his calculations.[70]

With the Earth humming at its electromagnetic resonant frequency, Tesla believed that power and messages could be picked up anywhere on the earth's surface by connecting a receiver to the ground. While energy moved from the magnifying transmitter to the receivers around the world via the stationary wave set up in the earth, Tesla anticipated the circuit in his system would be completed by having elevated terminals of some sort on both the transmitter and receivers. As we have seen, the elevated terminal at Wardenclyffe was designed to hold an enormous amount of electrical charge. Prior to going to Colorado, Tesla insisted that the return circuit involved a current moving through the atmosphere (see Chapter 12), but while he was working at Wardenclyffe he never offered a clear explanation as to what kind of return electrical connection he expected to make between the receiver and transmitter in order to complete the circuit. Drawing on ideas that Tesla later put forth concerning particle beams (see Chapter

16), some Tesla fans have speculated that he created a return circuit by using either X-rays or a laser to create a conducting path from the Wardenclyffe tower to the ionosphere, noting that there was a four-foot-wide hole in the top of the elevated terminal so that Tesla could have easily beamed the rays skyward.[71] Meanwhile, other Tesla specialists have been able to transmit power through the Earth for short distances using magnifying transmitters, but they have not been able to settle on an explanation for what happens in the return circuit that can be related to modern electrodynamic theory.[72]

As his fundamental patents suggest, Tesla initially thought that the receiver would need to be a tower similar in size to the transmitter (see Figure 12.11). However, at some point in his work at Wardenclyffe, he decided that a large receiving tower was not necessary and that messages and power could be picked up by using smaller devices. Instead, he now proposed that homes be fitted out with a ground connection and small elevated terminal so that they could receive the power necessary to illuminate his vacuum-tube lamps. He also designed a combination receiver and clock that would be powered by wireless energy and hence receive a precise time signal, noting that "[t]he idea of impressing upon the earth American time is fascinating and very likely to become popular."[73]

In many ways, Wardenclyffe was the fulfillment of Tesla's dreams. For nearly a decade he had been planning in his imagination a system for broadcasting power around the world, and now that system was taking shape in the real world. Tesla was effectively assembling the network of people, ideas, money, and resources needed to get ahead—and stay ahead—of Marconi.

But would the system at Wardenclyffe actually work? To some extent, I don't think Tesla was at all worried as he had full confidence in his abilities as an inventor; if he could imagine Wardenclyffe working, then it was sure to work. Moreover, by the end of 1901, he had all the trappings—the illusions—of success. As long as he lived like a millionaire at the Waldorf, had the support of J. P. Morgan, got ample press coverage, and was building an impressive station, then all would be right. The illusions confirmed the ideals that Tesla saw in his imagination.

CHAPTER FIFTEEN
THE DARK TOWER
(1901–1905)

To achieve a great result is one thing, to achieve it at the
right moment is another.

TESLA TO J. P. MORGAN, 13 OCTOBER 1904

BLINDSIDED BY MARCONI,
DOUBLE-CROSSED BY MARTIN

Through the fall of 1901, as Tesla supervised construction of his laboratory and tower at Wardenclyffe, he was confident that he was close to
success. As he gaily wrote to Katharine Johnson on 13 October 1901:

> 13 is my lucky number and so I know you will comply with my wish . . .
> [for you to] come to the Waldorf. And if you do—when I transmit my
> wireless messages across seas and continents you will get the finest bon-
> net ever made [even] if it breaks me. . . .
>
> I have ordered a simple lunch and you must come en masse. We
> must exhibit Hobson. . . . I know he likes me better than you.[1]

At the same time, Tesla assured Morgan that he was making progress. Summarizing his work for the Great Man in November, he confidently stated that his latest patents covered the production of "electrical effects of virtually unlimited power, not obtainable in any other ways heretofore known." Moreover, his patents covered techniques for highly efficient transmission. While other transmission methods suffered losses proportional to the square of the distance covered, Tesla claimed that his losses were significantly less and only in simple proportion to the distance. "This feature alone," reported Tesla, "bars all competition."[2]

But the competition, namely Marconi, was hardly deterred by such claims. As we saw in Chapter 12, Marconi had transmitted messages across the English Channel in March 1899, prompting Tesla to take up his experiments in Colorado Springs. Seven months later, Marconi had come to New York and used his apparatus to provide reports from the America's Cup yacht races, hoping to secure contracts from the New York newspapers or the U.S. Navy. Unsuccessful in getting a contract with the navy, Marconi returned to England and resumed work on increasing transmission distance as well as on developing a way to tune his transmitters and receivers so that they operated on a particular frequency. By early 1900 he could cover distances up to 185 miles, and he secured a British patent (No. 7777 of 1900) for a system of using specially wound jiggers (or coils) in his antenna circuits that permitted tuning.[3]

While Marconi publicly insisted that it was not yet possible to transmit messages wirelessly across the Atlantic, privately he had decided that he should try to achieve this goal as soon as possible. Marconi came to this decision because he was worried about the situation of his business. Despite its best efforts, Marconi's Wireless Telegraph and Signal Company had yet to sign a major contract with the British or American navy, the Post Office, or the marine insurance group, Lloyd's of London. Though Marconi stock was being bid up speculatively by London investors, the meager equipment sales nonetheless meant that the company was running out of capital. To solve these problems, Marconi argued that what was needed was a dramatic demonstration of the potential of his wireless system; if he could span the Atlantic, not only could his company establish a monopoly on ship-to-shore communications, it could also start competing with the profitable cable message business. Although the company's directors initially objected to his bold plan, Marconi was able to convince them that transatlantic communication was feasible, and in July 1900 the board of directors gave him the go-ahead.

To undertake the "great thing" (Marconi's term for transmitting across the Atlantic), Marconi quietly established new stations.[4] In England, he set up a station at Poldhu in Cornwall and in America he built a station on Cape Cod at South Wellfleet. At each station, he constructed a huge circular antenna comprised of twenty two-hundred-foot masts. While scouting out locations in Cape Cod in early 1901, Marconi heard rumors about Tesla's plans—perhaps that Tesla was getting support from Morgan—and this news prompted him to step up his efforts.[5]

To span the Atlantic, Marconi realized that he needed to scale up his system, not only in terms of the size of the antenna arrays but the amount of power employed by the transmitter. Up to 1900, Marconi had relied on small transmitters that used induction coils and batteries; to reach across the Atlantic, however, he needed a much more powerful transmitter. To increase the power, Marconi looked to the company's new scientific consultant, John Ambrose Fleming. The Pender Professor of Electrical Technology at University College, London, Fleming had met Tesla during his visit to London in 1892. Completely familiar with power engineering, Fleming designed a transmitter with a 25-kilowatt AC generator, 20,000-volt transformers, and high-tension capacitors. Not only was this electrical equipment similar in size to the apparatus Tesla had at Wardenclyffe, but Fleming arranged the circuit in a manner similar to what Tesla used with his magnifying transmitter in both Colorado and Wardenclyffe. Using this new system designed by Fleming, Marconi and his associates were able to generate at Poldhu foot-long sparks that were as thick as a man's wrist.[6]

Fleming installed this new transmitter in August 1901, but before it could be tested, gale-force winds blew down the antenna array in Poldhu. Marconi quickly replaced this array, but then another storm destroyed the antenna array on Cape Cod. Still determined to conduct a transatlantic test, Marconi, George Kemp, and P. W. Paget sailed to St. John's, Newfoundland, at the end of November. Marconi chose St. John's because it was the closest spot in North America to England (2,200 miles or 3,500 kilometers). Before leaving England, Marconi instructed the operators at Poldhu to transmit "SSS" in Morse code between 3:00 and 6:00 PM each day from 11 December on.

Marconi and his associates arrived in St. John's on 6 December 1901. Using an aerial suspended from a kite flying in a winter storm, both Marconi and Kemp heard the "SSS" signal on 12 and 13 December while using a sensitive telephone receiver. On 14 December, Marconi issued a press announcement, and his achievement was extensively covered in the Sunday newspapers on the following day.[7]

In the *New York Times*, news of Marconi's transatlantic signals dominated the front page. Calling his achievement "the most wonderful scientific development of recent times," the paper included a biography of Marconi. Since Tesla had "hinted at the possibility of 'telegraphing through the air and earth,'" the paper asked Tesla for his impressions. Tesla not only reminded the paper that he had discussed the possibility of wireless telegraphy years earlier but suggested that the transmission of power was much more important than being able to send a few short messages. To round out its coverage, the *New York Times* gave the last word to T. C. Martin, who, as editor of *Electrical World*, could put Marconi's achievement into context. Anxious to praise Marconi as the newest technological wunderkind, Martin noted that he was both surprised and pleased that Marconi had succeeded in spanning the Atlantic. With regard to his former friend Tesla, Martin pointed out that although Tesla had envisioned wireless telegraphy, he unfortunately had not been able to follow through and be the first to span the Atlantic. As Martin observed, "In a book which I published some eight years ago on Tesla's work is embodied one of his lectures in which he gives wireless telegraphy considerable attention. He expressed his belief in the matter so clearly that he made up my mind for me. I am only sorry, therefore, that Mr. Tesla, who has given the matter so much thought and experimentation, and to whose initiative so much of the work is due, should not also have been able to accomplish this wonderful feat."[8]

Over the next several weeks, skeptics questioned whether Marconi had actually heard the three dots that made up an "S" in Morse code. As rival wireless inventor Lee de Forest noted in his diary: "Signor Marconi has scored a shrewd coup. Whether or not the three dots he heard came from England or, like those Tesla heard, from Mars, if I am aught of a prophet we will hear of no more trans-Atlantic messages for some time. In this art, as in all other inventions, advance will be by slow growth and evolution, not by magnificent bounds from 100 to 2,000 miles."[9]

Since Hertzian waves were supposed to travel in straight lines like light waves, wondered other scientists, how did the waves follow the curvature of the Earth and not just fly off into space? (We now attribute this to the fact that radio waves bounce off the Heaviside-Kennelly layer in the ionosphere, but this layer was not discovered until 1924.) Could Marconi and Kemp, anxious to hear the three dots, have simply heard atmospheric crackling that they imagined to be the signal from Poldhu? And note that there were no independent witnesses to the event. The only two people to hear the signals were Marconi and his assistant Kemp; indeed, the world had to accept Marconi's word that he had heard a signal from across the Atlantic.[10]

But in spite of these doubts, Martin decided that Marconi was the man of the hour and that he should take "up the cudgels" on Marconi's behalf and show that Marconi was "not being a faker." To do so, Martin arranged for the young Italian to be the guest of honor at the annual dinner of the American Institute of Electrical Engineers (AIEE) on 13 January 1902. Having been president and toastmaster at many AIEE events, Martin easily convinced the leadership of the institute to agree to honor Marconi. However, since not everyone believed Marconi's claims, Martin found it difficult to get the engineering community to come out for the dinner, and he was obliged to ask Elihu Thomson to lend his support to the event. As word spread that Thomson favored recognizing Marconi, Martin was able to fill the three hundred seats in the Astor Gallery of the Waldorf-Astoria Hotel.[11]

To make it a truly memorable evening, Martin worked to ensure that everything highlighted Marconi's achievement. The menu featured sketches of Marconi antennas on two lighthouses, signaling "S" in three dots all the way across the ocean. At either end of the ballroom were hung two large tablets with "Poldhu" and "St. John's," in letters formed of electric lamps. Connecting the two tablets was a wire on which were inserted clusters of three lamps at intervals, designed to represent the three dots flashed across the Atlantic, and these lamps were flashed periodically during dinner.[12]

Although Tesla resided at the Waldorf, he could not bring himself to attend the dinner honoring Marconi and chose to be out of town that night. Noting his absence, toastmaster Martin read a letter in which Tesla congratulated Marconi but conspicuously left out any mention that a message had been sent across the Atlantic.

After reading a few more letters of congratulation, Martin invited Marconi to address the group. Marconi outlined the achievements of his system so far—seventy ships equipped with wireless and twenty shore stations in England. He described his experiments in Newfoundland, including the problems he had faced with flying kites in the wintry weather. After mentioning that he hoped that wireless telegraphy would allow for cheaper messages than the existing undersea cables, Marconi concluded by raising a glass and toasting the institute.

Marconi's speech was followed by remarks from Thomson and Professor Michael Pupin of Columbia University. Both men emphasized that although the proof for Marconi's achievement was limited, they accepted Marconi at his word since they knew and trusted him. As Pupin put it, "[I]n scientific work we never believe anything until we see a demonstration of it. I believe that Sig. Marconi has transmitted the famous three dots across the Atlantic, but I must say that I believe him because I know him personally. If I did not know him personally, I would not believe him, because the proof which Signor Marconi has furnished is not sufficiently strong from a purely scientific point of view; but knowing him personally as I do, I believe his statement." Pupin further offered an analysis of Marconi's work with electromagnetic waves, combining physics, mathematics, and engineering. In doing so, he included a swipe at Tesla for his ideas about transmitting signals through the earth:

> I also heard a man say, "Years ago I thought of transmitting wireless signals by the wobbling of the charge of the earth." Well, any one of us can think of schemes like that; any one of us who has had any experience in inventing can think of schemes like that as fast as you can write them down, for anybody knows you can transmit electrical waves to any distance, that is, mathematically and physically perhaps. But how about the engineering side of it? I said to this man, "Give me an engineering specification of your apparatus by means of which you intend to wobble the charge of the earth, and then I will believe you; not before."
>
> Now this is what Signor Marconi has done: He has written out a specification for setting up apparatus and wobbling the charge of the earth and transmitting the signals between wires.[13]

Thanks to Martin's efforts, the AIEE dinner helped establish Marconi as the inventor of wireless telegraphy in the public mind. Nevertheless, knowing that his transatlantic effort in Newfoundland was not sufficient proof, Marconi undertook a second demonstration to prove his system. In February 1902 he sailed from England to America on the passenger ship *Philadelphia*, which was fitted out with a Marconi wireless system. During the crossing, Marconi periodically received Morse code messages from Poldhu, and he invited the ship's captain and chief officer to listen to the messages and note on a nautical chart the time and location where they heard the messages. Much more than the signal received at Newfoundland, this witnessed chart from the *Philadelphia* documented that Marconi had a system that could transmit across the Atlantic.[14]

COUNTERING WITH "WORLD TELEGRAPHY"

With the newspapers filled with stories about Marconi, Tesla could not resist taking a poke at his rival in print. When a reporter from the *New York Sun* asked if there was any similarity between his system and that of Marconi, Tesla smiled and said, "I respect rigorously the rights of others, and when I give my system to the world I shall ask the entire technical profession to point to any feature of my system . . . which is not of my own creation. I admire skill and enterprise, and my best wishes for success accompany those who sell ready made shoes; but I myself prefer not to use them. They are cheap, but they raise corns and bunions."[15] In doing so, Tesla was taking the high ground in that he would let his fellow experts decide who was the original creator of the new wireless technology. At the same time, the metaphor of ready-made shoes is revealing: not only did Tesla have higher standards than Marconi (in real life, Tesla insisted on bespoke shoes), but he also suggested that Marconi's choice of selling cheaper shoes would result in problems (i.e., corns and bunions) for the user.

But while he could poke fun at Marconi in the press, Tesla realized that he needed to offer some explanation to Morgan. In early January 1902 Tesla wrote Morgan, beginning with an assessment of the equipment developed by Marconi with Fleming's help:

I have carefully examined the records and find that, in what the Marconi-Fleming Syndicate are now using, not a vestige of their old patented apparatus is left. . . . All the essential elements of these [new] arrangements . . . are broadly anticipated by my patents of 1896 and 1897. . . . They have adopted my resonating transmitter which magnifies enormously the currents conveyed, and in this connection my grounded receiving circuit or "multiplier," my transforming circuits at both ends, my "Tesla coil" and my modern type of the same, the "Tesla transformer," my system of dependent tuned circuits and numerous minor improvements. Nothing can be done just now, but as the water finds its level, so every one will get his due.

Rather than be defensive about Marconi, though, Tesla chose to be upbeat and bold. "I need not tell you," he continued,

that I have worked as hard as I have dared without collapsing. I have examined and rejected hundreds of experiments aiming at improvement in every way and at the attainment of the best result with the capital at command, and I am glad to say, that by slow and steady advances I have managed to contrive a machine, with which I shall produce an electrical disturbance of sufficient intensity to be perceptible over the whole earth. It will be feeble at some places and I fear, unsuitable for practical use, but I am sure that, when my apparatus is . . . in [a] condition to deliver its maximum energy—at the rate of one million horse-power—with the first throw of the switch, I shall send a greeting to the whole world, and for this great triumph I shall ever be grateful to you![16]

But more than just send out a signal, Tesla now introduced Morgan to a new business plan. Marconi might choose to cover yacht races and provide ship-to-shore communications, but Tesla dismissed these applications as unprofitable and hardly worthy of support from a Great Man. Indeed, as he later explained to Morgan, "When I discovered, rather accidentally, that others . . . were secretly employing [my apparatus], I found myself confronted with wholly unforeseen circumstances. . . . I could not develop the business slowly in a grocery-shop fashion. I could not report yacht races or signal incoming steamers. There was no money in this. This was no business for a man in your position and importance."[17]

Instead, Tesla proposed to Morgan a plan for a "World Telegraphy System" in which a number of transmitting stations would collect news and broadcast to customers via individual receivers. As he boasted to Morgan,

> The fundamental idea underlying this system is to employ a few power plants, preferably located near the large centers of civilization and each capable of transmitting a message to the remotest regions of the globe. These plants are to be connected by wires, cables and any other means with the civilized centers nearby, and as fast as they receive the news, they pour them [*sic*] into the ground, through which they spread instantly. The whole earth is like a brain, as it were, and the capacity of this system is infinite, for the energy received on every few square feet of ground is sufficient to operate an instrument, and the number of devices which can be so actuated is, for all practical purposes infinite. You see, Mr. Morgan, the revolutionary character of this idea, its civilizing potency, its tremendous money-making power.[18]

Although Tesla was certainly not thinking about the computers, software, and packet-switching that were necessary to create the World Wide Web, his fundamental idea that all news should be collected and disseminated around the world is suggestive of the beliefs that came to underlie the World Wide Web in the 1990s. "The World-Wide Web (W3)," noted media scholars Noah Wardrip-Fruin and Nick Montfort, "was developed to be a pool of human knowledge, which would allow collaborators in remote sites to share their ideas and all aspects of a common project."[19]

Tesla believed that he and Morgan would make money by manufacturing receivers, and he envisioned several versions. For instance, the receiver could be a printer that would produce a newspaper for customers at home, and in so doing his world telegraphy system would "do away not only with the cables but with the newspapers also, for how can journals as present [continue to] exist when everybody can have a cheap machine printing its own news?" At the same time, Tesla was also developing a receiver with a loudspeaker that would permit him to speak Morgan's "name in a telephone, and it will be repeated loud everywhere in the tone of my voice. This, I may tell you now, I have since long ago conceived as a means of acquitting myself fittingly of my debt to you."[20]

FIGURE 15.1. "Tesla's Wireless Transmitting Tower, 185 feet high, at Wardenclyffe, N. Y., from which the city of New York will be fed with electricity, and by means of which the camperout [sic], the yachtsman and summer resort visitor will be able to communicate instantly with friends at home."

From "Tesla's Tower," *New York American,* 22 May 1904 in TC 17:11.

But by far Tesla's most imaginative idea for a receiver was a hand-held device connected to a vertical wire on a short pole or even a lady's parasol so that it could pick up voice messages anywhere in the world (Figure 15.1). As Tesla promised in 1904, "An inexpensive receiver, not bigger than a watch, will enable him to listen anywhere, on land or sea, to a speech delivered, or music played in some other place, however, distant."[21] Here in the opening years of the twentieth century, we see Tesla conjuring up a vision of a device much like a transistor radio or cell phone, with the promise of providing instantaneous access to information anytime, anywhere.

In dreaming of a receiver that would be used by everyone, Tesla was an early harbinger of the consumer culture of the twentieth century. As engineers and inventors perfected the machinery for the mass production of goods, astute managers realized that the challenge would be to stimulate demand for the huge volume of goods produced. In other words, if they were going to take advantage of the economies of scale that came with mass production, industrialists would have to create products that would be desired and used by millions of consumers. A classic example of a mass-produced product for this new consumer culture was the Model T; as Henry Ford explained, the Model T was intended to be "a motor car for the great multitude . . . large enough for the family but small enough for the individual to run and care for . . . so low in price that no man making a good salary will be unable to own one."[22] Just as Ford envisioned everyone having a Model T, so Tesla believed that everyone would soon have one of his wireless receivers.

Viewed from the perspective of the twenty-first century, it may seem obvious that a consumer culture would grow out of mass production and that a significant portion of the global economy would depend on the mass consumption of products like cell phones, iPods, and laptop computers. Yet this consumer revolution was not entirely obvious in the opening years of the twentieth century and, indeed, must have looked alien to the leaders of industrial and producer culture. For instance, believing that money was to be made by selling inventions and machines to companies and not to individual consumers, Edison found it difficult after 1900 to understand how it was that the motion-picture industry was shifting away from the improvement of cameras and projectors to the development of new types of movies.[23] Likewise, Morgan must have had a hard time making sense of Tesla's vision of selling millions of receivers. Morgan had made his money in developing the railroads and the steel industry—the very heart of a producer culture. Much as he could never understand the rise of the automobile industry (and hence never invested in it), it is likely that he had little appreciation for Tesla's vision of world telegraphy. Tesla was unfortunately trying to use a consumer-culture argument to convince someone steeped in producer culture.

As exciting as world telegraphy was to him, Tesla did not neglect his vision of broadcasting power since "the transmission of energy in appreciable amounts, enabling at any point of the globe the positive operation of innumerable contrivances," would have a tremendous effect on the advancement of humanity. While the potential of broadcasting power could be demonstrated with the Wardenclyffe plant, large-scale transmission would require a bigger station, preferably located near a source of cheap electric power such as Niagara Falls. Of course Tesla would need more money, but once again, he was confident that the work could be done quickly. As he explained to Morgan, "To do this, a plant of five thousand horse-power would be needed. This power could be had at Niagara. . . . The preliminary work, the plans, estimates, options, etc.,—all the work on paper, would not involve an expenditure of more than twenty-five thousand dollars. By the time my present plant is completed and I have made demonstrations to your satisfaction, all this preparatory work could be done and, if you should elect so, before the next winter the large plant could be put in operation." Tesla felt strongly that a demonstration of the wireless transmission of power was the best way to respond to Marconi's bold appropriation of his techniques; as he wrote to Morgan some months later, "The only way to fully protect myself was to develop apparatus of such power as to enable me to control effectively the vibrations throughout the globe."[24]

Lest Morgan hesitate to support either world telegraphy or a new power transmission station at Niagara, Tesla closed his January 1902 letter with an egotistical flourish, reminding the financier that he was dealing with a genius who brought about technological revolutions: "Now, Mr. Morgan, am I backed by the greatest financier of all times? And shall I lose great triumphs and an immense fortune because I need a sum of money!! Is it not due to the honor of this country, that it be identified with this achievement[?] Have I not contributed to its greatness and prestige and have my inventions not exircised [*sic*] a revolutionary effect upon its industries[?] These are not my claims, Mr. Morgan, only my credentials."[25]

Tesla's business proposals landed on Morgan's desk just as he was entering another extremely busy period; as Henry Adams observed in April 1902, "Pierpont Morgan . . . is carrying loads that [would] stagger the strongest nerves." During the first months of 1902, Morgan was organizing the International Mercantile Marine, which brought together five transatlantic steamship lines to create a fleet of 120 ships; it was Morgan's hope that he could create rational order for transatlantic shipping just as he had done for the railroads. From ships, Morgan moved to agricultural equipment, and in the summer of 1902 he and his partners organized International Harvester by merging the

McCormick Harvesting Machine Company with four other firms. In addition, Morgan found that his newly formed United States Steel was undercapitalized, and he spent much of the year trying to raise funds for it.[26]

Morgan's biggest headache in 1902, however, came from Washington. The previous summer, Morgan had fought off a hostile takeover bid for the Northern Pacific Railroad, and to prevent future raids he organized the Northern Securities Company. As a holding company, Northern Securities brought together the Northern Pacific, the Great Northern Railroad, and the Chicago, Burlington, & Quincy so that these major trunk lines could coordinate operations. Because Northern Securities now controlled nearly all of the rail traffic from Minnesota to Washington State, its creation prompted a public outcry. In response, President Theodore Roosevelt decided to go after Northern Securities in order to demonstrate that he wasn't afraid to challenge big business.

Since these projects demanded most of his attention, Morgan does not seem to have been particularly concerned about Tesla. Indeed, the Great Man included Tesla as a guest at an extravagant luncheon honoring Prince Henry, the brother of Kaiser Wilhelm II.[27] In the meantime, Tesla pushed ahead on his own. In February 1902 he hired Lowenstein back to help with Wardenclyffe. To strengthen his patent portfolio, Tesla had his attorney Parker Page vigorously pursue an interference case with Reginald Fessenden concerning Tesla's patent applications for tuning, and Tesla provided testimony during the summer to support his case.[28] He also started planning a factory at Wardenclyffe since it was "now imperative to provide facilities for manufacturing a great number of receiving apparatus."[29]

Before leaving on his annual trip to Europe in April 1902, Morgan met with Tesla and told the Wizard that he did not want to get involved personally with building a new transmission station at Niagara or a new factory for manufacturing receivers. However, although he did not want to invest his own money, Morgan indicated that he would be willing to help Tesla raise money by reorganizing the Nikola Tesla Company and issuing new securities.[30]

In response to this "favorable answer received," Tesla spent the summer of 1902 working on two tasks. First, since Morgan would not support a new plant at Niagara, Tesla decided to push his system to see how much power he could generate. To pursue this task, Tesla and his secretary, Scherff, moved out to Wardenclyffe for the summer. "My efforts will be in large measure rewarded," he reported back to Morgan, "for by straining every part of my machinery to the utmost I shall be able to reach what I consider almost the maximum possible performance with the power available—a rate of energy delivery of 10 millions of horsepower—more than twice that of the entire Fall of Niagara. Thus the waves generated by my transmitter will be the greatest spontaneous manifestation of energy on Earth."[31]

Pleased with the potential results of pushing Wardenclyffe to the limit (and thereby putting Marconi in his place), Tesla turned to identifying new potential investors. Drawing on his contacts in the upper echelons of New York society, Tesla brought together a group of subscribers, "all people of high standing." Knowing that Morgan wished to continue to keep his involvement a secret, Tesla was careful not to mention the Great Man to this group.[32]

In late September 1902, Tesla and Morgan met to map out a plan to raise capital by creating a new company. Here Morgan was genuinely helping Tesla since the fundamental business of an investment banker is to organize companies and issue securities. Tesla's new company was to be capitalized at $10 million and would issue $5 million in bonds, $2.5 million in preferred stock, and $2.5 million in common stock. To secure the working capital needed to equip a new factory for manufacturing Tesla's inventions (presumably receivers), 50% of the bond and stock issue would be offered for sale to outside investors. Another 40% of the bonds and stocks would be given to Tesla and his old associates in the Nikola Tesla Company since the company had "incurred great expenses perfecting the inventions." For his part, Tesla planned to sell some of his bonds in order to repay the $150,000 advanced by Morgan, but he also intended to give one-quarter of his interest to an associate (probably Lowenstein) "in whose ability and integrity I believe, and who is to unite all of his energies with mine, to bring to the greatest possible success this undertaking, in which our honor will be engaged." The remaining 10% of the new bonds and shares was to remain unissued, but Morgan thought that he should get one-third of these remaining securities for the patents that had been assigned to him by Tesla. As it did with other stock issues, J. P. Morgan & Company would presumably make its money by taking a commission on the bonds and shares it sold to outside investors.[33]

With this plan in place, Tesla resumed his efforts to secure money from the New York social elite. Offering shares at $175 each, he approached several prominent women, including Mary Mapes Dodge, Mrs. E. F. Winslow, and Caroline Clausen Schwarz, the wife of toy-store magnate, F.A.O. Schwarz. Unfortunately, Tesla got few takers among the elite, who seemed to have regarded investing in his enterprise as a risky proposition. "I am tired of speaking to pusillanimous people," fumed Tesla, "who become scared when I ask them to invest $5000 and get the diarrhoea [*sic*] when I call for ten."[34]

Unable to attract investors but determined to push ahead at Wardenclyffe, Tesla raised $33,000 by selling personal property and borrowed another $10,000 from a bank in Port Jefferson, the next town over from Wardenclyffe. He also tapped Scherff for small loans, which eventually amounted to thousands. Nevertheless, the bills continued to mount—Tesla owed Westinghouse $30,000 for equipment, he had not paid the telephone company for running a special line out to the laboratory, and James Warden was suing him for not paying the property taxes. Worried that news of his financial difficulties would scare off potential investors, Tesla told Scherff to keep the reporters away.[35]

As his hopes for raising money by selling shares evaporated, Tesla blamed Morgan for his problems. Tesla felt that Morgan's various ventures had created turbulence on Wall Street and inflation generally. The stock market, responding to a nationwide coal strike, was very jittery during the fall of 1902, prompting Morgan to join with other bankers to create a $50 million fund to shore up the market in case of an emergency. In April 1903, the Federal Court of Appeals in St. Paul ruled that Morgan's Northern Securities Company was an illegal combination, and this announcement nearly caused a panic on the New York Stock Exchange. Writing a week after this crisis on Wall Street, Tesla complained, "Mr. Morgan, you have raised great waves in the industrial world and some have struck my little boat. Prices have gone up in consequence, twice, perhaps three times higher than they were and there were expensive delays, mostly as a result of activities you excited."[36]

Nevertheless, Tesla hoped the Great Man would bail him out. "Financially, I am in a dreadful fix," he admitted to Morgan in early July 1903. "But if I can complete this work, I can readily show that by my wireless system power can be transmitted in any amount, to any desired distance and with high economy. Of the three hundred horse-power developed by my oscillator on Long Island, two hundred and seventy-five,—perhaps a little more—can be recovered at the greatest distance in Australia." Knowing that the transmission of power through the Earth from New York to Australia sounded fantastic, Tesla assured Morgan that it was a gamble worth taking since this breakthrough would transform the world and Tesla possessed full control of the invention: "If I had told you such as this before, you would have fired me out of your office. Now you see, Mr. Morgan, what I work for. It means a great industrial revolution. It will be [the] one thing worthy of your attention, as I have always assured you. There is no incertitude about this, it is an absolute. My patents confer a monopoly. Will you help me or let my great work—almost complete—go to pots?"[37]

Although Morgan was immersed in reorganizing the International Merchant Marine trust (which also had run into political and financial difficulties), he agreed to meet with Tesla. However, following this conference, Morgan decided that he would no longer support Tesla. On 17 July 1903 he sent Tesla a terse note: "I have received your letter of 16th inst., and in reply would say that I should not feel disposed at present to make any further advances."[38]

Angry, Tesla expressed his frustration by cranking up the power at Wardenclyffe and hurling lightning bolts. As the *New York Sun* reported, Tesla's neighbors witnessed "all sorts of lightning . . . from the tall tower. . . . For a time the air was filled with blinding streaks of electricity which seemed to shoot off into the darkness on some mysterious errand. The display continued until after midnight." When asked to explain these flashes, Tesla replied, "It is true that some of them have had to do with wireless telegraphy" and that if the local people "had been awake instead of asleep, at other times [they] would have seen even stranger things. Some day, but not at this time, I shall make an announcement of something that I never once dreamed of."[39]

THE WIRELESS SPECULATIVE BUBBLE

So why did Morgan decide to stop supporting Tesla in 1903? If Morgan had provided the Wizard with perhaps another $100,000—the cost of one Old Master painting—Tesla could have tested his ideas, and the patents he had assigned to Morgan might have become highly valuable.[40] Morgan could have profitably sold or licensed these patents to someone else who could exploit this technology commercially.

Morgan certainly did not need an elaborate reason for refusing to continue to support Tesla. He had already sunk $150,000 into the project and Tesla had promised in late 1900 to span the Atlantic in six to eight months and the Pacific a year later. Two and a half years had now elapsed, Marconi had transmitted across the Atlantic, and Tesla had yet to provide any sort of demonstration of his system. Morgan could have easily concluded that Tesla was not a good risk.

The explanation most frequently offered for Morgan's decision to withdraw his support from Tesla was that Morgan had become concerned that Tesla had no plan for making money from wireless power and that he intended to give the power away for free. Perhaps the most colorful version of this story comes from Andrija Puharich, an inventor and physician who conducted research in parapsychology: "Now, I always got this second hand; you won't find it anywhere in print, but Jack O'Neill gave me this information as the official biographer of Nikola Tesla. He said that Bernard Baruch told J. P. Morgan, 'Look, this guy is going crazy. What he is doing is, he wants to give free electrical power to everybody and we can't put meters on that. We are just going to go broke supporting this guy.' And suddenly, overnight, Tesla's support was cut off, the work was never finished."[41] As we have seen, Tesla emphasized in his letters to Morgan how he would use Wardenclyffe for communications and that he intended to make money by manufacturing and selling receivers. Though Tesla was excited about the prospect of transmitting power, his letters reveal that he knew that this was the harder idea to sell to Morgan.

Instead, a more realistic explanation for Morgan's withdrawal of support comes from Tesla himself. When testifying in 1916 about his wireless work, Tesla recalled that he had "excited the interest of a great man" and that he had begun making preparations "for a very big undertaking." However, at the last minute, this patron had backed out, concerned that the wireless industry had entered a "phase of stock jobbing" or speculation. Consequently, the Great Man told Tesla that he "could not touch it with a 20-foot pole." As we have seen, Tesla referred to Morgan as the Great Man in order to hide Morgan's involvement.[42]

Indeed, there was a speculative bubble surrounding wireless in the first decade of the twentieth century. As promising reports of Marconi's demonstrations circulated in 1900 and 1901, shady promoters such as G. P. Gehring and Lancelot E. Pike turned from hawking questionable mining stocks to offering shares in new wireless companies. While Gehring bought up the rights to Amos Dolbear's telephone patents from the 1880s to create the American Wireless Telephone and Telegraph Company, Pike promised quick money by buying stock in one of the many subsidiaries of American Wireless. To give his scheme an air of credibility, Pike rented an elaborately furnished office in the same building in Manhattan that housed the headquarters of the United States Steel Corporation, and there any doubting investor could see the Dolbear instruments at work. Though Pike promised to establish wireless service between New York and Philadelphia, he never bothered to do so and instead absconded with investors' money.[43]

Pike's shenanigans were only the opening round in the mania surrounding wireless, and far more worrisome for Tesla must have been the companies formed around the inventions of Lee de Forest. The son of a Congregationalist minister in Alabama, de Forest grew up determined to overcome his modest background. Following family tradition, he attended Yale. While studying physics there, de Forest read Tesla's biography and dreamed of becoming Tesla's assistant. With help from a classmate, Ernest K. Adams (the son of Edward Dean Adams), de Forest secured an interview with Tesla in 1896, but Tesla turned him down. Undeterred, de Forest received his Ph.D. in 1899 (his dissertation was titled "The Reflection of Short Hertzian Waves from the Ends of Parallel Wires"), and he again applied for a job in Tesla's laboratory.[44]

After being turned down a second time by Tesla, de Forest started experimenting while working as a junior telephone engineer at Western Electric in Chicago. Since the filings in Marconi's coherer had to be continually reset by tapping, de Forest worked on a detector that reset itself automatically. With the help of two colleagues in Chicago, de Forest invented an electrolytic responder in 1901, and he set out for New York to demonstrate his new invention during the America's Cup races and find financial backers.

After being rejected by several investors, de Forest met Abraham S. White, a promoter who had made a fortune in real estate and hawking fireproof chemicals, in January 1902. White was clearly a master hustler, and one reporter described him thus: "White's hair and mustache were flaming red; his eyes of china blue. He wore patent-leather shoes, a silk hat, a flower in his buttonhole, a handsome gold watch chain, a pear-shaped pearl scarfpin and a diamond ring that was not too big. He smoked corkscrew-shaped cigars which he handed out freely, was never without a fat roll of $100 gold certificates, which he peeled off with the easy indifference of an actor handling stage money."[45]

White was quick to appreciate the potential of de Forest's responder and promptly organized the American DeForest Wireless Telegraph Company. Capitalized at $3 million, White served as president and de Forest as vice president and scientific director. To show off de Forest's invention, White built a penthouse laboratory with glass walls on the roof of a building at 17 State Street at the foot of Manhattan, a few blocks from Wall Street. Across New York harbor, a second station was erected at the Castleton Hotel on Staten Island. With potential investors watching, de Forest would send and receive messages from State Street to Staten Island, after which White would take the visitors to lunch. Over these meals, White would wax eloquent, sketching out for the investors a vision of

> wireless stations all along the Eastern seaboard and the Gulf of Mexico, across the continent from the snow-capped peaks of Alaska to Panama. Every ship that touched American shores was to pay the De Forest Company tribute; the stations were to compete with the telephone and telegraph. Wireless would span the Atlantic and Pacific and the cable would be supplanted.
>
> Subsidiary companies were to be established in Canada, England, Europe, Africa, the Orient, Australia, and South America. In reasonable time fifty such subsidiaries could be expected, all paying patent royalties to the American parent. Investors would buy millions of dollars of stock. . . .
>
> Warming up to his subject, White would take out a pencil and . . . calculate what the De Forest Company could earn in a year. Assuming only fifty ships equipped with De Forest instruments in a year, at

$5000 a ship, that was $250,000. Messages to and from these ships, another $250,000; transatlantic and transpacific messages, together, $4,000,000 more. Then there were the "10,000 islands in all the oceans"—what a blessing wireless would be to them! Add another $500,000. Total, $5,000,000—that is, "conservatively speaking."[46]

Enthralled by such prospects, investors snapped up de Forest stock, and White proceeded to form subsidiary after subsidiary and to reorganize the parent company over and over, always inviting the current stockholders to be the first to buy these new stock issues and thereby expand their holdings.[47] White allowed de Forest to set up a laboratory and spend some of the incoming cash on experiments. With this support, de Forest was able to demonstrate his equipment to both the U.S. Army and Navy in 1903 and secure contracts from both branches.[48]

However, White reserved much of the income for continuing his promotional efforts, which included a "wireless automobile" in February 1903. Equipped with de Forest transmitters, four of these vehicles cruised Wall Street, stopping periodically to collect stock prices from curb brokers that were then transmitted to the brokerage houses as well as the offices of the *Wall Street Journal* (see Figure 15.2).[49]

Glancing out his office window, Morgan could not have failed to notice the DeForest Wireless automobiles creating a scene. Equally, his staff would have undoubtedly kept him apprised of various speculative fads that came and went among investors. For Morgan, these fads were a nuisance since they revealed to the public and the government the risky side of Wall Street. Such fads were a problem not only because promoters such as Pike and White were raising more capital than necessary and then not investing it in developing the company's business but also because stockholders were at risk of losing everything when shares—purchased at premium prices—inevitably plummeted on the market. With the Roosevelt administration attacking Morgan's Northern Securities Company and complaints about the shaky structure of International Mercantile Marine, Morgan would have hardly been in the mood in July 1903 to get further involved in Tesla's wireless venture. Morgan simply could not afford the risk of having anything to do with these questionable activities in the emerging wireless industry. Given what White was doing with DeForest

FIGURE 15.2. De Forest Wireless Automobile operating in the New York financial district in 1903.
From "A Perambulating Wireless Telegraph Plant," *Electrical World and Engineer,* 28 Feb. 1903, p. 374.

Wireless, we should not be surprised that Morgan would not want to touch Tesla's venture with "a 20-foot pole."

Morgan withdrew his support of Tesla, then, not because he necessarily doubted Tesla but because he was disturbed by speculation in the wireless industry. In effect, what happened to Tesla at the critical moment in his wireless work was that he was penalized by the questionable behavior of other entrepreneurs in the industry. To be sure, there may have been technical problems with what Tesla was trying to do at Wardenclyffe (discussed later), but he never got the chance to tackle these issues fully since the actions of shysters like Pike and White denied him the capital he needed.

SCRAMBLING FOR CASH

Although Morgan was unwilling to invest any more of his own money in Tesla's venture, he was not opposed to letting others invest in Wardenclyffe, provided they put up new capital and he got what he thought was a reasonable portion of any stock issued by the new company.[50] Hence Tesla spent the next two years cultivating other investors and looking for new ways to raise money to complete Wardenclyffe.

In scrambling for cash Tesla was in a difficult spot, as the tide of public opinion was now running against him. Over the previous fifteen years, Tesla had been regarded as the great electrical wizard in the popular press. While professional engineers and scientists had deplored his frequent pronouncements in the tabloids and periodically criticized his ideas, their views did not seem to affect Tesla's celebrity. Now both the press and the scientific community turned against Tesla. As Laurence A. Hawkins, an engineer later associated with General Electric, wrote in 1903, "Ten years ago, if public opinion in this country had been requested to name the electrician of greatest promise, the answer would without doubt have been 'Nikola Tesla.' To-day his name provokes at best a regret that so great a promise should have been unfulfilled. In ten years the attitude of the scientific press has passed from admiring expectancy to good natured banter and at last [to] charitable silence." Hawkins went on to challenge Tesla's claims of having invented the AC motor, listed every unfulfilled prediction that Tesla had made in the 1890s, and provided a scorching critique of his 1901 *Century* article. For Hawkins, Tesla's downfall was ultimately caused by his weakness for publicity: "Not even the brilliancy of . . . his early work, not even the persistent efforts of powerful friends, moved by their commercial interests to magnify and exalt the value of his patented inventions, could avert the discredit to his reputation as a scientist brought upon himself by his wild struggles for notoriety. He has been condemned by his own extravagant boasts." Faced with this negative publicity, Tesla knew he had to take bold steps to restore his credibility. "My enemies have been successful in representing me as a poet and visionary," Tesla admitted to Morgan, such "that it is absolutely imperative for me to put out something commercial without delay."[51]

As a first measure, Tesla sought to raise cash by developing other inventions. In the summer of 1903, he formed the Tesla Electric and

Manufacturing Company in order to produce small Tesla coils for use in scientific laboratories and to power X-ray tubes. With the company capitalized at $5 million, Tesla found it difficult to attract investors and so this venture was put on hold. Tesla continued to hope that he could make money from this invention, and in 1905 he tried partnering with Pearce, an electrical instrument manufacturer in Brooklyn, to produce coils for under $50. When this scheme foundered, Tesla designed a small "ozonizer," a portable ozone generator that could be used to sanitize rooms since the ozone killed germs. In addition, he also began working on a new form of steam turbine (see Chapter 16).[52]

Since these inventions would take time to develop a cash flow, Tesla sought short-term financing through a loan from a bank in Serbia, assuming that his fame in his home country would carry the day. However, news of Tesla's patent litigation involving the "lion-capitalists" of America made the Serbian bankers nervous and they declined his request. In giving Tesla this bad news, his maternal uncle Petar Mandic wrote, "Dear Nikola! Don't be discouraged at all; you are, thank God, young; you don't have to bow to anyone or will not lose face."[53]

Determined not to lose face, Tesla turned again to John Jacob Astor, who had advanced money for developing his ideas for wireless lighting. However, still miffed that Tesla had spent the advance not on lighting but on research at Colorado Springs, Astor declined and wrote in October 1903, "While wishing you all possible luck, do not care to go into the company myself."[54]

Tesla next cultivated the financier Thomas Fortune Ryan, who was regarded as "the most adroit, suave and noiseless man" on Wall Street. Born in rural Virginia, Ryan had invested in rapid-transit companies in New York City as well as tobacco companies in Virginia. In 1898, Ryan merged his tobacco interests with James B. Duke's American Tobacco to create the Tobacco Trust. Tesla asked Ryan to invest $100,000 in Wardenclyffe, hoping that amount would "be sufficient to reach the first commercial results" and would "pave the way to other greater successes." Though interested, Ryan ultimately chose not to invest. "[A] Great many disappointments today!" Tesla remarked to Scherff in November 1903. "I wonder how long this is going to keep on."[55]

Undeterred by Ryan's refusal, Tesla followed up on his idea to wirelessly transmit power from Niagara and turned to his old business associate, William B. Rankine. Rankine had worked with Adams to establish Niagara power on the American side of the falls, but in 1892 he founded a second company, Canadian Niagara Power, to harness the Canadian Horseshoe Falls. After years of wrangling with the Canadian government, Rankine broke ground in 1901 on a 20,000-horsepower hydroelectric plant that began delivering power in early 1905. However, since there was no local market for this additional power, Rankine and Tesla began discussing the possibility of building a wireless station that could transmit up to 10,000 horsepower and presumably help Canadian Niagara Power reach new customers. Tesla thought this new station would "offer a great convenience to the whole world" since the power could be used to operate clocks and stock tickers, each of which would consume only one-tenth of a horsepower. Tesla estimated that a commercial-scale station would cost $2 million, and he asked Morgan to invest $500,000. However, it is not clear how Tesla or Rankine would have secured the remaining funds for this station.[56]

In the midst of this desperate search for funding, Tesla was cheered by visits with his friend Richmond Hobson. Having resigned from the navy and planning to run for Congress, Hobson spent much of 1903 on a nationwide speaking tour. Realizing that a credible candidate had to be a married man, Hobson had begun courting Grizelda Hull of Tuxedo Park, New York. Hobson's courtship with Miss Hull was stormy; on the one hand, she idolized Hobson as a great war hero, but on the other hand, she regarded him as an insincere social butterfly.[57] After seeing Miss Hull for a Christmas visit in 1903, Hobson was struggling with his feelings and so he went to visit his old friend. As he told Miss Hull,

> The day closed in homesickness till I went to see dear Tesla. He kissed me on the cheek, as once before and when I left him at one o'clock last night, I felt prepared and ready for another year and for future years. When at one time I referred to the great and trying obstacles that he was fighting against he said, "Hobson, I wish they were a thousand times greater, my only fear is that this world can not make obstacles as great as I need and long to meet."
>
> It made me feel ashamed of myself almost. You will never hear me refer to obstacles again, unless in a thought of impotence to get at them and in defiance of their magnitude.

And now au revoir, Grizelda, (won't you let me call you Grizelda?) I am shoving off from shore. The horizon is misty[,] I can not see the land beyond. But my soul is fortified. . . . I have been with you and with Tesla.[58]

Tesla, too, was fortified by his visit with Hobson (Figure 15.3). In early 1904 he produced an elaborate prospectus announcing "that in connection with the commercial introduction of my inventions, I shall render professional services in the general capacity of consulting electrician and engineer." Working as a consulting engineer, Tesla boasted to Morgan that he could easily earn $50,000 a year. The prospectus included a list of Tesla's patents, quotes from his lectures and articles, and a photograph of Wardenclyffe (Figure 15.4). Printed on vellum paper and tucked in an envelope bearing a large red wax seal with the initials "N.T.," this prospectus was regarded by the *Electrical World* as "a manifesto worthy of the original genius issuing it." "It was hard work to get it up," admitted Tesla to Scherff, but he hoped it would attract new investors. Tesla asked Robert and Katharine Johnson to give him "a list of people almost as prominent and influential as the Johnsons who desire to get into high society" so that he could send the prospectus to them. As splendid as this manifesto was, Tesla must have already been feeling discouraged, as he signed his note to the Johnsons, "Nikola Busted."[59]

And yet Tesla pressed on. To back up the prospectus, he gave several newspaper interviews and published a description of his work at Colorado Springs. Much to the Great Man's chagrin, Tesla openly acknowledged in this story that he was being supported by Morgan. He also announced his plans with the Canadian Niagara Power Company to distribute the power. Ever enthusiastic about the potential for wireless power and messages, Tesla concluded by proclaiming: "[W]hen the

FIGURE 15.3. Tesla in 1904. From NTM.

FIGURE 15.4. First page of Tesla's Prospectus, February 1904.
From "A Striking Tesla Manifesto," *Electrical World* 43:256 (6 Feb. 1904) in TC 16:159.

first plant is inaugurated and it is shown that a telegraphic message, almost as secret and non-interferable as a thought, can be transmitted to any terrestrial distance, the sound of the human voice, with all its intonations and inflections, faithfully and instantly reproduced at any other point of the globe, the energy of a waterfall made available for supplying light, heat or motive power, anywhere—on sea, or land, or high in the air—humanity will be like an ant heap stirred up with a stick: See the excitement coming!"[60]

Such proclamations, though, did not stir up investors. In the spring of 1904, Tesla met with Charles A. Coffin, president of General Electric, and his associates, observing that "If they refuse, they are simply snoozers." A few weeks later, Tesla made a pitch to John Sanford Barnes, another Wall Street financier and president of St. Paul and Pacific Railroad. A graduate of the Naval Academy at Annapolis, Barnes had served as a naval officer during the Civil War and gone on to write a book about torpedoes and submarine warfare.[61] Given Barnes's interest in naval technology, Tesla may have met this potential investor while he was working on his radio-controlled boat. To entice Barnes, Tesla had his attorney Parker Page prepare a legal analysis of his wireless patents, which emphasized that they gave broad control of this technology and were easily worth $5 million. In spite of this patent analysis, Barnes passed up the opportunity to invest in Wardenclyffe. "*I swear*," growled Tesla to Scherff, "if I ever get out of this hole nobody will catch me without cash!"[62]

Through 1905, Tesla continued to seek out major investors for Wardenclyffe, including the banker Jacob Schiff, who had participated against Morgan in the raid on the Northern Pacific in 1901.[63] To be sure, all the individuals Tesla approached had the financial means to support him, but why did they turn him down?

Tesla looked like a bad risk because of the way he had structured the original deal with Morgan. (Recall that Tesla, not Morgan, had drafted the original agreement.) In return for the $150,000, Tesla had assigned Morgan a 51% share of his wireless patents. While Tesla could certainly assign his remaining 49% of the patents to new investors and create a new company, the new company could not exercise its patent rights unless Morgan agreed to cooperate. (In order to gain a monopoly, the patents could only be developed by one company, not two competing companies.) Morgan consistently assured Tesla that he would cooperate provided that the new investors put up fresh capital and that he got a reasonable portion of any stock issued by the new company. Again, Morgan was not opposed to seeing Tesla's inventions developed—he simply did not want to invest any more of his own money. However, to new investors this looked like a lousy deal since they would be taking all of the risk (i.e., putting in new capital and developing the company) while Morgan would enjoy substantial gains thanks to his early investment of only $150,000. Why, the investors must have wondered, should they work so hard only to line Morgan's pocket? Over and over, potential investors asked Tesla, "If this is a good thing, why does not Morgan see you through?"[64] Hence, no matter how hard he pitched the potential of wireless power, Tesla could not convince new investors that the long-term returns outweighed the short-term risks of setting up a new company.

TURNING ON MORGAN

As Tesla struggled to persuade new investors to join the Wardenclyffe venture, he bombarded Morgan with letters, alternating between demanding and pleading for more money. At times arrogant and other times groveling, Tesla admitted that these letters to the Great Man were often "written in moments of despondency when the pain [was] too hard to bear."[65]

Tesla's arguments in these letters took various forms. Sometimes he promised Morgan fantastic returns on his investment; as he explained in September 1903, "If you will give me your earnest support in this you can have a greater income than Rockefeller from his oil-wells. And you will have at your mercy the cutthroats who are trying to undo your work and get your royal mantle. I only need to complete this plant, Mr. Morgan, the rest will take care of itself." In other letters, he sought to convince Morgan by claiming that his wireless inventions would revolutionize daily life:

> I have never attempted, Mr. Morgan, to tell you even a hundredth of what can be readily accomplished by the use of certain principles I have discovered. If you will imagine that I have found the stone of the philosophers, you will not be far from the truth. They will cause a revolution so great that almost all value and all human relations will be profoundly [upended]. These new developments do not concern any country in particular, but the whole world and they are in line with your own efforts. The commercial possibilities they offer are simply

infinite, and you are the only man today who possesses the genius and power to compel the universal adoption of these ideas and that is why I approached you two years ago.[66]

Failing that, Tesla sought Morgan's sympathy. Hearing that Morgan was going to meet with the Archbishop of Canterbury, Tesla wrote a long letter in October 1904 that concluded: "Since a year, Mr. Morgan, there has been hardly a night when my pillow is not bathed in tears, but you must not think me a weak man for that. I am perfectly sure to finish my task, come what may. I am only sorry that after mastering the difficulties which seemed insuperable, and acquiring special knowledge which I now alone possess, and which, if applied effectively, would advance the world a century, I must see my work delayed."[67]

Morgan responded to this overture with a curt "No," and Tesla began to show his anger. Knowing that Morgan viewed himself as a devout Anglican, Tesla burst out:

> You are a man like Bismarck. Great but uncontrollable. I wrote purposefully last week hoping that your recent association [with the archbishop] might have rendered you more susceptible to a softer influence. But you are no Christian at all, you are a fanatic musoulman [i.e., a Muslim]. Once you say no, come what may, it is no.
>
> May the gravitation repel instead of attract, may right become wrong, every consideration no matter what it may be, must founder on the rock of your brutal resolve. . . .
>
> You let me struggle on, weakened by shrewd enemies, disheartened by doubting friends, financially exhausted, trying to overcome obstacles which you yourself have piled up before me.[68]

As Tesla's anger grew, it came to be mixed up with superstition, saints, and salvation. Writing on 14 December 1904, Tesla told Morgan,

> Owing to a habit contracted long ago in defiance of superstition, I prefer to make important communications on Fridays and the 13th of each month, but my house is afire and I have not an hour to waste.
>
> I knew that you would refuse. What chance have I to land the biggest Wall Street monster with soul's spider thread!

Your letter reached me just on the day of my patron saint—the greatest of all—St. Nikola. There was a silent agreement between St. Nikola and myself that we would stick to each other. He did well for a time, but during the last three years he has forgotten me,—as you have.

You say that you have fulfilled your contract with me. *You have not.*

I came to enlist your genius and power, not because of money. You should know that I have honored you in so doing as much as I have honored myself. You are a big man, but your work is wrought in passing form, mine is immortal. I came to you with the greatest invention of all times. I have more creations named after me than any man that has gone before not excepting Archimedes and Galileo—the giants of invention. Six thousand million dollars are invested in enterprises based on my discoveries in the United States to-day. I could draw on you at sight for a million dollars if you were the Pierpont Morgan of old.

When we entered our contract I furnished: 1) patent rights; 2) my ability as an engineer and electrician; 3) my good will. You were to furnish 1) money; 2) your business ability; 3) your good will. I assigned patent rights to you which in the worst case are worth ten times your cash investment. You advanced the money, true, but even this first clause of our contract was violated. There was a delay of two months in furnishing the last $50,000—a delay which was fatal.

I complied conscientiously with the second and third obligation[s]. You ignored yours deliberately. Not only this, you discredited me.

There is only one way to do [*sic*], Mr. Morgan. Give me the money to finish a great work, which will advance the world a century and reflect honor on all that come after you. Or else make me a present and let me work out my salvation.[69]

As the months went by and Tesla was unable to either attract new investors or convince Morgan, Tesla succumbed to the ultimate delusion: that he was *the* inventor of *the most important* invention of *all time*. As he proclaimed to Morgan in February 1905:

Let me tell you once more. I have perfected the greatest invention of all time—the transmission of electrical energy without wires to any distance, a work which has consumed 10 years of my life. It is the long sought stone of the philosophers. I need to complete the plant I have constructed and in one bound, humanity will advance centuries.

THE DARK TOWER / 361

> I am the *only man* on this earth *to-day* who has the peculiar knowl-
> edge and ability to achieve this wonder and another one may not come
> in a hundred years. There has been a long and painful delay. My nerves
> are not of iron, and all this knowledge and ability may be lost to the
> world. Help me to complete this work or else remove the obstacles in
> my path.[70]

Despite his anger, Tesla never gave up hoping that Morgan would
support Wardenclyffe. In the summer of 1905, Tesla told Scherff that
he had a hunch Morgan would contact him when he returned from his
annual trip to Europe, but of course the Great Man did not call. Even
as late as 1911 Tesla held out hope; writing to John Hays Hammond Jr.
about a joint venture for a radio-controlled boat, Tesla said, "I have
already interested a gentleman who signs himself J.P.M. in part of my
wireless inventions and my friend Astor is now waiting for the completion of my plant to go into the wireless power transmission business
which should be a colossal success."[71]

IS THE EARTH LIKE A WATER BALLOON
OR THE OCEAN?

As Tesla struggled to find new investors and to persuade Morgan to
support him, new problems kept cropping up. Along with the Port
Jefferson Bank demanding loan payments, James Warden took legal
action for nonpayment of the mortgage on the Wardenclyffe property. Meanwhile, a Wardenclyffe employee named Clark sued for back
wages. In May 1905, after seventeen years (the regular life of a patent),
Tesla's AC motor patents expired; even though Tesla was no longer
receiving royalties from them, he may have felt some regret since these
patents represented a major contribution to society and industry. "The
obstacles in my way," confessed Tesla to Scherff, "are a regular hydra.
Just as soon as I chop off a head, two new ones grow."[72]

As he wrestled with these problems, Tesla also received the upset-
ting news that Hobson was going to marry Grizelda. "Do you know,
my dear Tesla," wrote Hobson, "you are the very first person, outside of my family that I thought of and while the ceremonies will be

simple, I wish to feel you present in standing close to me on this occasion so full of meaning in my life. Indeed, I would not feel the occasion complete without you. You occupy one of the deepest chambers in my heart."[73]

While Tesla was outwardly happy for Hobson and served as an usher at the wedding, inside he must have been deeply disappointed, for this marriage meant that Hobson had chosen Grizelda over him. Although we will never know for sure whether Tesla and Hobson had a physical relationship, it is undeniable that they were very close emotionally. Just as Szigeti had left him fifteen years earlier, now Tesla must have felt abandoned by Hobson. Tesla and Hobson remained friends, and in the 1930s they would periodically get together to see a movie and talk for hours.[74]

Perhaps to ease this emotional blow, Tesla threw himself into his work. During the summer of 1905, he and Scherff worked on making a strong electrical connection between the magnifying transmitter and the Earth at the bottom of the 120-foot well beneath the Wardenclyffe tower. This ground connection was the key to the entire system since it was there that Tesla would deliver electrical energy into the earth's crust in order to set up stationary waves and broadcast power around the world.

As we have seen, Tesla planned to get a grip on the Earth by using sixteen pipes radiating out from the bottom of the well. To push these pipes three hundred feet into the ground, he devised special machinery that used compressed air. However, this machinery presented all sorts of problems, and letters between Tesla and Scherff indicate that they struggled to come up with reliable valves and had to redesign parts of the machinery.

These letters reveal Tesla wavering between optimism and great anxiety. With each small change, he was hopeful that "we shall get excellent results this time and lay the foundation for a great success." Here we see a familiar trait—that Tesla could see hope in the smallest pieces of evidence. But now there was anxiety; as he confided to Scherff, "The troubles and dangers are at their height. . . . The Wardenclyffe specters are hounding me day and night. . . . When will it end?"[75]

But the problem was not simply the pipes; the real specter was how the Earth actually responds when electrical energy is pumped into it.

FIGURE 15.5. Tesla's vision of the earth as filled with an incompressible fluid. From NT, "Famous Scientific Illusions," *Electrical Experimenter,* Feb. 1919, pp. 692–694 ff.

Based on his Colorado experiments, Tesla assumed that the Earth behaved electrically as if it were filled with an incompressible fluid; if one pumped this fluid into the Earth on one side, it would come gushing out of one-way valves on the other side (Figure 15.5). Think of the Earth here as a balloon filled with water; if one pumped water into one side at the resonant frequency of the balloon, then water would come squirting out the valves on the other side. Hence if the Earth was filled with an incompressible electrical fluid, it would indeed be possible to transmit electric power through the Earth with minimal losses. Tesla would be right.[76]

But what if this was not the case? What if the Earth behaved instead as if it were filled with a *compressible* fluid? In this situation, then, Tesla might pump electrical energy into the ground and set up electrical waves, but these waves would dissipate and eventually disappear. As an illustration, consider what happens when a stone is dropped in the water at the edge of the ocean. Waves ripple out from the point at which the rock hits the water, but since the ocean is a compressible fluid, these waves do not travel continuously across the ocean to the other side. Unfortunately for Tesla, from an electrical standpoint, the Earth behaves as if it is filled with compressible fluid; it is more like an ocean than a water balloon. Hence it is doubtful that electric power can be transmitted through the Earth in the way Tesla envisioned it.

To be sure, there are individuals today investigating how Tesla's ideas for wireless power may be made to work. One approach is to postulate that Tesla was generating not only ordinary electromagnetic waves but other forms of electromagnetic radiation as well. In particular, some investigators argue that Maxwell's equations predict that a moving electrical charge produces a second form of radiation popularly known as scalar waves that are longitudinal (just as Tesla emphasized), do not dissipate, and travel faster than the speed of light. However, most physicists regard scalar waves as being experimentally unproven and hence not part of accepted theory.[77] While future researchers may prove that Tesla was on to something with wireless power, I am siding with the physicists and take the view that there was a technical problem at Wardenclyffe, a disjuncture between what Tesla thought should happen and how the Earth actually functions.

DESCENT INTO DARKNESS

Tesla never gave up on his ideas about transmitting power through the earth, but the fact that he could not get his ideas to square with reality during his time at Wardenclyffe was deeply disturbing. Tesla was supremely confident in his own powers of discovery and invention; "my ideas are always rational," he said in the 1930s, "because I am an exceptionally accurate instrument of reception." Indeed, as he told a reporter in 1904, he would go crazy if he doubted his abilities.[78] Moreover, because of his Orthodox religious heritage and his faith in the rationality of science, Tesla firmly believed that there were fundamental truths that could be discovered about the natural world. Just as he had identified the ideal of the rotating magnetic field and manifested it in the AC motor, Tesla believed that what he imagined about transmitting power through the Earth must be true; both were products of his mind. In this sense, Tesla was like the Nobel Prize–winning mathematician John Nash; when asked how a rational mathematician could possibly believe that extraterrestrials were sending him messages, Nash calmly explained that "the ideas I had about supernatural beings came to me the same way that my mathematical ideas did. So I took them seriously."[79]

Hence, when Tesla could not get Wardenclyffe to work the way he wanted, he must have been confronted with a serious dilemma: either he was wrong or nature was wrong. Unable to accept either alternative, Tesla suffered a nervous breakdown. As he reported in his autobiography, "No subject to which I have ever devoted myself has called for such concentration of mind and strained to so dangerous a degree the finest fibers of my brain as the system of which the Magnifying Transmitter is the foundation. . . . Despite my rare physical endurance at that period the abused nerves finally rebelled and I suffered *a complete collapse*, just as the consummation of the long and difficult task was almost in sight [emphasis added]."[80]

While Tesla was angry and depressed through 1904, the full breakdown came in the fall of 1905. In September, Tesla's business associate, Rankine, died suddenly at age forty-seven; Rankine had helped organize the Nikola Tesla Company in 1895 and had been instrumental in Tesla's attempted negotiations with Canadian Niagara Power. In October, feeling overwhelmed by the problems at Wardenclyffe, Tesla confessed to Scherff that "The troubles are so many that I am eager to see what solution the good Lord has provided for. This time he will have to send Santa Claus with a full bundle."[81] In November, Tesla thought that he had convinced Carnegie's former partner, Henry Clay Frick, to invest in wireless power, but Frick, Tesla, and Morgan were never able to agree to terms.[82] In December, Tesla admitted that he had been dangerously ill for the previous month and that it was as severe as the bout of cholera he had suffered as a youth. On Christmas Eve, T. C. Martin dropped him a note: "I have been sorry to hear of your recent illness— well concealed from your friends and the public—and I am also very glad to hear of your recovery. Please stay well and strong."[83]

Tesla's mental collapse continued well into 1906; even in April, Scherff wrote to his boss, "I have received your letter and am very glad to know that you are vanquishing your illness. I have scarcely seen you so out of sorts as last Sunday, and I was frightened."[84]

Throughout his life, Tesla was deeply interested in how the mind worked, and he subsequently described his breakdown to the poet and journalist George Sylvester Viereck, who published a popular book about Freudian psychoanalytical theory in the 1920s.[85] During his breakdown, Tesla explained to Viereck, he wrestled in his dreams with traumatic events from his life. Many of these dreams involved his mother, but they began with a recollection of the death of his brother Dane or Daniel:

> On my slow return to the normal state of mind I experienced an exquisitely painful longing after something undefinable. During the day I worked as usual and this feeling, though it persisted, was much less pronounced, but when I retired the night, with its monstrous amplifications, made the suffering very acute until it dawned upon me that my torture was due to a consuming desire to see my mother.
>
> Thought of her led me to review my past life beginning with the earliest impression of my childhood and I was dismayed to find that I could not recall clearly even her features except in one scene. It was a dismal night with rain falling in torrents. My brother, a youth of eighteen and [an] intellectual giant, had died. My mother came to my room, took me in her arms and whispered almost inaudibly: "Come and kiss Daniel." I pressed my mouth against the ice cold lips of my brother knowing only that something dreadful had happened. My mother put me again to bed and lingering a little said with tears streaming: "God gave me one at midnight and at midnight he took away the other one." The remembrance was like an oasis in the wilderness kept alive by some strange prank of the brain in the midst of oblivion.
>
> My recollections came slowly gaining in clearness and after weeks of thinking the images appeared sharply defined and in a fullness of light which astonished me. Uncovering more and more of my past life I came to review my American experiences. In the meantime my craving had become almost unbearable and every night my pillows were wet from tears.[86]

In the dreams that followed, Tesla relived the stress of lecturing in London and Paris in 1892, his subsequent journey home to see his dying mother, and the vision he experienced at the moment of her passing (see Chapter 8). As he told Viereck, at the moment that he recalled the vision of his mother floating up to heaven on a cloud,

> a feeling of absolute certitude swept over me that my Mother was dead and, sure enough, a maid came running who brought the message. This knowledge gave me a terrific shock and suddenly I became aware that I was—in New York! My Mother had died years before but I had

forgotten it! How could this happen I asked myself horrified and bitterness, pain and shame overwhelmed me. My sufferings had been real though the events were but imaginary reflections of previous occurrences. What I experienced was not the awakening from a dream but the restoration of a particular department of my consciousness.

As always, Tesla refused to attribute these experiences to any psychic or spiritual causes and insisted they were the result of overwork and external stimuli. As he explained to Viereck,

> I am proving constantly, by every thought and action of mine, that I am nothing more than an automaton responding to external stimuli and passing through an infinitude of different existence[s], from the cradle to the grave.
>
> The explanation of these mental phenomena is, after all, very simple. Through long concentration on a special subject certain fibers in my brain, for want of blood supply and exercise, were benumbed and could no longer respond properly to outside influences. With the diversion of my thoughts they were gradually vivified and finally brought back to their normal condition. . . . The desire to see my Mother was due to my examination of some artistic fabrics woven by herself which had awakened in me tender memories shortly before I began to concentrate.

"The practical lesson of all this," Tesla said to Viereck in closing, "is to beware of concentration and be content with mediocre achievement."[87] Sadly, Tesla seems to have taken this lesson to heart, for after his breakdown in 1905 he never again attempted a project as ambitious as wireless power at Wardenclyffe. While he lived another thirty-eight years, his career as a bold innovator had come to an end.

CHAPTER SIXTEEN
UISIONARY TO THE END
[1905–1943]

Life is a moderately good play with
a badly written third act.

TRUMAN CAPOTE

Tesla lived well into the twentieth century and passed away at the age of eighty-seven in 1943. He continued to invent, but as John G. Trump, an MIT professor, observed after reviewing Tesla's papers in 1943, "his thoughts and efforts during at least the past fifteen years were primarily of a speculative, philosophical and somewhat promotional character—often concerned with the production and wireless transmission of power—but did not include new sound workable principles or methods for realising such results."[1] Ideal and illusion continued to shape Tesla's creative approach to the very end.

BLADELESS TURBINES

As he recovered from his nervous breakdown in 1906, Tesla hoped that he could resume work at Wardenclyffe. To raise the necessary funds, he shifted his creative efforts from electricity to mechanical engineering. In doing so, he took up his old dream of flying.

Tesla had dreamed of flying since he was a boy, and one of the applications he planned to pursue once he had perfected his wireless power system was to transmit power to aircraft. As he explained in 1911, "Twenty years ago I believed that I would be the first man to fly; that I was on track of accomplishing what no one else was anywhere near reaching. . . . My idea was a flying machine propelled by an electric motor, with power supplied by stations on the earth."[2]

Because he was concentrating on an electrically powered aircraft, Tesla paid little attention to how inventors such as the Wright brothers were utilizing lightweight gasoline engines, which, in part, allowed them to fly their first airplane in 1903.[3] Realizing that both automobiles and aircraft would require lighter and more powerful engines, Tesla now turned to investigating bladeless turbines.

Tesla conceived of the idea of a bladeless turbine by drawing an analogy with the rotating magnetic field in his AC motor. Just as the rotating field "dragged" the rotor along in his motor, Tesla thought it should be possible to have a fluid such as steam or compressed air drag a series of disks fastened to the turbine's shaft (Figure 16.1). By positioning the disks closely together and placing them at right angles to the flow, Tesla found that he could use the viscosity of the fluid to spin the disk stack. All fluids have the property of viscosity, and fluids like molasses have high viscosity while gases like air have low values. Whatever their viscosity, all fluids "stick" to solid surfaces; that is, the molecules of the fluid directly in contact with the surface move at the velocity of the solid surface. At the same time, molecules that are farther away from the surface are slowed by viscous interaction with the molecules near the surface. This results in a transition layer between the "no-slip" surface and the "freestream" velocity that is known as a boundary layer.

Tesla found that he could take advantage of the "viscous shear" of the boundary layer—that tension between the molecules freely flowing and those already sticking to the surface—to transfer energy from the flowing fluid to the disk stack, eliminating the need for complex blades. Instead, by carefully adjusting the gap between the disks so that it matched the characteristics of the viscosity and velocity of the fluid used, Tesla hoped to create an efficient engine. In Tesla's design, the fluid flow entered the periphery of the turbine and exited at the

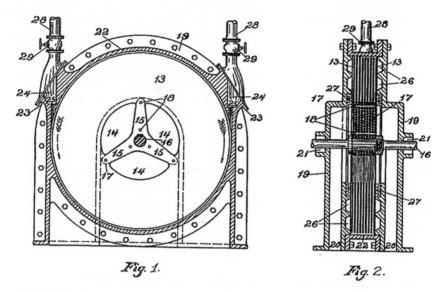

FIGURE 16.1. Tesla Turbine. In this design, steam or compressed air would be introduced to either the valve on the top left or top right and thus cause the disks to turn in one direction or the other.

    From NT, "Turbine," US Patent No. 1,061,206 (filed 21 Oct. 1909, granted 6 May 1913).

center shaft. As the fluid spiraled toward the center, energy was extracted from the flow to drag the disks and cause the shaft to turn. Tesla further found that by reversing the flow so that fluid entered at the center and exited at the periphery, his turbine would also function as a pump or blower.[4]

    Like his other inventions, the bladeless turbine was based on an ideal: that two basic properties of any fluid—viscosity and adhesion—could be used to create the perfect engine. Using only simple disks instead of the complex blades found in the axial turbines invented by Charles Parsons and Gustaf de Laval, Tesla believed that his turbine would be cheaper to build and maintain. But even more important, he was confident that his design would deliver more horsepower per pound of machine, allowing it to be used extensively in automobiles and aircraft. "I have accomplished what mechanical engineers have been dreaming about ever since the invention of steam power," crowed Tesla in 1911. "That is the perfect rotary engine."[5]

    But just as it took Tesla years to go from the ideal of a rotating magnetic field to a functioning AC motor, the perfection of the bladeless turbine involved a great deal of careful engineering. Tesla soon found himself involved in testing different configurations and materials for the disks in his turbine. Julius C. Czito (the son of Tesla's longtime assistant Kolman Czito) built the first prototype in 1906 with eight disks, each six inches (15.2 centimeters) in diameter. The machine weighed less than 10 pounds (4.5 kilograms) and developed 30 horsepower. Tesla soon discovered that the rotor attained such high speeds—up to 35,000 revolutions per minute (rpm)—that the metal disks were stretched out of shape. In 1910 Czito built a larger model for Tesla with twelve-inch (30.5 centimeters) disks and when they limited this machine to 10,000 rpm, it developed 100 horsepower. In 1911 they built a third prototype with disks 9.75 inches (24.8 centimeters) in diameter. Again they reduced the rpm to 9,000 and found the power increased to 110 horsepower. Impressed with the amount of power produced relative to the size of the prototype, Tesla told reporters that his turbine represented "a powerhouse in a hat."[6]

Following his usual strategy of patent-promote-sell, Tesla initially hoped that he would be able to sell his turbine patents to a manufacturer and use the money to finish Wardenclyffe. In March 1909, Tesla pitched his turbine to John Jacob Astor; once again, Astor refused to invest. Consequently, he set up the Tesla Propulsion Company with Joseph Hoadley and Walter H. Knight. Hoadley was associated with Alabama Consolidated Coal and Iron, which planned to install Tesla pumps or air blowers for use in its blast furnaces. Tesla then filed two patent applications—one for a pump, another for a turbine in October 1909. Confident of the potential success of this invention, Tesla rented an office suite in the new Metropolitan Life Tower on Madison Square, then the tallest building in the world.[7]

To show off the potential of this new invention, Tesla arranged to demonstrate his turbine at the Waterside Power Station of the New York Edison Company in 1911–12. For these tests, Tesla built two turbines with 18-inch (45.7 centimeters) disks, and each engine developed 200 horsepower at 9,000 rpm. These two turbines were installed on a single base with their shafts connected by a torque spring (Figure 16.2). When steam was fed to the turbines, each would turn in an opposite direction and the torque spring would measure the power developed as the two turbines pushed against each other. Here the illusion that would illustrate the ideal was that the two engines would engage in a tug-of-war.

FIGURE 16.2. Tesla's turbine test apparatus at the Edison Waterside Station, New York in 1912. Note the light-colored torsion spring connecting the two engines.

From Frank Parker Stockbridge, "The Tesla Turbine," *World's Work* 23 (1911–12): 543–48.

However, the engineers watching the test did not understand what Tesla was doing with the torque spring and they expected to see the turbines turning; when they did not see engine shafts turning, they concluded that the test was a failure. This time, the illusion backfired.[8]

When J. P. Morgan passed away in 1913, Tesla attended the Great Man's funeral, and two months later he approached Morgan's son Jack for support. Still dreaming of Wardenclyffe, Tesla hoped that Jack would invest in wireless power; however, when Jack showed no interest in that project, Tesla pitched his turbine plans. Willing to take a small risk, Jack loaned Tesla $20,000 in four $5,000 installments. Using this money, Tesla tried selling his turbine idea to Sigmund Bergmann, an old colleague of Edison's who had set up a large manufacturing operation in Germany. Determined to keep up appearances,

FIGURE 16.3. Tesla in his office in the Woolworth Building, circa 1916.
From Deutsches Museum.

Tesla moved into new offices in the Woolworth Building, which had replaced the Metropolitan Life Tower as the tallest building in the world (Figure 16.3). Unfortunately World War I broke out in 1914 and Tesla was unable to consummate the deal with Bergmann in Germany. At the same time, Jack Morgan lost interest in the project and became deeply involved in helping the French and British finance the war.[9]

Over the next ten years, Tesla worked on his turbine with engineers at Pyle National in Chicago, Allis-Chalmers in Milwaukee, and the Budd Company in Philadelphia. They were unable to overcome the problem that Tesla had noticed early on: speeds over 10,000 rpm placed incredible stresses on the thin disks in the turbine, causing them to deform. Although Tesla searched for better steel alloys to use in his design, he was never able to find sufficiently strong materials.

Moreover, Tesla's turbine design seems to have fallen between two separate areas of industry. On the one hand, to develop it properly,

Tesla needed the assistance of engineers at firms like Allis-Chalmers or General Electric where they specialized in making rotating machinery, but their expertise was primarily in the area of axial turbines and not necessarily relevant to building a better boundary-layer turbine such as Tesla's. On the other hand, the major markets for a lightweight Tesla turbine would have been the automobile and aviation industries, but these companies were focusing on the development of high-performance piston engines and had little interest in tangling with a turbine engine.

However, when used as a pump, Tesla's design works remarkably well, and today the DiscFlo Corporation of Santee, California, manufactures pumps based on Tesla's ideas. In addition, Phoenix Navigation and Guidance Inc. (PNGinc), in Munising, Michigan, is experimenting with disk turbines using advanced materials such as carbon-fiber, titanium-impregnated plastic and Kevlar reinforcement in their disks. And there is a devoted group of amateurs who continue to work with Tesla's ideas and share their results through the Tesla Engine Builders Association.[10]

BANKRUPTCY AND OTHER DISAPPOINTMENTS

Unable to find investors for his turbine after Jack Morgan, Tesla's finances again went into a tailspin. He was forced to give up his office in the Woolworth Building for more modest space at 5 West 40th Street. In 1916, New York City took him to court to collect $935 in back taxes, and Tesla had to admit that his income was only $350 to $400 a month, which was barely enough to cover his expenses. "How do you live?" the judge asked Tesla.

> "Mostly on credit," he replied. "I have a bill at the Waldorf that I have not paid for several years."
> "Are there other judgments against you?"
> "Scores of them."
> "Does anybody owe you any money?"
> "No sir."

"Have you any jewelry?"

"No Sir; jewelry I abhor."

Tesla explained that he was still president and treasurer of the Nikola Tesla Company but that 90% of the company's stock had been pledged between 1898 and 1902 to bankers, creditors, and friends. Although the company once had a portfolio of two hundred patents, most had expired. To keep the company going, Tesla had named two former employees, Fritz Lowenstein and Diaz Brutrago, as directors. After learning that he owned neither real estate nor an automobile, the court appointed a receiver to manage his affairs.[11]

In the midst of these financial troubles, Tesla had two incidents with major scientific awards. In November 1915, the *New York Times* published an early report that Tesla and Edison were going to share that year's Nobel Prize in physics; since Marconi had already won the prize in 1909, it seemed entirely plausible that the two wizards would now be sharing the prize. Although he had not yet been officially notified, Tesla told the paper that "I have concluded that the honor has been conferred on me in acknowledgement of a discovery announced a short time ago which concerns the transmission of electrical energy without wires." Unfortunately the *New York Times* was wrong and the 1915 Nobel Prize in Physics was awarded to William H. Bragg and his son W. L. Bragg. Disappointed, Tesla rationalized his loss in a letter to his friend Robert Underwood Johnson: "In a thousand years, there will be many recipients of the Nobel Prize, but I have not less than four dozens of my creations identified with my name in the technical literature. These are honors real and permanent, which are bestowed, not by a few who are apt to err, but by the whole world which seldom makes a mistake."[12]

A year later Tesla received happier news when he learned that the American Institute of Electrical Engineering (AIEE) wished to present him with its highest honor, the Edison Medal. As noted earlier (Chapter 12), Tesla's relationship with the institute may have been strained because members feared negative publicity if stories of Tesla's sexual proclivities became public, with the likely result that Tesla seldom participated in institute affairs after serving as vice president in 1892–93. Though the award was announced in December 1916, the medal was not formally presented to Tesla until May 1917, and the delay may well have been caused by Tesla's reluctance to accept the award for an organization from which he felt socially ostracized. As he bitterly wrote to B. A. Behrend, the senior Westinghouse engineer who had nominated him: "You propose to honor me with a medal which I could pin upon my coat and strut for a vain hour before the members and guests of your Institute. You would bestow an outward semblance of honoring me but you would decorate my body and continue to let starve, for failure to supply recognition, my mind and its creative products which have supplied the foundation upon which the major portion of your Institute exists. And when you would through the vacuous pantomine [*sic*] of honoring Tesla you would not be honoring Tesla but Edison who has previously shared unearned glory from every previous recipient of this medal."

Behrend was able to persuade Tesla to accept the medal, but Tesla continued to have mixed feelings. In May 1917 he attended the banquet at the Engineers' Club and was charming with his colleagues, but moments before the awards ceremony he disappeared. Frantic, Behrend searched everywhere for Tesla, only to find him in Bryant Park across the street from the club. There, Tesla was busy feeding the pigeons. When he was done, Tesla went with Behrend to the ceremony where he gave an inspiring speech recounting his early life and describing his creative approach.[13]

WIRELESS LITIGATION AND MINOR INVENTIONS

In the mid-1910s Tesla was as poor as he had been in the mid-1880s when he was forced to dig ditches (see Chapter 4). Yet, just as he did thirty years earlier, Tesla turned to minor inventions to get himself out of the financial hole.

First, he sought ways to earn something on his wireless patents. In 1903, anxious that the Germans had a strong company to challenge British Marconi, Kaiser Wilhelm II had encouraged all of the German wireless companies to come together as a single company, Gesellschaft für drahtlose Telegraphie System Telefunken. Further, to compete with Marconi in the American market, Telefunken had organized a subsidiary, the Atlantic Communication Company, around 1911, and this company retained Tesla as a consultant; doing so made sense since years before one of Telefunken's leading German researchers, Adolph Slaby, had publicly called Tesla the "father of the wireless" in order to needle Marconi.[14]

When World War I broke out, the British navy immediately cut all of the undersea telegraph cables from Germany, and the only link between Germany and the United States were the stations Atlantic Communication had built at Sayville on Long Island and the station at Tuckerton, New Jersey, set up by another German company, HOMAG. Determined to force these stations to be closed so that it would have full control over the information flowing from Europe to America about the war, the British government asked American Marconi to sue Atlantic Communication for patent infringement in 1914. Both sides—the British and German governments, Marconi and Telefunken—recognized the stakes of this legal battle and brought in their superstars; while Marconi traveled to New York, Telefunken sent two physicists, Jonathan Zenneck and Ferdinand Braun (who had shared the 1909 Nobel Prize with Marconi). In addition, Atlantic Communication retained the leading American patent counsel, Frederick P. Fish, to lead the defense team and it asked Tesla to serve as an expert witness. Consequently, from 1915 to 1917 Telefunken paid Tesla about $1,000 a month. Relying on Tesla, Braun, and Zenneck, Atlantic Communication made a strong case against Marconi, and in May 1915 American Marconi asked for a postponement. The rationale given by the Marconi lawyers was that Italy had entered the war and that the Italians needed Marconi home in order to help with the war effort; one wonders if American Marconi also saw the suit as a losing proposition.[15]

Emboldened by these legal developments, Tesla launched his own lawsuit against Marconi for patent infringement in August 1915. In this case, Tesla challenged the U.S. patent issued to Marconi in 1904; Tesla gave a substantial deposition in 1916 recounting in detail his wireless work. According to Tesla expert Gary Peterson, "Nothing significant resulted from this [case] until 1916 when the Marconi Wireless Telegraph Company of America itself sued the United States for alleged damages resulting from the use of wireless during WWI." This case,

American Marconi v. the United States, took decades to wind through the legal system but resulted in a 1935 ruling by the United States Court of Claims that invalidated the fundamental Marconi patent because it was anticipated by Tesla and other early inventors. This ruling was affirmed by the U.S. Supreme Court in 1943, and based on this ruling, many people feel that Tesla finally gained at least a legal victory over Marconi.[16]

But since all this litigation did not produce any immediate income in the 1910s, Tesla used his turbine research to spin off several inventions. While performing tests on pumps, he had learned that as he decreased the space between the disks and the wall of the pump, the relationship between the fluid's momentum and its speed changed from being a square to a linear function.[17] Drawing on this insight, Tesla patented improvements in automobile speedometers, frequency meters, and flow meters.[18] Around 1918, Tesla licensed these patents to the Waltham Watch Company.[19] Drawing on its manufacturing capability in making precision clocks and instruments, Waltham introduced a line of "scientifically built speedometers" that were installed in luxury automobiles such as the Pierce-Arrow, Lincoln, and Rolls-Royce. In promoting its speedometers, Waltham Watch sometimes included Tesla's name in advertisements.[20]

In addition to developing a speedometer and other measuring instruments, Tesla used what he learned about capillary forces and surface tension in his turbine work to develop a new process for refining metals. Around 1930 he produced a report titled "Process of De-Gassifying, Refining, and Purifying Metals," which he circulated to several companies including United States Steel. Tesla appears to have licensed this process to the copper-mining giant American Smelting and Refining Company (today ASARCO).[21]

BIRTHDAY PARTIES FOR A RECLUSE

Through the 1920s, Tesla lived off these modest royalties. Nevertheless, he continued to encounter financial difficulties; for instance, after hiring an attorney, Ralph J. Hawkins, to help with some legal work, Tesla neglected to pay him fees amounting to $913, and Hawkins was obliged to take Tesla to court in June 1925. Warned by Hugo Gernsbac, the editor of *Electrical Experimenter*, that it might be highly embarrassing if became known that Tesla was nearly penniless, the Westinghouse Company reluctantly agreed in 1934 to put Tesla on the payroll as a "consulting engineer" and pay him a monthly pension of $125.[22]

To supplement this income, Tesla wrote articles for popular magazines, and his autobiography appeared in several installments in Gernsback's *Electrical Experimenter* in 1919. Ever the visionary, Tesla loved to speculate on new applications of electricity and radio. For instance, in 1917 he described a scheme for detecting ships by training a powerful ray of short wave impulses at objects and then picking up the reflection of the ray on a fluorescent screen; thus he anticipated radar that was subsequently developed in the 1930s using microwave electronics. One early radar pioneer, Émile Girardeau, was inspired by Tesla and recalled that his first system in France was "conceived according to the principles stated by Tesla."[23]

Shortly after World War I, Tesla was approached by a representative of the newly formed Soviet Union. V. I. Lenin believed that electrification was essential for the success of communism, and one of his slogans was "Communism is the Soviet power plus electrification of the whole country." Nationwide electrification figured prominently in GOELRO, a loose ten-year plan of industrial transformation that Lenin introduced in 1920.[24] Hence Lenin was apparently interested in employing Tesla's wireless power system to distribute electricity across the vast distances of the Soviet Union. As Tesla reported in 1919, "only recently an odd looking gentleman called on me with the object of enlisting my services in the construction of world transmitters in some distant land. 'We have no money,' he said, 'but carloads of solid gold, and we will give you a liberal amount.' I told him that I wanted to see first what will be done with my inventions in America, and this ended the interview."[25]

Tesla never abandoned his dream of broadcasting power from Wardenclyffe, and for many years he struggled to hold off his creditors and retain control of the property. In 1904 Tesla had mortgaged Wardenclyffe to George C. Boldt, the owner of the Waldorf-Astoria, so that he could continue living at the hotel. In 1917 creditors tore down the tower for scrap metal, and in 1921 the courts awarded the property to the Waldorf-Astoria as restitution for Tesla's long-overdue hotel bill.[26]

These disappointments depressed Tesla, and he became a recluse, spending much of his time walking the streets of Manhattan and feeding pigeons. He particularly liked the pigeons in Bryant Park behind the New York Public Library, and today one end of the park (at Sixth Avenue and West 40th Street) is officially designated "Nikola Tesla Corner." Tesla continued to live in hotel rooms, moving from one hotel to another when he could no longer pay the bill and after complaints that he was keeping too many pigeons in his room.[27]

To mark Tesla's seventy-fifth birthday in 1931, Kenneth Swezey, a young science writer, organized a special party for the Wizard. Swezey asked seventy prominent scientists and engineers from around the world to send letters of congratulations, which he presented to Tesla in a special volume. Included in the volume were messages from Albert Einstein, Sir Oliver Lodge, Robert A. Millikan, Lee de Forest, and Vannevar Bush. The letters were reprinted in Yugoslavia and prompted the establishment of the Nikola Tesla Institution in Belgrade. *Time* magazine ran a cover story in which the aging inventor held forth on his plans to disprove Einstein's theory of relativity, on his belief that splitting atoms released no energy, and the importance of interplanetary communications.[28]

Enjoying the warmth of this publicity, Tesla subsequently held press conferences each year on his birthday (Figure 16.4). At the 1932 conference, he announced that he had a new motor that would work on cosmic rays. On his seventy-ninth birthday, he recounted how he had developed a pocket-sized mechanical oscillator that could destroy the Empire State Building (see Chapter 10). For his eightieth, Tesla informed reporters that he wiggled his toes several hundred times before bed as a way of toning up his body so that he would live for 135 years.[29] And at the 1937 birthday celebration, Tesla was presented with gold medals by the governments of Yugoslavia and Czechoslovakia.[30] A reporter from the *New York World-Telegram*, though, perhaps best summarized these annual events when he described the 1935 fete: "Twenty-odd newspapermen came away from his Hotel New Yorker birthday party yesterday, which lasted six hours, feeling hesitantly that something was wrong either with the old man's mind or else their own, for Dr. Tesla was serene in an old-fashioned Prince Albert and courtly in a way that seems to have gone out of this world."[31]

FIGURE 16.4. Tesla at his birthday interviews, 1935.
From http://www.pbs.org/tesla/ll/ww_teslab_pop.html.

A PARTICLE BEAM WEAPON AND GLOBAL INTRIGUE

While Tesla's birthday predictions seemed simultaneously mundane
and far-fetched to reporters, the old illusionist succeeded in getting
their attention at his 1934 birthday press conference. During that interview, Tesla announced that he was perfecting a particle beam weapon.
As the *New York Times* explained, Tesla now claimed that he could
"send concentrated beams of particles through the free air, of such tremendous energy that they will bring down a fleet of 10,000 enemy airplanes at a distance of 250 miles from a defending nation's border and
will cause armies of millions to drop dead in their tracks." Invoking

the same theme that he had used in 1898 with his radio-controlled boat (see Chapter 12), Tesla promised that this new invention would abolish war because this death beam "would surround each country like an invisible Chinese wall, only a million times more impenetrable. It would make every nation impregnable against attack by airplanes or by large invading armies."[32]

Did Tesla actually design such a weapon? For many years, engineers and Tesla fans did not know for sure, but in 1984 a paper surfaced that circulated among Tesla specialists and was later confirmed as genuine by the Tesla Museum in Belgrade. Titled "The New Art of Projecting Concentrated Non-dispersive Energy through Natural Media," this paper outlined a system for accelerating minute tungsten or mercury particles to very high speeds. Still not convinced of the efficacy of Hertzian waves, Tesla insisted on using particles, not rays: "I want to state explicitly that this invention of mine does not contemplate the use of any so-called 'death rays.' Rays are not applicable because they cannot be produced in requisite quantities and diminish rapidly in intensity with distance."[33] In addition, Tesla undoubtedly wanted to distance himself from free-wheeling characters like Harry Grindell Matthews, who told English newspapers in 1924 that he had developed a death ray that could bring down airplanes but which he refused to demonstrate to skeptical British government officials. In particular, the Air Ministry wanted Grindell Matthews to prove that his ray could deliver enough power to kill the pilot in an enemy plane by making his blood boil.[34]

Rather than using rays, then, Tesla's plan was to accelerate tiny mercury particles to a velocity forty-eight times the speed of sound. To energize these particles, Tesla proposed an electrostatic generator similar to Robert Van de Graaff's design, but in place of a charge-carrying belt he would use a circulating stream of desiccated air propelled by a Tesla pump or blower through hermetically sealed ductwork (Figure 16.5). This stream of air would pass two discharge points where it would be ionized by high-voltage direct current. The ions would then be carried up by the airstream where the charge would accumulate in a large spherical terminal similar to one used on the top of the Wardenclyffe tower. To increase the electrical capacity of the spherical terminal, it was studded with evacuated glass bulbs, each of which contained an umbrella-shaped electrode. "I am confident," wrote

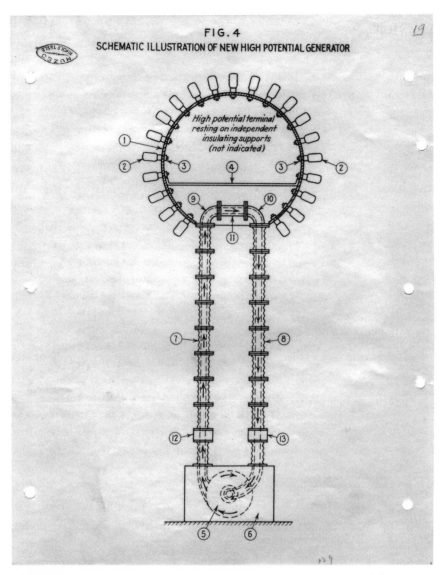

FIGURE 16.5. Tesla's plan for high potential generator to be used as a particle beam weapon.

Key:

5	Tesla turbine or blower
7 and 8	ducts for air
12 and 13	points where air flow would be ionized by high-voltage direct current
1	sphere charged to high voltage
2	evacuated glass bulbs to increase sphere's electrical capacity

Source: NT, "The New Art of Projecting Concentrated Non-dispersive Energy Through Natural Media," Fig. 4. From NTM.

Tesla, "that as much as one hundred million volts will be reached with such a transmitter providing a tool of inestimable value for practical purposes as well as scientific research."[35]

Inside the sphere, Tesla would maintain a high vacuum into which he would introduce millions of tiny mercury particles. Although Tesla did not specify exactly how it would happen, these particles would be charged to the same high voltage of the whole sphere and then accelerated out of the sphere through a specially designed projector (Figure 16.6). The projector would shoot a single row of highly charged particles that would deliver prodigious amounts of energy over a great distance. Again, Tesla provided no explanation as to how he would keep the particles in the beam from scattering as they left the projector and headed toward a target.

In certain ways, Tesla's design for this beam weapon reflected the state of the art in high-voltage engineering. When evaluated in 1943 by John Trump, who had earned his Ph.D. at MIT with van de Graaff, Trump noted, "The proposed scheme bears some relation to present means for producing high-energy cathode rays by the cooperative use of a high-voltage electrostatic generator and an evacuated electron acceleration tube."

Though the basic principle of using electrostatic forces to accelerate particles is sound—it is used in multiple devices ranging from the picture tubes in older televisions to the gigantic particle accelerators employed in nuclear physics—Tesla's description is not sufficiently detailed to conclude that he could have actually built such a weapon. To quote Trump again, "Tesla's disclosures . . . would not enable the construction of workable combinations of generator and tube even of limited power, though the general elements of such a combination are succinctly described."

In particular, modern particle accelerators use massive amounts of energy to impart high velocities to subatomic particles such as electrons and protons, and it would take even more energy to accelerate the sort of macroscopic particles that Tesla had in mind in order to have them travel any distance. As Trump concluded, "It is well known . . . that such devices, while of scientific and medical interest, are incapable of the transmission of large amounts of power in non-dispersed beams over long distances."[36]

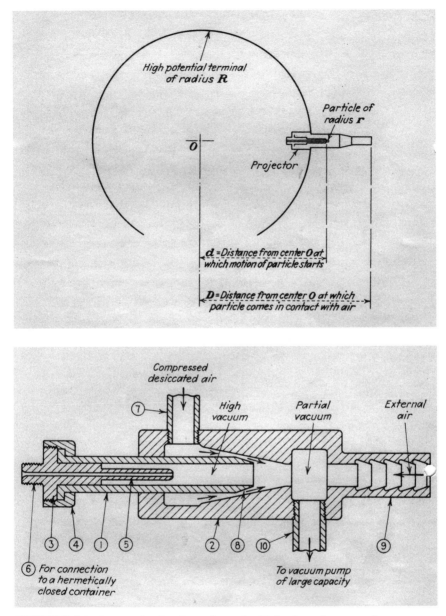

FIGURE 16.6. Diagrams showing the projector that Tesla planned to use to shoot a stream of highly-charged mercury particles from his beam weapon.

Source: NT, "The New Art of Projecting Concentrated Non-dispersive Energy Through Natural Media," Figs. 1 and 5. From NTM.

Indeed, Paul Nahin argues that in order for Tesla's particle beam weapon to satisfy the British Air Ministry's requirement of making even one pilot's blood boil, the beam would have to deliver over 36,960 watts of power. To destroy an army of one million soldiers—as Tesla claimed his weapon could do—one would need 7.4 nuclear reactors each delivering 5,000 megawatts. Clearly the power required for this weapon raises questions about its feasibility.[37]

In spite of its impracticality, Tesla's beam weapon generated a great deal of publicity when it was announced in 1934. Sensing an opportunity, Tesla contacted several governments and let a shady character, Titus deBobula, draw up architectural plans for what the new tower might look like. A Hungarian architect, deBobula met Tesla in New York in the 1890s when Tesla "took the youth under his protection" and helped him book passage back to Hungary. DeBobula subsequently returned to America, married an heiress, and set up an architectural practice. However, Tesla and deBobula soon parted in 1935 when deBobula tried to borrow money from Tesla and to get him involved in a deal selling arms to Paraguay. Nevertheless, some Tesla fans believe that deBobula helped Tesla set up a secret laboratory underneath the 59th Street bridge in Manhattan.[38]

Tesla pitched the particle beam weapon to Jack Morgan, who had previously loaned him money to develop the turbine. As Tesla explained to Jack in November 1934:

> The flying machine has completely demoralized the world, so much that in some cities, as London and Paris, people are in moral fear from aerial bombing. The new means I have perfected afford absolute protection against this and other forms of attack.
>
> You know how your father assisted me in the development of my wireless system. He did not get any returns but I am convinced that if he were living he would be gratified by the knowledge that my inventions are universally applied. I still gratefully remember your own support although the war deprived me of the success I had achieved. . . .
>
> These new discoveries, which I have carried out experimentally on a limited scale, have created a profound impression. One of the most pressing problems seems to be the protection of London and I am writing to some influential friends in England hoping that my plan will be adopted without delay. The Russians are very anxious to render their

borders safe against Japanese invasion and I have made them a proposal which is being seriously considered. I have many admirers there especially on account of the introduction of my alternating system to an extent unprecedented. Some years ago Lenin made me twice in succession very tempting offers to come to Russia but I could not tear myself from my laboratory work.

Words cannot express how much I am aching for the same facilities which I then had at my disposal and for the opportunity of squaring my account with your father's estate and yourself. I am no longer a dreamer but a practical man of great experience gained in long and bitter trials. If I had now twenty five thousand dollars to secure my property and make convincing demonstrations I could acquire in a short time colossal wealth. Would you be willing to advance me this sum if I pledged to you these inventions?[39]

Jack Morgan declined to underwrite Tesla's work on the particle beam weapon, but Tesla did not hesitate to use his particle beam in his ongoing financial skirmishes with the management of New York hotels. When the managers at the Governor Clinton Hotel demanded that he pay his $400 bill, Tesla offered a working model of his weapon to them as collateral. Since the model was supposedly worth $10,000, the managers accepted Tesla's offer and the old gentleman signed a promissory note. However, when Tesla delivered the model to the hotel clerk, he gravely warned that the box would detonate if opened by an unauthorized person; sufficiently frightened, the staff hid the model in the hotel's backroom vault.[40]

In the meantime, several different countries expressed concern over and interest in his particle beam weapon. Hearing that Tesla might offer the weapon to the League of Nations in Geneva, Breckinridge Long, the U.S. ambassador to Italy, warned that "if Tesla should give the secret to Geneva, it would be in the hands of half a dozen governments in Europe and they would be using the beam instead of guns to fight one another. If the United States Government should obtain control of it, no other government would obtain it and the American government could act as a guardian."[41] A career diplomat, Long had worked with Woodrow Wilson to help establish the League, but the State Department ignored his advice and assumed responsibility for Tesla's weapon.

Tesla next entered into negotiations with the Soviets. In April 1935 he signed an agreement with the Amtorg Trading Corporation. Founded by Armand Hammer in 1924, Amtorg ostensibly coordinated trade between the United States and the USSR while collecting intelligence on American science and technology for the Soviet military. In return for $25,000, Tesla agreed "to supply plans, specification, and complete information on a method and apparatus for producing high voltages up to 50 million volts, for producing very small particles in a tube open to air, for increasing the charge of the particles to the full voltage of the high potential terminal, and for projecting the particles to distances of a hundred miles or more. The maximum speed of the particles was specified as not less than 350 miles per second." Soviet scientists and engineers studied Tesla's plans and corresponded with him further, but we do not know whether the Soviets tested such a device in the 1930s.[42]

While Tesla entered into an agreement with the Soviets, he also offered his weapon to the British government under Neville Chamberlain for $30 million. In the mid-1930s, Chamberlain sought to create a stable Europe by adopting a conciliatory stance toward Adolf Hitler and Germany, and Chamberlain may have hoped that by acquiring (or even appearing to acquire) Tesla's weapon he could deter any further aggressive moves by Hitler. Hence Tesla corresponded with the British government in 1936 and 1937, but in January 1938 the British politely declined to pursue the matter further.[43]

During the period in which he carried on these global negotiations, Tesla claimed that spies tried to steal his plans for the particle beam weapon. His room had been broken into and his papers examined, but the thieves left empty-handed. The reason for this, explained the old man to O'Neill, was that he had not committed any of the key details to paper but carried them around in his head.[44]

Following the outbreak of World War II in 1939, Tesla made another attempt to interest the U.S. government in funding his particle beam weapon. On his eighty-fourth birthday in 1940, he announced that he was ready to work with Washington to introduce a new beam weapon, his "teleforce," which was not based on his wireless power inventions but on a new principle. As the *Baltimore Sun* reported,

> The beam, he said, would be only one hundred-millionth of a square centimeter in diameter and could be generated from a special plant

that would cost no more than $2,000,000 and would take only about three months to construct. A dozen such plants, located at strategic positions along the coast, he said, would be enough to defend the country against all possible aerial attack.

The beam would melt any [airplane's] engine, whether Diesel or gasoline driven, and would also ignite any explosives aboard. No possible defense against it could be devised, as it would be all-penetrating, he declared.

Tesla informed the reporters that he was ready to go to work for the government at once and that he was in good health. However, should the government take up his offer, he warned, "I would have to insist on one condition—I would not suffer interference from any experts. They would have to trust me." Not surprisingly, Tesla did not hear back from federal officials. Although J. Edgar Hoover at the FBI received a tip about Tesla's latest plan, Hoover appears to have been indifferent to the old Wizard's predictions.[45]

A QUIET DEATH AND A FINAL ILLUSION

Despite Tesla's claims that he was feeling well, by 1940 his health was in decline. Three years earlier, while taking his daily walk around Manhattan, he had been hit by a taxi. Tesla refused medical treatment for his injuries and he never fully recovered. Tesla had always been particular about what he ate, but now he insisted on a strict diet, first of boiled vegetables, then only warm milk. Staff at the Hotel New Yorker described him as having "a vigorous temperament and emphatic ideas about personal health," which included keeping everyone at least three feet away to avoid their germs. Worried about his health, the Nikola Tesla Institute in Yugoslavia arranged a monthly stipend of $600 beginning in 1939.[46]

By 1942, Tesla spent much of his time in bed, mentally active but physically weak. He was suffering from some senility; in July he sent a telegraph messenger boy to deliver $100 to Mark Twain (who had died in 1913) at 35 South Fifth Avenue, the address of his old laboratory.[47] Tesla refused to see most visitors, but at the behest of his nephew Sava Kosanović he met with King Peter II, the exiled ruler of Yugoslavia. Tesla also spent some time discussing his particle beam and other inventions with a young man from Kansas named Bloyce Fitzgerald. Fitzgerald had studied electrical engineering and had corresponded with Tesla since 1935 about his own anti-tank gun; some Tesla fans think that Fitzgerald was personally appointed by President Roosevelt to look after the ailing inventor.[48]

Tesla died quietly in his sleep on the night of 7 January 1943, and the cause listed on his death certificate was coronary thrombosis or heart attack. His funeral took place on 12 January at the Cathedral of St. John the Divine in New York, and like the funerals of his parents, the service was presided over by several priests prominent in the Serbian Orthodox Church. It was attended by a number of scientific dignitaries and two thousand mourners. Among those at the funeral was fellow radio pioneer Edwin Howard Armstrong, who commented, "The world, I think, will wait a long time for Nikola Tesla's equal in achievement and imagination."[49]

In the days that followed, condolences were received from the president and Eleanor Roosevelt, Vice President Henry A. Wallace, and several Nobel Prize winners. New York City mayor Fiorello LaGuardia offered a eulogy on the radio. Although it is not traditionally done in the Serbian Orthodox Church, Kosanović decided that Tesla's remains should be cremated.[50]

In the weeks following Tesla's death, the U.S. government became momentarily worried about Tesla's particle beam and hence the contents of his papers, setting off years of speculation that there was indeed a beam weapon and a government cover-up.[51] In the 1980s, under the Freedom of Information Act, the FBI released 250 pages of documents related to Tesla, and Tesla aficionados have combed these papers looking for a conspiracy. My reading of these documents suggests a curious story of greed and bureaucratic politics but no conspiracy.

On the morning after Tesla died, Kosanović went to Tesla's room at the Hotel New Yorker, ostensibly to see if his uncle had left a will. Then director of the Eastern and Central European Planning Board for the Balkans, Kosanović assumed that, as Tesla's nephew, he was the rightful heir to his uncle's papers and belongings. Kosanović was accompanied by his assistant, Charlotte Muzar, Bogoljub Jevtic, Boris Furlan, Kenneth Swezey, and radio historian George Clark. Since the safe was locked, Kosanović called in a locksmith to open it and change the combination. In the presence of three assistant managers from the hotel, Kosanović and his group inspected the contents of the safe and found Tesla's Edison Medal and a key ring. Kosanović removed a few photographs and the 1931 volume of birthday letters and then shut the safe again.[52]

While this was happening, Abraham N. Spanel informed Percy E. Foxworth in the FBI's New York office that the bureau needed to check on Kosanović and Tesla's papers. An entrepreneur, Spanel was born in Russia and immigrated to the United States when he was ten. In the 1930s, he built up the International Latex Corporation in Dover, Delaware, to manufacture Playtex girdles, brassieres, infant wear, and rubber gloves. A liberal Democrat, Spanel had close ties to Vice President Henry A. Wallace, and Spanel was later accused by conservative syndicated columnist Westbrook Pegler of being a communist. Spanel warned Foxworth that Tesla supposedly had an intense dislike for his nephew, that Kosanović would make Tesla's plans for a beam weapon available to the enemy, and that one of Vice President Wallace's aides had told him that the government was vitally interested in Tesla's papers.[53]

Spanel knew about Tesla's weapon plans because he was acquainted with Bloyce Fitzgerald. Early in the war, Fitzgerald had lined up a deal to sell his anti-tank gun to Remington Arms, but without telling Fitzgerald, Spanel convinced Remington to renege on the deal. Instead, Spanel arranged for Fitzgerald to develop the gun with the Higgins Boat Company in New Orleans in return for Spanel getting 80% of the profits. So why did Spanel go to the FBI about the Tesla plans? My guess (and it's only a guess) is that Spanel did not want Kosanović to get the plans in the hope that Fitzgerald would gain access to them so that he and Fitzgerald could then develop a weapon that they could sell to a military contractor.[54]

Spanel's concerns paralleled what the FBI's New York office had already been told by Fitzgerald on 7 January 1943, and so Foxworth sent a teletype to the Office of the Director of the FBI in Washington asking what steps should be taken. D. Milton Ladd, one of J. Edgar Hoover's two assistant directors, took the view that Kosanović should not be trusted and suggested that the New York office contact the New York State Attorney to see if Kosanović could be taken into custody on a burglary charge. However, in the meantime, another FBI agent, L.M.C. Smith, contacted the Office of the Alien Property Custodian (OAPC); this agency was interested in the case because even though Tesla was an American citizen, Kosanović was not, and hence the papers were now alien property and could be confiscated by the government if necessary. As far as the FBI was concerned, letting the OAPC take over was a perfect solution—the papers would not fall in Kosanović's hands (whom the FBI did not like) and the FBI did not have to get involved. As the other assistant FBI director, Edward A. Tamm, concluded, "L.M.C. Smith is handling this with [the] Alien Property Custodian so there appears to be no need for us to mess around in it."[55]

Consequently, Walter C. Gorsuch of the OAPC went to the Hotel New Yorker and seized all of Tesla's property in his bedroom and the adjoining storeroom. Consisting of two truckloads of material, this property was taken to the Manhattan Storage Company where there were already eighty barrels and bundles that Tesla deposited nine to ten years earlier. To determine whether there was anything vital to the war effort, the OAPC called in John G. Trump to examine the papers. Working on radar at MIT's Radiation Laboratory and an expert on high-voltage generators, Trump was well suited to undertake the review.[56] In addition to Trump, the OAPC allowed one of the top agents from Naval Intelligence, Willis De Vere George, to be present along with two enlisted men.[57] On 26–27 January 1943, Trump reviewed Tesla's files while the navy personnel microfilmed the papers that they found interesting.

After two days of rummaging through the papers at Manhattan Storage and finding nothing of interest, Trump decided that they should investigate the box that Tesla had left at the Governor Clinton Hotel. Arriving at the hotel, Trump and his group were escorted to the back storeroom where the managers brought out the box; recalling Tesla's dire warnings that the box would explode if opened by an unauthorized individual, the managers promptly left the storeroom. The agents accompanying Trump also pulled back, leaving him to open the box alone.

As Trump recalled, the box was wrapped in brown paper and tied with string. As he cut the string with his pocketknife, Trump thought that it was a lovely day and that he ought to be outdoors enjoying the weather. Removing the paper, he saw that the box was made of polished wood, typical of that used for the storage of a scientific instrument. Taking a deep breath, Trump lifted the hinged lid, only to discover that it contained an old resistance box, the kind used to measure resistance using the Wheatstone bridge technique. From beyond the grave, Tesla had pulled off one final illusion.[58]

On the basis of his review of the materials, Trump reported to OAPC that "it is my considered opinion that there exist among Dr. Tesla's papers and possessions no scientific notes, descriptions of hitherto unrevealed methods or devices, or actual apparatus which could be of significant value to this country or which constitute a hazard in unfriendly hands. I can therefore see no technical or military reason why further custody of the property should be retained." Consequently, the OAPC released the papers to Kosanović.[59]

Deeply embroiled in Yugoslav politics—at the outbreak of the war, Kosanović was loyal to King Peter II, but at some point he switched his allegiance to the new leader, Marshall Tito—Kosanović was in no position to do anything with the Tesla materials.[60] However, no sooner had the OAPC released the papers to Kosanović than the New York State Department of Taxation put them under seal until Tesla's back taxes were paid. Over the next seven years, the materials stayed at the Manhattan Storage Company while Charlotte Muzar paid the monthly rent.[61]

In June 1946, Kosanović returned to America as the Yugoslav ambassador. With funds from either the Yugoslav government or the Tesla Institute, Kosanović hired a lawyer, paid Tesla's remaining bills and back taxes, and arranged for the papers to be shipped from New York to a new museum being created in Belgrade. The papers arrived in the fall of 1951, and in February 1957 Muzar carried Tesla's ashes back to his home country. There they were placed in the Tesla Museum in a spherical urn—Tesla's favorite geometrical shape.[62]

Although Tesla's papers eventually went to Yugoslavia, that did not stop government scientists from being interested in them. In October 1945, the FBI received a request from Bloyce Fitzgerald, who, while only a private in the army, was supposedly heading up a top-secret project at Wright Field in Dayton, Ohio. The project was to perfect Tesla's "death ray," which the Army regarded as "the only possible defense against the offensive use by another nation of the Atomic Bomb."

Fitzgerald wanted access to either the Tesla papers at the Manhattan Storage Company or the microfilm made by Naval Intelligence; the FBI declined to assist him, saying in true bureaucratic fashion that it did not know anything about a microfilm that might be held by another agency. According to one letter in the FOIA file, scientists from the navy and the Office Strategic Services may have spent a month going over the microfilm, presumably in the 1940s. Over the next forty years, numerous individuals contacted the FBI about Tesla's papers, and the bureau's standard reply was that it did not have them. As government scientists became interested in the 1970s in Tesla's fireballs as a way of containing nuclear fusion, they again asked the FBI about the Tesla microfilm; curiously, no one seems to know what had become of the microfilm. In my mind, the disappearance of the microfilm does not make a conspiracy and "prove" that the Tesla papers contained plans for a secret weapon; as one of my professors used to say, "Never assume a conspiracy when mere ineptitude will suffice as an explanation."[63]

During the cold war, both the United States and the Soviet Union investigated particle beam weapons at various times in the belief that they could be used to disable incoming nuclear missiles. Most notably, in 1977 the retired head of Air Force Intelligence, General George J. Keegan, claimed in *Aviation Week & Space Technology* that the Soviets were building a large-scale charged particle beam at the Semipalatinsk Test Site in northeast Kazakhstan.[64] Although Keegan's claims were roundly denied by President Jimmy Carter and scientific experts, fear of a possible "death beam gap" provided the political impetus for a significant expansion of American research into space-based beam weapons.

Under the direction of the Defense Advanced Research Projects Agency (DARPA), work began on the ALPHA chemical laser project in 1978, the TALON GOLD targeting system in 1979, and the Large Optics Demonstration Experiment (LODE) in 1980. These programs formed the basis for the Strategic Defense Initiative (SDI) that Ronald Reagan announced publicly in 1983. In the 1980s, the Department of Defense added several programs to SDI including X-ray and chemical lasers as well as the neutral particle beam weapon; reminiscent of Tesla's "Chinese Wall" claims, SDI weapons were intended to create a "curtain" that would destroy incoming enemy missiles. While this work was under way, government scientists again asked the FBI about Tesla's papers, particularly since they believed that the Soviets had studied Tesla materials in Belgrade and that a review of any Tesla-related papers in the United States would provide insight into the Soviet beam weapons program.[65]

After the fall of the Soviet Union, American weapons experts visited the Semipalatinsk Test Site, only to discover that the Soviets had not been working on a beam weapon at all but on nuclear-powered rockets. This new information from Semipalatinsk prompted John Pike of the Federation of American Scientists to call Keegan's claims of a Soviet particle beam weapon "one of the major intelligence failures of the Cold War."[66] Since, in principle, a beam weapon is possible, the U.S. military may be continuing to investigate them; however, just as in Tesla's time, these weapons will require enormous amounts of energy and pose formidable problems in terms of keeping the beam from dissipating.[67]

EPILOGUE

I misunderstood Tesla. I think we all misunderstood
Tesla. We thought he was a dreamer and visionary. He
did dream and his dreams came true, he did have visions
but they were of a real future, not an imaginary one.

JOHN STONE STONE, 1915

In the decades since his death, Tesla has enjoyed a curious legacy.
On one hand, he is acknowledged by engineers for his contributions
to alternating current (AC), and in 1956 "Tesla" was adopted as the
name for the unit of measure for the flux density of magnetic fields.
On the other hand, thanks to the many colorful predictions he made
about his inventions, Tesla has become a figure in popular culture.
In these closing pages, I wish to reflect not only on Tesla as a cultural
figure but, more important, on what we can learn from him about the
invention process and the role that innovation plays in the economy.
In doing so, we are completing the task that Brisbane posed that summer night in 1894 when he interviewed Tesla: that we should seek "to
discover this great new electrician thoroughly; to interest Americans

in [Tesla's] personality so that they may study his future achievements with proper care."[1]

TESLA IN POPULAR CULTURE

Unlike his contemporaries Edison and Marconi, Tesla never made it into the history books of the second half of the twentieth century. To be sure, his invention of the AC motor and his pioneering work with electromagnetic waves certainly warrant a spot for him in the general American historical narrative. Tesla is absent from the history books, in part, because he never created a major eponymous corporation to manufacture his inventions; until the recent creation of Tesla Motors, there has been no equivalent of Marconi Cable and Wireless or Consolidated Edison. Rather than seek profits from manufacture, as we have seen, Tesla preferred to pursue a strategy of patent-promote-sell. Hence there was no large company like General Electric (who sees Edison as their founder) or the Radio Corporation of America (which was based on Marconi's patents) that needed to put Tesla forward as a "founding father." Westinghouse might have used Tesla in this manner, but Tesla's ties were primarily with George Westinghouse personally and not with the managers who ran the company for as long as it lasted in the twentieth century.

But another reason that Tesla is not in the history books of the late twentieth century is that in cold war America he was not a useful figure. Unlike Thomas Edison or the Wright brothers, Tesla was not born in America and hence could not represent "Yankee ingenuity"— the popular notion that Americans by nature were practical and technologically creative. Moreover, since most people assumed that Tesla was a mystic who did not use theoretical science to develop his inventions, he could hardly be invoked by those apostles of modernity who believed that technology grew out of research performed by scientists in university and corporate laboratories. And unlike Edison or Henry Ford, both of whom were seen as ordinary folk who understood the needs of the common man and in response created mass-produced goods such as automobiles, electric lights, and motion pictures, Tesla seemed effete, elitist, and, well, eccentric.[2] For the cold war establishment, then, Tesla was best ignored—a curious outsider who could be forgotten.[3]

But it was exactly his outsider status—his mystical qualities, his impractical streak, his rejection by establishment figures like Edison and Morgan—that made Tesla a hero of the counterculture. How could they not love all of Tesla's wonderful claims about free wireless power, talking to Mars, a race of robots, abolishing war, and death rays? Beginning in the 1950s, quirky individuals embraced Tesla, playing up his interest in interplanetary communications. For instance, Arthur H. Matthews claimed that he had worked with Tesla and that the Wizard had taught him how to build a "Teslascope" in order to communicate with aliens on other planets. Matthews also reported that Tesla had been born on Venus and that he knew this because "Venusians" had told him so during a visit to his home in Canada. Not to be outdone, Margaret Storm published a book printed in green ink that stated that Tesla had come by spaceship from Venus, accompanied by his "Twin-Ray," the white dove. Storm, too, had her own special radio set to contact extraterrestrials.[4]

During the energy crisis of the 1970s, Tesla became a hero of the free-energy movement. Adherents in this movement believe that there is advanced technology—often based on Tesla's ideas—that has not yet been adequately defined by conventional physics that can be used to access energy in the universe. Free-energy researchers take seriously Tesla's slogan that humans should be able to utilize energy by "attaching their machines to the very wheelwork of nature."[5] Some members of this movement believe that a corporate conspiracy—dating back to Edison and J. P. Morgan—is what prevents improved energy technology from being developed or adopted. One example of Tesla-inspired energy research was Robert Golka, who attempted from 1970 to 1988 to re-create Tesla's Colorado experiments in order to generate artificial ball lightning. Along with other scientists, Golka believed that fireballs could be used to develop techniques for containing high-energy plasma and usher in nuclear fusion.[6]

To draw together free-energy researchers and other individuals interested in Tesla, scientists and community leaders in Colorado Springs organized the first Tesla Symposium in 1984. Over the next fourteen years, this group held annual meetings, published proceedings, created a museum, and established the International Tesla Society. At one point the society boasted seven thousand members worldwide, but as a result of internal political struggles, it went bankrupt in 1998. In the decade since, Tesla aficionados have found homes in the Tesla Engine Builders Association, the Tesla Memorial Society of New York, and Cameron B. Prince's Tesla Universe website.[7]

Tesla has also been attractive to believers in New Age philosophy. As an alternative subculture, New Age thinking has been described as "drawing on both Eastern and Western spiritual and metaphysical traditions and infusing them with influences from self-help and motivational psychology, holistic health, parapsychology, consciousness research and quantum physics."[8] Given his personal charisma (tall, dark, handsome, and mysterious), his efforts to develop his willpower as a boy, and his supposed interest in psychic phenomena (which I think ran counter to his materialist view of cognition), it is not surprising that Tesla crops up across New Age sources and practices. As we have seen, his story is very much one of cultivating and perfecting his "intense and wild nature" (see Chapter 12), in which he learned how to invent drawing on both his rational faculties as well as his imagination; indeed, much of Tesla's tale can be interpreted as spiritual enlightenment and personal development.[9]

One manifestation of Tesla and New Age beliefs are the Purple Plates used by individuals to channel positive energy to address physical and emotional ailments. The plates were introduced by Ralph Bergstresser, who knew Tesla during the last six months of his life and subsequently developed the plates based on one of Tesla's patents.[10] Reflecting a mix of scientific and spiritual beliefs, one supplier of Purple Plates, the Swiss Tesla Institute, describes how these aluminum plates possess a lattice structure designed "to oscillate at a frequency that resonates with the frequency of the earth's own electromagnetic field that permeates the cavity between the earth's surface and the ionsphere. This field is known as the Schumann Resonance field and . . . is considered to be associated with the biofield (Chi or Qi, Prana, Orgon, etc.) containing all the essential information and energy necessary for the proper functioning, growth and evolution of any life form."[11]

Much of the New Age interest in Tesla reflects the fact that not everyone is comfortable with the rationality of modern life, particularly

the assumption that technological innovation is driven by the inexo-
rable logic of either the marketplace or science. The standard explanation today for why a new technology is introduced is that it is either
a response to market demand or the result of scientific discoveries.
Although both of these factors contribute to technology, they are not
necessarily meaningful to everyone; some people would like to believe
that new technology should also reflect the values, dreams, and wishes
of a culture. Technology is so important in our lives that, for some
people, it cannot simply be boiled down to the impersonal forces of
the marketplace or the laboratory.

For these folks, Tesla is a welcome alternative. He reveals that there
is more to technology than relentless economic or scientific rationality. Although educated in engineering, Tesla refused to be guided exclusively by the dictates of either science or the marketplace. Instead
his inventions came from within himself, and through his inventions
he sought to order his own life and the larger world around him. In
this sense, Tesla was like an artist or a poet, and his contemporaries
certainly described him at times as such. Thus, for people who do not
want to see the world in exclusively rational terms, Tesla is their man.
To quote *Hamlet*: "There are more things in heaven and earth, Horatio,
than are dreamt of in your philosophy" (Act 1, Scene 5).

I am not suggesting that New Agers are setting up a dichotomy between the material and the spiritual, that technologists have no soul.[12]
Rather, what some individuals hunger for is a culture in which the
spiritual and the material are intertwined. Tesla, as a cultural hero,
neatly eliminates this dichotomy: he created great technology by
being in touch with his inner self, and he reveals that one can weave
together the spiritual and the material.

Some corporations have recognized how Tesla represents a blend-
ing of the spiritual and material. This is the genius behind the marketing of the electric roadster manufactured by Tesla Motors: like their
namesake, they have figured out how to produce an automobile that
simultaneously speaks to both the spirit, in the sense of being envi-
ronmentally sound, as well as to people's materialistic side in the sense
of being fast, cool, and high-tech. One catchy slogan on the company's website proclaims "Zero emissions, zero guilt." As one Tesla car
fan wrote on a Tesla Motors' blog in 2009, "I am tired of the Green
revolution being all bean sprouts and bicycles. Hell, yeah, Tesla is

Green with a bite. . . . [M]y philosophy [is that] you don't have to wear
Birkenstocks and eat nuts and berries and sit cross legged for hours
and meditate. . . . A cleaner greener world doesn't mean we need to
devolve back to the middle ages."[13]

As a proponent of extraterrestrial life, hero of free energy, and New
Age saint, Tesla has proven to be an intriguing character for popular
culture. He brings together the spiritual and the material, he challenges the corporate establishment, and he invites flights of imagination to new technology and new worlds. For all these reasons, though
Tesla may be absent from the history books, he can be found everywhere in popular culture: as a hard-rock band, in Hollywood movies
such as *The Prestige* (2006), in novels such as *The Invention of Everything
Else* (2008), in plays and operas such as Jeff Stanley's *Tesla's Letters*,
and even a character in video games such as Capcom Entertainment's
"Dark Void Saga." As a spokesperson for Tesla Motors, observed,
"You know you've gone into mainstream pop glory when you're in a
videogame aimed at 18-year-old boys."[14]

TESLA AND DISRUPTIVE INNOVATION

But more than being an attractive figure for American pop culture,
Tesla offers insight into the nature of technological innovation from
which we can draw lessons relevant for today and tomorrow.

At first glance, this may seem like a surprising claim since it would
be easy to dismiss Tesla as being sui generis—that he was a highly eccentric inventor who worked over a hundred years ago—and so what
could inventors, engineers, and entrepreneurs possibly learn from
him? Innovation in the twenty-first century is all about a rapidly expanding scientific frontier, large collaborative teams, venture capital,
and global markets. What could a flighty visionary like Tesla teach us
that is relevant to the modern economy?

To respond to this challenge, we need to recall Schumpeter's idea
that economies grow as a result of two kinds of innovation (see Chapter 2). On one hand, there are the creative responses of entrepreneurs
and inventors who introduce new products, processes, and services
and in so doing dramatically change everyday life and reorder the

industrial world; as Clayton Christensen has suggested, these can be called disruptive innovations.[15] On the other hand, there are the adaptive responses of managers and engineers who undertake the steady and incremental work of establishing the corporate structures, manufacturing procedures, and marketing plans that allow products and services to be produced and consumed. Clearly the success of any economy depends on getting the right mix of disruptive and adaptive innovations.

Tesla introduced two disruptive innovations that changed the American economy at the end of the nineteenth and the start of the twentieth century. His AC motor in the late 1880s made it desirable for electrical utilities to shift from DC to AC so that they could provide not only lighting service but power that could be used by industry and consumers. By expanding the services that electrical utilities could provide, AC power allowed them to increase the size of their systems, pursue economies of scale, and drive down the cost of electricity over the long term. Equally important, Tesla introduced the idea of polyphase power; by distributing AC using either two or three phases, electrical utilities found that they were able to efficiently transmit more power over longer distances. In a nutshell, Tesla's AC inventions were essential to making electricity a service that could be mass-produced and mass-distributed; his inventions set the stage for the ways in which we produce and consume electricity today. For all these reasons modern versions of Tesla's AC motors can be found running household appliances, powering industrial machinery, and even keeping the hard disks of laptop computers spinning.

With wireless power, Tesla saw himself as launching a second electrical revolution; with this technology, he was going to pull an end-run around the existing wired networks being used for power, telephony, and telegraphy. As we have seen, Tesla was unable to perfect his system because he could not raise additional funds after he lost Morgan's support and because he could not come up with a way to introduce electrical energy into the earth's crust; consequently, Marconi's ideas about wireless telegraphy were pursued by companies in Europe and America. However, one should also keep in mind that while Marconi worked out point-to-point wireless telegraphy, a host of other inventors—Lee de Forest, Reginald Fessenden, John Stone Stone, and Edwin H. Armstrong—contributed the innovations needed to create broadcast radio. Radio was not invented by Marconi; it was the result of an evolutionary process involving a variety of people, including Tesla.[16]

Tesla played a crucial role in the evolution of radio in two ways. First, he came up with key ideas and devices—the importance of grounding transmitters and receivers, adjusting capacitance and inductance in order to tune devices, and the Tesla coil—all of which were borrowed and adapted by other radio pioneers. As his rival A.P.M. Fleming observed in 1943, Tesla "produced almost incidentally a whole succession of apparatus which was employed successfully by other workers striving for less ambitious ends. Had his primary concentration allowed him to pay more attention to the tools which his inventive genius improvised freely, then the great influence he had on radio development would have been obvious to all."[17] Radios, televisions, and cellular phones today employ variations of Tesla's ideas about tuning circuits to resonate to particular frequencies.

Second, Tesla was important to the evolution of radio because he inspired—indeed galvanized—his rivals into action. Tesla was one of the first inventors to investigate electromagnetic waves, and as his contemporary John Stone Stone emphasized, Tesla "did more to excite interest and create an intelligent understanding of these phenomena in the years 1891–1893 than any one else."[18] As we have seen, Tesla was a genuine inspiration to de Forest (though de Forest's subsequent actions harmed Tesla's efforts), and Marconi and his associates kept an eye on what Tesla was doing. At the very least, Tesla's rivals in wireless had to figure out ways to get around his patents, whether by creating alternatives or by fighting him in court.

TESLA AND THE PROCESS OF INVENTION

Just as it is useful to look at what Tesla invented—AC motors and wireless power—it is even more important to look at how he invented such disruptive technologies. Both facets of his creative style—ideal and illusion—offer insights for inventors, engineers, and entrepreneurs wishing to develop similarly bold innovations. In engineering and business schools, students are taught how to objectively analyze the performance of machines and systems, estimate consumer demand, and create new devices on the basis of what they measure—to create what Schumpeter called adaptive innovations. In contrast, what do we actually know—as historians, entrepreneurs, or policymakers— about the other of kind innovation, Schumpeter's creative innovation or Christensen's disruptive innovation? Where do these innovations come from? Are disruptive innovations simply caused by mysterious, unknowable forces like genius and luck? How does one harness disruptive technology so that it has a positive impact on a company, the economy, and society at large? These are the questions that Tesla's story answers.

Tesla's great strength was that he was willing to think like a maverick. With his motor, for instance, while most other investigators worried about changing the direction of the magnetic poles in the rotor, Tesla instead figured out how to create a rotating magnetic field in the stator (see Chapter 2). If everyone knocks at the front door, Tesla is suggesting, then one way forward is to go around the house and see if there is a back door.

To find that back door, though, one needs to cultivate an expansive imagination. One needs to be willing to go on extended flights of fancy, to conjure up not just shadows but entire machines and whole societies. Not all of us are able to access our imagination to this level and create images with the visual clarity that Tesla did—let's face it, he did have an unusual childhood in which he had to develop his imagination—but we need to cultivate, or at least tolerate, broad explorations via the imagination. If we don't take chances in our imagination, how can we even begin to find the maverick ideas or ideals?

Yet imagination is only half of the cognitive process as described by Tesla: it was the "wild" side of an inventor's nature, but there was also an "intense" side as well (see Chapter 13). In spite of the vast popular literature celebrating Eureka moments, Tesla found that an insight, intuition, or hunches had to be refined in the mind through rigorous thought and analysis. Tesla spoke to this refining stage when he described his creative process in 1921:

> Here in brief, is my own method: After experiencing a desire to invent a particular thing, I may go on for months or years with the idea in the back of my head. Whenever I feel like it, I roam around in my

imagination and think about the problem without any deliberate concentration. This is a period of incubation.

Then follows the period of direct effort. *I choose carefully the possible
solutions of the problem I am considering, and gradually center my mind on
a narrowed field of investigation* [emphasis added]. Now, when I deliberately think of the problem in its specific features, I may begin to feel
that I am going to get the solution. And the wonderful thing is that if
I do feel this way, *then I know I have really solved the problem and shall get
what I am after.*[19]

One small but visible clue of Tesla's intense analysis is that at his
South Fifth Avenue laboratory there was "a small black board which
hangs on the wall and bears evidence of hard usage. The black is
worn from this board in several spots, and the rest of it is covered
with figures and cabalistic signs."[20] Tesla wasn't drawing fantasies on
that board but using what mathematics he knew to hone and sharpen
an ideal as it took shape in his mind. For Tesla, then, an ideal did
not appear suddenly full-blown but was the result of two cognitive
activities: roaming around the imagination and carefully examining
possible solutions. Without these two activities working together,
in exquisite tension, I suspect that Tesla would not have invented
anything.

In our thinking about innovation and design today, we tend to di-
chotomize imagination and analysis. We assume that inventors work
primarily with the imagination and that engineers rely on rigorous analytical techniques drawn from science and mathematics. Seldom do
we admit that, for the development of disruptive technologies, both
activities are needed in the right measure. One needs to dream but one
needs to critically assess one's dreams in terms of scientific theory as
well as available materials and techniques.[21]

While inventors must keep imagination and analysis in exquisite
tension in their own minds, backers and patrons also help them maintain a useful balance between the two. Tesla was extremely fortunate
in his AC work that he had Charles Peck and Alfred Brown as his backers. These men encouraged Tesla to pursue his ideas for an AC motor,
but at the same time they provided the young inventor with valuable
feedback about what might actually work and what would appeal to
the businessmen who would invest in or buy his patents. Peck and

Brown were the hard flint against which Tesla's steel struck, from which the sparks of genius could ignite the fire of a breakthrough.

For Tesla, one of the great "what-ifs" is what might have happened if Peck had not died in 1890. With Peck's business savvy, would Tesla have had the guidance needed to shape his wireless power ideas into a strong set of patents that in turn could have been developed into a workable product or service? Would radio have developed more along the lines of lighting and power applications as opposed to communications? Unfortunately Tesla's later backers, Edward Dean Adams and J. P. Morgan, were not in a position to work as closely with him in shaping his inventions. Without someone like Peck to provide a mix of encouragement and critical advice, Tesla became intoxicated by the beauty of his ideal for wireless power and was not inclined to adjust his ideal to suit practical and business considerations.

The lesson to be drawn from Tesla and Peck is that we need to understand and appreciate how inventors and entrepreneurs forge relationships that foster a balance between imagination and analysis: the businessperson "grounds" the inventor's dreams in existing business practices and expectations but at the same time the inventor "inspires" the businessperson to see new possibilities in the technology. Alexander Graham Bell had this sort of relationship with his father-in-law, Gardiner Hubbard, as did Thomas Edison with William Orton of Western Union, but we should ask the same questions about the relationship between steam-engine pioneers James Watt and Matthew Boulton or Steve Wozniak and Steve Jobs with Mike Markula in the early days of Apple Computer.[22] For ideas to become disruptive technology, inventors must balance imagination and analysis not only in their own minds but also in relationships with their backers.

In thinking about how inventors interact with their backers, it's now time to turn from ideals to illusions. My impression is that even for an inventor like Tesla it can be very difficult to fully grasp an ideal; at any given moment, one can visualize in the mind's eye some facets of the ideal but not necessarily all of them. Inventors learn how to work around this difficulty and, indeed, to even take advantage of the ambiguity in their mental models as they generate alternative designs.[23] But in sharing an ideal with others, inventors have to confront this problem head-on: if they cannot fully access the ideal, how can they convey it to friends, patrons, patent examiners, and customers?

In the case of Tesla, we have seen that he resorted to using images, metaphors, and stories—what I have called "illusions" throughout this book. Illusions are not deceptions but approximations. Illusions are how an ideal escapes the mind of one individual and gets into the mind of another person. No ideal, no idea, no invention goes anywhere unless one is willing to tell a story about it, a story that another person finds interesting and persuasive.[24]

One major turning point in Tesla's life was converting his tin-can motor into the egg of Columbus demonstration apparatus in 1887. The tin-can motor thrilled Tesla because it was the physical realization of his ideal, but this motor meant nothing to Peck and Brown. For them to see something of the ideal, Tesla had to clothe it in the Columbus story and create a compelling vision for them. With the egg apparatus, Tesla created an illusion of possibility for his backers; yes, this motor could be something that the electrical industry will want. In turn, Peck and Brown worked with Tesla to use patents, reports in the technical press, his lecture, and laboratory demonstrations to create more illusions, to effectively link Tesla's motor with the needs and goals in the electrical industry at the time. Through this skillful promotion, Peck, Brown, and Tesla generated excitement about the motor, so much so that they convinced Westinghouse to sign a lucrative contract.

To a very real extent, promotion as a business strategy is all about the shrewd use of illusion. The promoter must get people excited, willing to invest, but not suspicious. The invention must fall in a good spot on the spectrum between feasible and improbable. If it's entirely feasible, then the invention may not be all that novel and hence not worth very much; equally, if the promoter overstates the case and the invention looks too good to be true, then investors may regard it as too risky and not invest. Hence getting the illusions right—exciting but feasible—was a real challenge for Tesla and his backers; as they seek investors, contemporary inventors and entrepreneurs face the same challenge.

Within this context, it is obvious that inventors need to be not only monitoring what is happening in the technical realm but also keeping a finger on the cultural pulse: What excites people? What are the burning issues of the day? What needs or dreams do people have to which an invention might be connected? From this perspective, it makes sense that Tesla became friends with Robert Underwood

Johnson and Mark Twain, for they were both carefully observing how American society was taking shape as it entered the twentieth century. Here I am reminded of one letter Tesla sent in 1899 from Colorado to Scherff in New York, asking what sort of stories were circulating in the daily papers; feeling isolated in the mountains, Tesla was hungry for the cultural themes he needed to shape illusions.[25]

For an inventor and his backers, illusions can be quite tricky since they need to figure out how big an illusion should be. Should they tie an invention to the specific and immediate needs of an industry?[26] Should they claim that an invention will revolutionize the whole industry? Or should they invoke broader cultural wishes? In the case of the motor, Tesla and his backers stayed close to the needs and expectations of the electrical industry and succeeded in landing the Westinghouse contract. As Tesla matured and was influenced by his friends, however, he framed progressively grander illusions for his inventions—his radio-controlled boat would abolish war, his wireless power system would revolutionize communications and the whole of society, his turbine was "the perfect rotary engine."

One way of looking at these grandiose claims is that Tesla simply let success go to his head. Another view is that he allowed himself to be unduly influenced by his friends in the press such as T. C. Martin and Johnson.[27] But we can also ask a counterfactual question: if Tesla had kept his illusions modest, would anyone have paid any attention to his inventions in the mid-1890s? To some extent, Tesla was responding to his era's "yellow journalism." As the large New York tabloids vied for circulation in the 1890s, they sought out stories with oversized claims, and with each successive retelling the claims had to be further exaggerated.[28] The scale of Tesla's illusions is thus both a product of his personality as well as the way that popular culture was then taking shape.

TESLA AND THE CREATIVE URGE

Ideal and illusion thus tell us a great deal about *how* an inventor invents, about the process of creating disruptive technologies. But we might also reasonably inquire about *why* an inventor invents. This is, after all, a biography, and you are entitled to ask about motivation.

I am going to focus here not on what motivates creativity in all sorts of individuals but on the specific case that Tesla represents: why do individuals go to the trouble of creating disruptive technologies? To develop a disruptive technology, one has to take significant chances, to pursue new devices and business practices, and though the personal and financial rewards are sometimes great, there are no guarantees of success. It is very hard to tell at an early stage which disruptive technology will take off and what constitutes the key innovation that an inventor or company may need to control in order to profit from the disruption. Consequently, it is not surprising that lots of talented engineers and entrepreneurs choose the safer path of developing technology in response to existing market demand; there, one can calculate the odds and convert the chances into manageable risks. More than once, Tesla told interviewers that he could have taken the safer path and made a fortune developing several inventions for which there was an immediate demand, but he chose to go after the grander and more difficult challenges.

One explanation that is frequently put forward to explain disruptive innovations in a variety of fields is what might be termed the "outsider argument." Individuals who are outside the established social, political, and economic hierarchy choose to innovate sometimes in order to gain access to the hierarchy, sometimes to challenge the status quo, and sometimes to do both. As outsiders, these creative folks do not see things in the same way as people inside the hierarchy and they have little to risk and everything to gain by innovating.

One can certainly apply the outsider argument to Tesla. While the electrical engineering community in late nineteenth-century America was made up mostly of native-born Protestant men of English, German, or Dutch ancestry, Tesla was a Slavic immigrant—a Serb—and his Orthodox religious background was distinctly different from the prevalent American Protestantism. Equally, we have clues that his heterosexual peers were uncomfortable with Tesla's attraction to men. And as we have seen, Tesla eschewed the emerging professional norms of writing up one's research in scientific papers for the more spectacular practices of bold public demonstrations and lively newspaper interviews. One indication of how Tesla's contemporaries perceived him as an outsider is the story that, shortly after hiring Tesla, Edison asked if he really was a cannibal.[29]

The outsider argument is helpful but it does not fully capture what was going on with Tesla. Yes, Tesla was an outsider seeking a place in the electrical engineering profession and respectable New York society, but by 1894 when he was interviewed at Delmonico's, he had achieved status and respectability. What made him keep going after that? Have we fully captured what made him tick?

Through his inventions, Tesla sought to reorder the world around him. With his AC motor, Tesla fervently believed in the ideal of a rotating magnetic field, even if that ideal required going from two to four or even six wires, changing the frequency in AC systems from 110 to 60 cycles, and even building entirely new systems for generating and distributing two- or three-phase power. For Tesla, the ideal of the rotating magnetic field was so compelling that rather than having him adapt his ideal to the world, the world should be reordered to make space for his ideal. In like manner, Tesla expected that people would find his wireless power scheme so obviously wonderful that they would shift from the existing wired systems to his new ideal.

So how did Tesla develop this desire to reorder the world to suit his ideals? Why would he even believe that it was possible? I think this desire stems from the way he developed his imagination as a boy in order to cope with his fears. From childhood, Tesla was confronted with a frightening and disorderly world on several levels: as Serbs, his family and relatives were strangers in the strange land of the Krajina, the Austrian Military Frontier; his family was traumatized by the loss of the oldest son, Dane; and Tesla himself suffered from scary images and nightmares. As we have seen, Tesla coped with this disorder by developing his willpower and applying it to his imagination. Rather than being controlled by fearful images, Tesla as a boy learned to channel his imagination in ways that permitted him to overcome the nightmares and go on pleasant mental journeys.

As he gained mastery over his imagination, Tesla applied this new talent to invention with significant emotional consequences, which brought him to another major turning point in his life. Since he could fly in his imagination, he wondered why he could not fly in real life; thus when he was twelve he experimented with creating a flying machine powered by a vacuum pump. Nothing thrilled Tesla more than seeing that homemade pump work, as even a slight motion confirmed that what he had imagined could come true in the material world, that the imaginary and material realms were indeed connected. In other words, if Tesla could imagine an orderly world, then it ought to be possible to make that order manifest in the material realm. Here was a powerful incentive to invent.

The discovery that order in the imagination could map onto material practices was reinforced by what Tesla absorbed about the logos from the Orthodox faith of his father, that the material universe is not only orderly but that everything in it—natural and man-made—has an underlying principle that can be discovered by individuals. From his religious background, then, Tesla learned that there were ideals that he should uncover and apply to the material world. Throughout his life, he strove to refine all of his faculties—mental, physical, and spiritual—so that he was as perfect an instrument as possible for detecting those ideals.

This brings us back to Schumpeter's notion of objective and subjective rationality. The typical engineer and manager use objective rationality in that they rely on measurements from the outside world: order is "out there" and they just need to find the pattern. In contrast, for the inventor and entrepreneur, order comes from within and they seek to apply this order to the outside world. New ideas, new disruptive technology come about when inventors and entrepreneurs seek to impose that internal order on the outside world. For subjective thinkers, the arrow runs from the inside to the outside, exactly the opposite of how objective thinkers see the world. Tesla's admirer Kenneth Swezey appreciated the difference between the objective and subjective when he wrote to Tesla in 1924 that "I have seen the recognized engineers scoff at your ideas, and hint at an unbalanced state of mind—but a mind certainly has to be somewhat unbalanced to overcome the momentum of flighty enthusiasm or the inertia of destructive conventionalism. These fellows are so well balanced that they can spin around and around and around at such a uniform and unwavering rate, that an eccentric movement—the energy and power behind the birth of a new satellite [i.e., an invention]—becomes impossible."[30]

So why did Tesla seek to impose his internal ideals on the outside world? Why take up this struggle? Though it is undoubtedly different for every inventor and entrepreneur, Tesla sought to reorder the world as a means of compensating for the disorder he felt inside. While Tesla possessed a great confidence in his "intense and wild nature"—his analytical and imaginative powers—he may have felt disordered inside throughout his adult life. As we have seen, he could be highly charming and sociable one day, withdrawn and taciturn the next. Equally, Tesla had periods of great energy and enthusiasm followed by spells of depression. And however one chooses to interpret Tesla's attraction to men, his sexual feelings may very well have contributed to his internal confusion. Hence Tesla was driven to invent, to impose his ideals on the material world in response to feeling disordered inside. If he could get the outside world to align with the ideals that came from his mind, he would again have some evidence of meaning in the universe.

With wireless power, Tesla was especially driven to impose his vision on the material world. He truly believed that he had discovered the penultimate ideal or, as he once told J. P. Morgan, the philosopher's stone. Based on the confirmatory evidence he gathered while in Colorado, he was utterly convinced that his system would work. In the years following, Tesla stopped worrying about the validity of his ideal and instead focused on getting the illusions right. As long as he lived like a millionaire at the Waldorf, had the support of J. P. Morgan, received ample press coverage, and worked on an impressive station at Wardenclyffe, all would be right. In the sense that a magician worries about creating the right illusion in the mind of the audience, so, too, had Tesla become something of a magician. Wardenclyffe had to work because he was never wrong; what he saw in his mind's eye had to map onto the world. But with Wardenclyffe, illusions got ahead of the ideal, and Tesla suffered a nervous breakdown when he was unable to grasp the disjuncture between how he thought his system should work versus how the Earth actually responds.

For Tesla, his ability to uncover an ideal and to find the illusions necessary to convey that ideal to others were his great strengths but also his greatest weaknesses. With the AC motor and other inventions, Tesla successfully balanced ideal and illusion in his mind and in his interactions with society. Tragically, with wireless power, Tesla became intoxicated by the beauty of his ideal but distracted by his illusions; he did not find a balance between the two. As Tesla once remarked, "Our virtues and our failings are inseparable, like force and matter. When they separate, man is no more."[31]

At the height of the Wardenclyffe adventure, Tesla wrote Morgan that "[m]y enemies have been successful in representing me as a poet and visionary."[32] In doing so, his enemies were accusing him of not solving the business and technical problems related to this project, and perhaps they were right. But we should not let their complaints keep us from seeing what Tesla teaches us about disruptive technology. Radically new technology comes from within, from a willingness to discern ideals and to connect them to the needs and wishes of society. Tesla reminds us that like poets, technologists need to think hard but dream boldly. Only by doing both will we be able to use technology, as Tesla did, to create a little bit of heaven here on earth.

NOTE ON SOURCES

One of the great pleasures of writing about inventors is that one has the opportunity to work with a wide range of materials, everything from love letters and diaries to notebooks, sketches, and models to business and legal records. To create the portrait of Tesla presented in these pages I have woven these materials together, and here I want to highlight the most useful sources so that future Tesla scholars will be able to have a running start for their investigations.

The El Dorado for future Tesla study is the Nikola Tesla Museum in Belgrade, Serbia. As I recounted in Chapter 16, all of the material confiscated by the Office of Alien Property was given in 1951 to the people of Yugoslavia, who in turn created this museum. The collections include not only 160,000 pages of correspondence and technical notes but also Tesla's clothing, personal effects, and a few scientific artifacts. For a detailed description of the museum's collections, consult Marija Sesic, ed., *Nikola Tesla Museum, 1952–2003* (Beograd: Nikola Tesla Museum, 2003). During two visits to the Tesla Museum (1998 and 2006), I was able to work with Tesla's scrapbooks, published legal testimony, and personal library. Over the years, the museum has graciously shared documents and photographs with me, and I hope that the museum will be able to provide access in the future to more of its manuscript holdings.

To complement what was available from the Tesla Museum, I consulted a variety of archives in the United States and Europe, and these are listed at the end of this section. I found much valuable material in the papers of Leland Anderson and Kenneth Swezey, both of whom spent decades locating Tesla-related material in the course of writing articles about the Wizard; both gentlemen hoped to write biographies of Tesla. In the 1960s, the Library of Congress hoped to secure a complete microfilm of the Tesla Papers that had gone to Belgrade, but only seven test reels were made by the Yugoslavian archivists. These reels, however, offer a rich trove of letters between Tesla and George Westinghouse, J. P. Morgan, Robert Underwood Johnson, George Scherff, and Mark Twain. The collection at Columbia University includes a large number of additional letters exchanged between Tesla, Johnson, and Scherff. Another remarkable collection are the letters sent by Tesla to Edward Dean Adams during the planning of the Niagara power system in 1893, and these are part of the Western New York Collection at National Grid. For information on Tesla's particle beam weapon, I relied on the FBI file that was released in the 1980s under the Freedom of Information Act; the version I used can be downloaded at http://www.scribd.com/ under "Freedom of Information Act file for Nikola Tesla, Federal Bureau of Investigation." Complementing my research in these manuscript collections, the Henry Ford Museum and the Science Museum in London hold a number of Tesla-related motors.

In writing about nineteenth-century inventors, one can fortunately rely on patent-related materials to glean insight into the creative process. In particular, I was very lucky to find Tesla's 1902 testimony about his work on alternating-current motors in the Tesla Museum. In addition, Leland Anderson published Tesla's testimony related to a patent interference case with Reginald Fessenden, as well as Tesla's 1916 testimony in a case with Marconi. Citations to these three testimonies can be found in the notes. Copies of Tesla's U.S. patents can be readily downloaded from Google Patents, but Jim Glenn published a complete set (New York: Barnes & Noble, 1994). If one wishes to examine the legal minutiae of Tesla's patents, one should consult John T. Ratzlaff's *Dr. Nikola Tesla Selected Patent Wrappers from the National Archives*, 4 vols. (Millbrae, CA: Tesla Book Company, 1980).

Over the years, the Tesla Museum has published several volumes of primary-source material for Tesla. Of these, the most valuable is Tesla's *Colorado Springs Notes* (2008), and the museum also produced facsimile editions of earlier notebooks from his work in Strasbourg and at the Edison Machine Works; again, full citations for these notebooks can be found in the notes. Another important primary source is Tesla's autobiography, which first appeared in 1919 but has since been widely reprinted and published on the World Wide Web. The version I used was edited by Ben Johnston (Williston, VT: Hart Brothers, 1982).

Next to manuscript sources and patent testimony, the most valuable resource for my research was *The Tesla Collection: A 23 Volume Full Text Periodical/Newspaper Bibliography* (New York: The Tesla Project, 1998) compiled by Iwona Vujovic, with support and encouragement from Dr. Ljubo Vujovic. As the title suggests, this collection contains thousands of articles related to Tesla, culled from dozens of American newspapers, magazines, and technical journals. It covers the years 1886 to 1920 and, as near as I can tell, is comprehensive; there are very few published items on Tesla that do not turn up in these twenty-three volumes. The availability of this published collection made an enormous difference in my research. Also helpful in terms of searching out Tesla-related materials was *Dr. Nikola Tesla Bibliography*, edited by John T. Ratzlaff and Leland I. Anderson (Palo Alto, CA: Ragusan Press, 1979).

Like any other biographer, I have relied on published books about Tesla for sources and clues. As noted in Chapter 11, the first biography of Tesla appeared in 1894 and was prepared by his editor and friend, Thomas Commerford Martin (*The Invention, Researches, and Writings of Nikola Tesla* [repr., New York: Barnes & Noble, 1995]). This first biography includes Tesla's famous lectures on the AC motor and early wireless work. In 1943, John O'Neill released a second biography, *Prodigal Genius*; O'Neill was the science editor for the *New York Tribune* and a personal friend of Tesla. O'Neill chose to portray Tesla as a Nietzschean superman with incredible mystical powers. While useful, *Prodigal Genius* is exasperating because, like many books on Tesla, it lacks footnotes. Margaret Cheney wrote two biographies, *Tesla: Man Out of Time* (1986) and with Robert Uth, *Tesla: Master of Lightning* (1999); both are good popular treatments of Tesla, and the latter is richly illustrated. In 1996, Marc J. Seifer brought out *Wizard: The Life and Times of Nikola Tesla* (New York: Birch Lane Press). Seifer uncovered new Tesla letters in various archives and helpfully published many of them verbatim in his text. In *Wizard*, Seifer concentrates on dramatically reconstructing Tesla's relationships with prominent individuals such as Robert and Katharine Johnson, Mark Twain, and John Jacob Astor, J. P. Morgan. Dan Mrkich's *Nikola Tesla: The European Years* (Ottawa:

Commoners' Publishing, 2004) provides fresh information about Tesla's formative years but needs to be used carefully because it lacks footnotes. Finally, one should be warned that there are a host of additional popular Tesla biographies and novels which are frequently based on O'Neill's biography and which continue to propagate many misleading ideas about Tesla.

ARCHIVAL COLLECTIONS

American Philosophical Society, Philadelphia, PA
    Elihu Thomson Papers
Bakken Museum, Minneapolis, MN
    Miscellaneous manuscript collection
Columbia University Library, Rare Books and Manuscript Library, New York, NY
    Nikola Tesla Papers
Deutsches Museum, Munich, Germany
    Jonathan Zenneck Papers
Federal Bureau of Investigation, Washington, DC
    Freedom of Information Act file on Nikola Tesla, available at
        http://scribid.com
Harvard Business School, Baker Business Library, Boston, MA
    R. G. Dun & Co. Collection
    Henry Villard Papers
Houghton Library, Harvard University, Cambridge, MA
    Corinne Roosevelt Robinson Papers
    Henry Villard Papers
Henry Ford Museum, Dearborn, MI
    Westinghouse Motor Collection
Benson Ford Research Center, Henry Ford Museum, Dearborn, MI
    Tesla to George S. Viereck, 17 Dec. 1934
    Trade Catalog Collection
Institute of Electrical and Electronic Engineers, Archives, Piscataway, NJ
    Biographical Files
Institution of Electrical Engineers, Archives, London, UK
    A. P. Trotter Reminiscences
Library of Congress, Manuscript Division, Washington, DC
    Elmer Gertz Papers
    Richmond P. Hobson Papers
    Nikola Tesla Microfilm

National Museum of American History, Archives Center, Smithsonian Institution, Washington, DC:
  George Clark Radioana Collection
  Lloyd Espenshied Papers
  Kenneth Swezey Papers
National Museum of American History, Dibner Library, Smithsonian Institution, Washington, DC
  Manuscript Collections
New-York Historical Society, New York, NY
  Edward P. Mitchell Papers
  Naval History Collection
New York Public Library, Manuscripts and Archives Division, New York, NY
  The Century Magazine Collection
  Personal Miscellaneous Collection
Nikola Tesla Museum, Belgrade Serbia
  Library of Nikola Tesla
  Scrapbooks of Clippings
Oxford University, Bodelian Library, Oxford, UK
  Marconi Archives
Science Museum, London, UK
  Tesla Motor, ca. 1888
University of Virginia, Special Collections Library, Charlottesville, VA
  Clifton Barrett Library
Western Pennsylvania Historical Society, Pittsburgh, PA
  Leland Anderson Papers [Anderson Collection]

ABBREVIATIONS AND SOURCES

1892 Lecture	Tesla, Nikola. "Experiments with Alternate Currents of High Potential and High Frequency," in Thomas Commerford Martin (TCM)
1904 Essay	Tesla, Nikola. "Transmission of Electric Energy Without Wires," *Electrical World and Engineer,* 5 March 1904, pp. 429–31 in *The Tesla Collection* (TC) 16:166–68.
1915 Autobiographical Sketch	Tesla, Nikola. "An Autobiographical Sketch," *Scientific American,* 5 June 1915, 537 and 576–77. Reprinted in *Nikola Tesla: Lectures, Patents, Articles* (Beograd: Nikola Tesla Museum, 1956), pp. A-195–99.
Edison Medal Speech	Tesla, Nikola. Speech upon Receiving Edison Medal, 1917. Box 6, fol. 21, Kenneth Swezey Papers (KSP).
Fessenden Interference	Nikola Tesla, Testimony in Patent Interference 21,701, *NT vs. Fessenden,* in Leland I. Anderson, ed., *Nikola Tesla: Guided Weapons & Computer Technology,* (Breckenridge, Colo: Twenty-First Century Books, 1998). Martin, *The Inventions, Researches, and Writings of Nikola Tesla* (New York: The Electrical Engineer, 1894; reprinted Barnes & Noble, 1995), 198–293.
Mrkich, "NT Father."	Mrkich, Dan. "Nikola Tesla's Father—Milutin Tesla (1819–1879)." American Srbobran. Pittsburgh, March 2001. http://www.serbnatlfed.org/Archives/Tesla/tesla-father.htm.
EDA	Edward Dean Adams.

GW	George Westinghouse.
JJA	John Jacob Astor.
JPM	J. P. Morgan.
KJ	Katharine Johnson.
KSP	Kenneth Swezey Papers, Archives Center, National Museum of American History, Smithsonian Institution, Washington, D.C.
LC	Library of Congress.
NT	Nikola Tesla.
NT, CSN	Tesla, Nikola. *From Colorado Springs to Long Island: Research Notes, Colorado Springs, 1899–1900, New York, 1900–1901* (Beograd: Nikola Tesla Museum [NTM], 2008).
NT, Motor Testimony	Tesla, Nikola. Testimony in *Complaint's Record on Final Hearing,* Vol. 1: Testimony, *Westinghouse vs. Mutual Life Insurance Co. and H. C. Mandeville* [1903], Item NT 77, Nikola Tesla Museum, Belgrade, Serbia.
NT, *My Inventions*	Tesla, Nikola. *My Inventions: The Autobiography of Nikola Tesla,* ed. B. Johnston (Williston, Vt.: Hart Brothers, 1982).
NT, Problem of Increasing Human Energy	Tesla, Nikola. "The Problem of Increasing Human Energy," *The Century Magazine* (June 1900), pp. 175-211 in *The Tesla Collection* [TC] 15:19–55. Also available at http://www.pbs.org/tesla/res/res_art09.html.
NT, Radio Testimony	Tesla, Nikola. *Nikola Tesla on His Work with Alternating Currents and their Application to wireless Telegraphy, Telephony, and Transmission of Power: An Extended Interview,* ed. Leland Anderson (Breckenridge, Colo.: Twenty First Century Books, 2002).
NTM	Nikola Tesla Museum, Beograd, Serbia.
NY Herald, 1893	"Scientists Honor Nikola Tesla," *New York Herald,* 23 April 1893, in *The Tesla Collection* (TC), 6:91–93.
RUJ	Robert Underwood Johnson.
TAE	Thomas A. Edison.
TC	*The Tesla Collection: A 23-Volume Full-Text Periodical/Newspaper Bibliography*, Iwona Vujovic, comp., (New York: Tesla Project, 1998).
TCM	Thomas Commerford Martin.
Tesla biography, 1890	Thomas Commerford Martin [TCM] "Electrical World Portraits. XII. Nikola Tesla." *Electrical World* 15:106 (15 Feb. 1890) in *The Tesla Collection* [TC] 2:42.

NOTES

A list of abbreviations used in the notes can be found on pages 421–22.

INTRODUCTION

1. Oliver Carlson, *Brisbane: A Candid Biography* (New York: Stackpole Sons, 1937).

2. Arthur Brisbane, "Our Foremost Electrician," *New York World,* 22 July 1894 in TC 9:44–48. Unless otherwise stated, all quotes in this section are from the article.

3. Quoted in Lately Thomas, *Delmonico's: A Century of Splendor* (Boston: Houghton Mifflin, 1967), 244.

4. Laurence A. Hawkins, "Nikola Tesla, His Work, and Unfulfilled Promises," *Electrical Engineer* 30:99–108 in TC 16:111–20, on 114.

5. Ibid., 99.

6. Tesla's popularity is illustrated by the existence of the Tesla Memorial Society of New York. From 1984 to 1999, there was a second organization, the International Tesla Society, that held annual conferences and published proceedings. There are also two hobbyist groups dedicated to experimenting with his inventions, the Tesla Coil Builders Association and the Tesla Engine Builders Association, as well as dozens of websites devoted to Tesla.

7. Nick Francesco, "Who Is Nikola Tesla?" Nick's Personal Web Site, http://nickf .com/tesla.php.

8. Thomas P. Hughes, *Networks of Power: Electrification in Western Society, 1880–1930* (Baltimore: Johns Hopkins University Press, 1982); David E. Nye, *Electrifying America: Social Meanings of a New Technology* (Cambridge, MA: MIT Press, 1990); Harold C. Passer, *The Electrical Manufacturers, 1875–1900* (Cambridge, MA: Harvard University Press, 1953); Jill Jonnes, *Empires of Light: Edison, Tesla, Westinghouse, and the Race to Electrify the World* (New York: Random House, 2003).

9. This theme was developed by John J. O'Neill in *Prodigal Genius: The Life of Nikola Tesla* (New York: Ives Washburn, 1944).

10. TCM, *The Inventions, Researches, and Writings of Nikola Tesla* (New York: The Electrical Engineer, 1894; repr., Barnes & Noble, 1995); Slavko *Boksan, Nikola Tesla und sein Werk*

(Vienna: Deutscher Verlag für Jugend und Volk, 1932); Inez Hunt and Wanetta Draper, *Lightning in His Hand: The Life Story of Nikola Tesla* (Hawthorne, CA: Omni Publications, 1964); Margaret Cheney, *Tesla: Man Out of Time* (New York: Prentice-Hall, 1981); Marc J. Seifer, *Wizard: The Life and Times of Nikola Tesla* (New York: Birch Lane Press, 1996); and Margaret Cheney and Robert Uth, *Tesla: Master of Lightning* (New York: Barnes & Noble, 1999).

11. W. Bernard Carlson, "Invention, History, and Culture," in *Science, Technology, and Society*, ed. S. Restivo (New York: Oxford University Press, 2005); W. Bernard Carlson, *Innovation as a Social Process: Elihu Thomson and the Rise of General Electric, 1870–1900* (New York: Cambridge University Press, 1991); and W. Bernard Carlson and Michael E. Gorman, "Understanding Invention as a Cognitive Process: The Case of Thomas Edison and Early Motion Pictures, 1888–1891," *Social Studies of Science* 20 (August 1990): 387–430.

12. In asserting that nature or an invention is not simply "out there" waiting for an inventor to come along and discover it, I am drawing on ideas from the sociology of scientific knowledge. See Bruno Latour, *Science in Action: How to Follow Scientists and Engineers through Society* (Cambridge, MA: Harvard University Press, 1987) and Harry Collins and Trevor Pinch, *The Golem: What Everyone Should Know about Science* (New York: Cambridge University Press, 1993).

13. W. Bernard Carlson, "The Telephone as a Political Instrument: Gardiner Hubbard and the Political Construction of the Telephone, 1875–1880," in *Technologies of Power: Essays in Honor of Thomas Parke Hughes and Agatha Chipley Hughes*, ed. M. Allen and G. Hecht (Cambridge, MA: MIT Press, 2001), 25–55; Claude S. Fischer, *America Calling: A Social History of the Telephone to 1940* (Berkeley: University of California Press, 1992); Wiebe E. Bijker et al., eds., *The Social Construction of Technological Systems* (Cambridge, MA: MIT Press, 1987).

14. Thomas P. Hughes, *American Genesis: A Century of Invention and Technological Enthusiasm, 1870–1970* (New York: Viking, 1989), 53–95.

15. Plato, *The Republic*, trans. Desmond Lee, 2nd ed. (New York: Penguin, 1974), 316–25.

16. "Tesla—Inspiration," Notecard, KSP.

17. NT, Edison Medal Speech. This long quote has been broken into two paragraphs to ease reading.

18. Bishop Kallistos Ware, *The Orthodox Way* (Crestwood, NY: St. Vladimir's Seminary Press, 1995), 32–33.

19. Geoffrey N. Cantor, *Michael Faraday, Sandemanian and Scientist: A Study of Science and Religion in the Nineteenth Century* (Basingstoke, Hampshire: Macmillan, 1993); Colin Russell, *Michael Faraday: Physics and Faith* (New York: Oxford University Press, 2000).

20. Joseph A. Schumpeter, "The Meaning of Rationality in the Social Sciences," in *The Economics and Sociology of Capitalism,* ed. Richard Swedberg (Princeton: Princeton University Press, 1991), 316–38.

21. Michael E. Gorman et al., "Alexander Graham Bell, Elisha Gray, and the Speaking Telegraph: A Cognitive Comparison," *History of Technology* 15 (1993): 1–56; W. Bernard Carlson, "Invention and Evolution: The Case of Edison's Sketches of the Telephone," in *Technological Innovation as an Evolutionary Process*, ed. J. Ziman (New York: Cambridge University Press, 2000), 137–58.

22. As an example, consider how David Bowie played Tesla in the movie *The Prestige* (2006). Bowie did not completely disguise himself as Tesla but rather selected a few key traits that would convey to the audience that he was Tesla. My thinking here has also been shaped by Kenneth Silverman, *Houdini!!! The Career of Ehrich Weiss* (New York: HarperCollins, 1996).

23. Arthur C. Clarke, *Profiles of the Future: An Inquiry into the Limits of the Possible*, rev. ed. (New York: Harper & Row, 1973), 21.

24. David Lindsay, *Madness in the Making: The Triumphant Rise and Untimely Fall of America's Show Inventors* (New York: Kodansha, 1997).

CHAPTER ONE
AN IDEAL CHILDHOOD

1. Schumpeter, "Rationality in the Social Sciences."

2. Nikola Tesla, "My Inventions," *Electrical Experimenter*, May–October 1919; reprinted as NT, *My Inventions*. The page numbers used in the notes refer to the 1982 Johnson edition.

3. Tim Judah, *The Serbs: History, Myth, and the Destruction of Yugoslavia* (New Haven: Yale University Press, 1997), 5.

4. http://en.wikipedia.org/wiki/Lika.

5. http://en.wikipedia.org/wiki/Military Frontier.

6. *NY Herald*, 1893, 92. See also Notecard on Kosanović's criticism of O'Neill's mss., KSP.

7. http://en.wikipedia.org/wiki/Illyrian provinces.

8. *NY Herald*, 1893.

9. Cheney and Uth, *Master of Lightning*, 5.

10. Mrkich, "NT Father." See also [Dan] Mrkich, *Nikola Tesla: The European Years* (Ottawa: Commoners' Publishing, 2004), 52–53.

11. The French established twenty-five *gymnasia* in the Illyrian provinces; see http://en.wikipedia.org/wiki/Illyrian_provinces.

12. Mrkich, "NT Father."

13. Mrkich, *Tesla: The European Years*, 53. This church was burned down in 1941 and restored in the 1980s, only to be destroyed again during the fighting in 1992 (Mrkich, "NT Father"). It was rebuilt by the Croatian government in 2006.

14. NT, *My Inventions*, 28; Mrkich, *Tesla: The European Years*, 62.

15. NT, "A Story of Youth Told by Age," http://www.pbs.org/tesla/ll/story_youth.html.

16. After Tesla's death, this book passed to his nephew, Sava Kosanović, who was Yugoslavia's ambassador to the United States. In 1950, Kosanović presented this rare volume to President Harry Truman and it now resides in the Truman Presidential Library in Independence, Missouri. See George C. Jerkovich, "An Unusual Treasure: Library's Serbian Book of Liturgy Found to Be a Rarity," *Whistlestop: Harry S. Truman Library Newsletter* 5, no. 4 (Fall 1977) and Mrkich, *Tesla: The European Years*, 67.

17. Mrkich, *Tesla: The European Years*, 54.

18. Then as now, the Serbian Orthodox Church follows the Julian calendar known as Old Style; in the Western or Gregorian calendar, Tesla would have been born on 20–21 July 1856. See *Nikola Tesla: Lecture before the New York Academy of Sciences, April 6, 1897*, ed. Leland I. Anderson (Breckenridge, CO: Twenty-First Century Books, 1994), ix.

19. Mrkich, *Tesla: The European Years*, 55–56.

20. NT, *My Inventions*, 46.

21. NT, "Story of Youth Told by Age."

22. NT, *My Inventions*, 30; *NY Herald*, 1893.

23. NT, *My Inventions*, 45.

24. Ibid., 46.

25. Ibid., 36, 31–32.

26. Ibid., 28.

27. During the Croatian War of Independence in the 1990s, Gospić suffered greatly. The town was held by rebel Croatian government forces while the Serb forces of the Republic of Serbian Krajina occupied positions directly to the east and often bombarded the town from there. During the U.S. Operation Storm, control of the area finally shifted to the rebel Croatian government in August 1995. See http://en.wikipedia.org/wiki/Gospić.

28. Mrkich, "NT Father"; NT, *My Inventions*, 30.

29. NT, *My Inventions*, 46–47.

30. Ibid., 28; Robert V. Bruce, *Bell: Alexander Graham Bell and the Conquest of Solitude* (Boston: Little, Brown, 1973), 66–69.

31. NT, *My Inventions*, 30.

32. Ibid., 36.

33. Ibid., 47.

34. Ibid., 36.

35. Albert Tezla, *Hungarian Authors: A Bibliographical Handbook* (Cambridge, MA: Belknap Press of Harvard University Press, 1970), 256–62; "Hungarian Novels," *Foreign Monthly Review and Continental Literary Journal* (1839): 203, viewed on GoogleBooks; *A History of Hungarian Literature: From the Earliest Times to the mid-1970's*, chapter 10, "Social Criticism and the Novel in the Age of Reform," http://mek.niif.hu/02000/02042/html/23.html.

36. NT, *My Inventions*, 36–37.

37. Ibid., 34, 32–33.

38. M. K. Wisehart, "Making Your Imagination Work for You," *American Magazine* 91 (April 1921): 13ff., on 60.

39. NT, "Problem of Increasing Human Energy," 184. Tesla put forward his theory that humans are "meat machines" in conversation with his biographer John O'Neill; see *Prodigal Genius*, 261–62.

40. NT, *My Inventions*, 47–48.

41. Ibid., 48.

42. Ibid., 51.

43. Along with the fire engine, one wonders if Tesla had also read about the atmospheric steam engine invented by Thomas Newcomen in 1712 since this engine also used both a vacuum and air pressure to create motion.

44. O'Neill, *Prodigal Genius*, 26.

45. NT, *My Inventions*, 51–52.

46. Ibid., 52.

47. See NT, *My Inventions*, 53. Elsewhere Tesla suggested that he read Twain's *Tom Sawyer* while he was recovering from cholera in 1873; see NT to Watson Davis, 27 March 1938, Notecard, KSP.

48. *NY Herald*, 1893; NT, *My Inventions*, 53–54.

49. NT, Edison Medal Speech, 8.

50. NT, *My Inventions*, 54.

51. Mrkich, *Tesla: The European Years*, 73–74; O'Neill, *Prodigal Genius*, 36; NT, *My Inventions*, 55.

52. NT, *My Inventions*, 33; Wisehart, "Making Your Imagination Work for You," 60.

53. Ware, *The Orthodox Way*, 32–33.

54. Mrkich, *Tesla: The European Years*, 8–9. The document cited by Mrkich is dated 22 September 1876 but Tesla started his studies at Joanneum in the fall of 1875, so this document may be for the second year of Tesla's scholarship.

55. *Nikola Tesla Museum, 1952–2003* (Belgrade: NTM, 2006), 108.

CHAPTER TWO
DREAMING OF MOTORS

1. Josef W. Wohinz, Hg., *Die Technik in Graz: Aus Tradition für Innovation* (Wien: Böhlau,1999).

2. Tesla biography, 1890.

3. Copy of Tesla's course transcript, Box 7, Folder 13, KSP.

4. 1915 Autobiographical Sketch.

5. Michael Brian Schiffer, *Draw the Lightning Down: Benjamin Franklin and Electrical Technology in the Age of Enlightenment* (Berkeley: University of California Press, 2003).

6. Michael Brian Schiffer, *Power Struggles: Scientific Authority, and the Creation of Practical Electricity before Edison* (Cambridge, MA: MIT Press, 2008), 49–74.

7. According to Tesla's transcript (cited in note 3), he took two courses with Pöschl in 1876–77.

8. During the nineteenth century, inventors developed two forms of electric illumination, arc lighting and incandescent lighting. In an arc light, a strong current is passed through two carbon rods; as the rods are drawn slightly apart, a spark jumps the gap between the rods and provides a brilliant light. See Carlson, *Innovation as a Social Process*, 80–82.

9. "The First Transmission of Power by Electricity," *Electrical World* 6 (12 December 1885): 239–40; Silvanus P. Thompson, *Dynamo-Electric Machinery*, 3rd ed. (London: E&FN Spon, 1888), 13.

10. 1915 Autobiographical Sketch, 537.

11. NT, *My Inventions*, 57.

12. Ibid.; *NY Herald*, 1893.

13. Tesla biography, 1890; "The Westinghouse Company Secures the Tesla Motor," *Electrical Review* (NY), 11 August 1888 in TC 1:83.

14. NT, *My Inventions*, 59.

15. Carlson, *Innovation as a Social Process*, 88–95.

16. *NY Herald*, 1893, 92.

17. NT, *My Inventions*, 56.

18. Mrkich, *Tesla: The European Years*, 10; NT, *My Inventions*, 37; O'Neill, *Prodigal Genius*, 43.

19. Mrkich, *Tesla: The European Years*, 10–11. Mrkich notes that, hoping to extricate himself from the obligations of his military scholarship, Tesla first wrote the *Queen Bee* in October 1876 to ask for a scholarship.

20. Mrkich, *Tesla: The European Years*, 16; O'Neill, *Prodigal Genius*, 44.

21. NT, *My Inventions*, 37; Mrkich, *Tesla: The European Years,* 17.

22. Tesla's police record from Maribor, March 1879, in Mrkich, *Tesla: The European Years*, 18.

23. Mrkich, *Tesla: The European Years*, 76.

24. NT, *My Inventions*, 37.

25. Mrkich, *Tesla: The European Years*, 92, 98.

26. Ibid., 77; Daniel Mayer, "Nikola Tesla in Prague in 1880—Some Details from Tesla's Life, Until Now Unpublished," *Tesla III Millennium: Fifth Annual Conference Proceedings* (Beograd, 1996), VI67–VI69; Seifer, *Wizard*, 19.

27. 1915 Autobiographical Sketch, 537.

28. NT, *My Inventions*, 59.

29. 1915 Autobiographical Sketch, 537.

30. Tesla biography, 1890.

31. Mrkich, *Tesla: The European Years*, 100.

32. For biographical information on Tivadar Puskás, see http://www.budpocket guide.com/TouristInfo/famous/Famous_Hungarians10.asp.

33. Edward Johnson to Uriah Painter, 17 December 1877, *The Papers of Thomas A. Edison*, ed. R. A. Rosenberg et al. (Baltimore: Johns Hopkins University Press), 3:676–79; Paul Israel, *Edison: A Life of Invention* (New York: John W. Wiley, 1998), 148–49.

34. Puskás was not the inventor of the telephone exchange but one of several entrepreneurs to see its potential. The first telephone exchange was established in May 1877 by E. T. Holmes in Boston when he added telephones to his existing burglar alarm network. See Carlson, "The Telephone as a Political Instrument," 25–55, on 42.

35. Quotes are from NT, *My Inventions*, 59 and Tesla biography, 1890. It is difficult to establish the exact sequence of events in Budapest, as Tesla offered several versions in his recollections. The sequence followed here comes from the Tesla biography, 1890 and 1915 Autobiographical Sketch, which is that Tesla first worked at the telegraph office and then had his Eureka moment. Alternatively, *My Inventions* (59, 65) suggests that Tesla fell ill waiting for the position with Puskás's telephone company and only went to work as a draftsman in the government telegraph office after he had his Eureka moment in the park.

36. Tesla biography, 1890; NT, Motor Testimony, 191.

37. NT, *My Inventions*, 59.

38. 1915 Autobiographical Sketch, 537; NT, Motor Testimony, 321.

39. NT, *My Inventions*, 60–61.

40. In his 1903 patent testimony, Tesla made no mention of drawing diagrams in the sand for Szigeti in Budapest in 1882. However, he did recall that he drew diagrams in the dirt when he explained his motor to several colleagues at the Edison Lamp Works outside Paris around 1883.

41. NT, *My Inventions*, 61.

42. Szigeti testified in 1889 that Tesla told him about a polyphase motor in Paris in May 1882, but he said nothing about a Eureka moment in Budapest. See Szigeti, 1889 deposition.

43. To be sure, there is a strong resemblance between the copper disk in Arago's device and the disk rotors used by Tesla in his first motors, and this resemblance led one textbook author to conclude that "Tesla's motor is a form of Arago's experiment . . . in which a revolving magnet draws round after it a conducting disc." See G. C. Foster and E. Atkinson, *Elementary Treatise on Electricity and Magnetism* (London: Longman, Green, 1896), 490.

44. Silvanus P. Thompson, *Polyphase Electric Currents and Alternate-Current Motors* (Spon, 1895), 422–25.

45. Ibid., 447.

46. Ibid., 437–40.

47. 1915 Autobiographical Sketch, 576.

48. Indeed, as far as I can tell, the concept of alternating currents being in or out of phase with each other did not appear in the engineering literature until an 1883 paper by John Hopkinson in which he discussed the problem of using several ac generators on a single circuit; see "Some Points in Electric Lighting," in *Original Papers by the Late John Hopkinson*, ed. Bertram Hopkinson (Cambridge: Cambridge University Press, 1901), 1:57–83, on 67–69.

49. 1915 Autobiographical Sketch, A198.

50. Clayton M. Christensen, *The Innovator's Dilemma: When New Technologies Cause Great Firms to Fail* (Boston: Harvard Business School Press, 1997).

51. Thomas K. McCraw, *Prophet of Innovation: Joseph Schumpeter and Creative Destruction* (Cambridge, MA: Belknap Press of Harvard University Press, 2007).

52. Schumpeter, "Rationality in the Social Sciences," 329–30.

CHAPTER THREE
LEARNING BY DOING

1. NT, *My Inventions*, 65.

2. See Osana Mario, "Historische Betrachtungen uber Teslas Erfindungen des Mehrphasenmotors und der Radiotechnick um die Jahrhundertwende," in *Nikola Tesla-Kongress für Wechsel- und Drehstromtechnik*, proceedings of a conference held at the Technical Museum in Vienna, 6–13 September 1953 (Vienna: Springer-Verlag, 1953), 6–9, on 7. Mario heard the story from his professor at the Vienna University of Technology, Johann Sahulka, who in turn had learned the story when he met Tesla at the 1893 Chicago World's Fair. Neither Mario nor Sahulka could determine whether Tesla worked at Ganz or was just a visitor.

3. "Foundry Museum, Budapest," http://sulinet.hu/oroksegtar/data/kulturalis_ertekek_a_vilagban/Visegradi_orszagok_technikai_2/pages/angol/003_mo_muszaki_muem_ii.htm; "Ganz Works," http://www.omikk.bme.hu/archivum/angol/htm/ganz.htm.

4. Carlson, *Innovation as a Social Process*, 88–91.

5. Tesla later explained that the two different currents would have been the result of the two coils having different inductances, but it is doubtful that in 1882 Tesla fully understood the concept of inductance. What is important here is that he recognized that somehow two different alternating currents were producing a rotating magnetic field. See Mario, "Historische Betrachtungen," 7.

6. In his patent testimony, Tesla never mentioned this story about the ring transformer at Ganz because, in all likelihood, he was anxious not to suggest to his opponents that he had learned anything about AC from the engineers at Ganz, lest it then be suggested that he had simply appropriated the idea of using AC to create a rotating magnetic field from them. In my view—thanks to his ideal of a rotating magnetic field—Tesla was probably one of the few people who could have looked at the ball spinning on top of the ring transformer and envisioned a practical motor.

7. "1890 Biographical Sketch."

8. NT, Motor Testimony, 186.

9. Walter L. Welch, *Charles Batchelor: Edison's Chief Partner* (Syracuse: Syracuse University Press, 1972), 50.

10. NT, Motor Testimony, 195.

11. Paul Israel, *From Machine Shop to Industrial Laboratory: Telegraphy and the Changing Context of American Invention, 1830-1920* (Baltimore: Johns Hopkins University Press).

12. NT, Motor Testimony, 187-88, 195, 306.

13. Szigeti, 1889 deposition, A398.

14. Tesla described his invention to David F. Cunningham, Milton F. Adams, Charles M. Hennis, and James F. Hipple; see NT, Motor Testimony, 189-90, 274-75. Szigeti testified that Tesla described a similar eight-wire generator-motor system in Paris in 1882. See Szigeti, 1889 deposition, A398-A401.

15. Frederick Dalzell, *Engineering Invention: Frank J. Sprague and the U.S. Electrical Industry* (Cambridge, MA: MIT Press, 2009).

16. This company was probably not organized because Cunningham was only a minor figure in the French Edison companies and hardly in a position to raise capital. See NT, *My Inventions*, 66; NT, Motor Testimony, 274-75; Seifer, *Wizard*, 29; Francis Jehl, *Menlo Park Reminiscences* (Dearborn, MI: Edison Institute, 1938), 2:680, 682.

17. NT, *My Inventions*, 66-67.

18. Julius Euting, *A Descriptive Guide to the City of Strassburg and Its Cathedral*, 7th ed. (Strassburg: Karl J. Trübner, n.d.), 84-85.

19. Wilhelm I visited Strasbourg in September 1879 and his biographies do not mention an explosion. See Paul Wiegler, *William the First: His Life and Times*, trans. C. Vesey (Boston: Houghton-Mifflin, 1929), 377 and Edouard Simon, *The Emperor William and His Reign* (London: Remington, 1888), 2:189.

20. NT, Motor Testimony, 185-86; NT, *My Inventions*, 67. For the notebook Tesla kept while in Strasbourg, see NT, *Tagebuch Aus Strasburg, 1883-1884* (Beograd: NTM, 2002); this notebook (pp. 249-50) shows that Szigeti was on the payroll at Strasbourg from October 1883 to February 1884.

21. Alfred Ritter von Urbanitzky, *Electricity in the Service of Man*, ed. R. Wormell (London: Cassell & Co., 1886), 548-51.

22. NT, Motor Testimony, 188; von Urbanitzky, *Electricity in the Service of Man*, 296-99; Thompson, *Dynamo-Electric Machinery*, 267-68.

23. Szigeti, 1889 deposition, A400.

24. NT, Motor Testimony, 181, 192.

25. Szigeti, 1889 deposition, A400.

26. NT, Motor Testimony, 220, 184; NT, "Electric Magnetic Motor," U.S. Patent 424,036 (filed 20 May 1889, granted 25 March 1890), especially figure 3; TCM, *Inventions, Researches, and Writings*, 69.

27. NT, Motor Testimony, 182; Benjamin Silliman, *Principles of Physics, or Natural Philosophy*, 2nd ed. (Philadelphia: Theodore Bliss, 1863), 608.

28. NT, *My Inventions*, 67; NT, Motor Testimony, 177-82, 284; Szigeti, 1889 deposition, A400.

29. NT, Edison Medal Speech.

30. NT, Motor Testimony, 190; NT, *My Inventions*, 67-68.

31. NT, Motor Testimony, 186. In *Prodigal Genius*, p. 60, O'Neill attributes this letter of introduction to Batchelor, but Batchelor was already back in New York. Instead Barbara Puskás reported to me that she had seen a letter like this written by Tivadar Puskás in her family's archives.

32. Notes in Box 1, KSP; Walter Chambers, "Tesla Too Busy to Be Honored at Radio Show," 25 September 1929, p. 26, KSP.

33. NT, *My Inventions*, 71.

34. On the *Oregon*, see http://en.wikipedia.org/wiki/SS_Oregon_%281883%29.

35. NT, *My Inventions*, 71–72; NT, *Notebook from the Edison Machine Works* (Belgrade: NTM, 2003), 12.

36. NT, Motor Testimony, 186, 195; NT, *My Inventions*, 72; Alfred O. Tate, *Edison's Open Door* (New York: E. P. Dutton, 1938), 149.

37. See NT, Motor Testimony, 195; 1915 Autobiographical Sketch, A199.

38. TAE, "Arc Lamp," U.S. Patent 438,303 (filed 10 June 1884; granted 14 October 1890); NT, Motor Testimony, 193.

39. Carlson, *Innovation as a Social Process*, 211–16.

40. Israel, *Edison*, 221–25; Passer, *The Electrical Manufacturers*, 34–38; "Arc and Incandescent Interests Combined," *Electrical World* 5 (9 May 1885): 188. Not aware of Edison's dealings with AEM, Tesla thought that his arc-lighting system was abandoned because of a deal Edison struck with the Edison Illuminating Company; see NT, Motor Testimony, 193.

41. "The Edison System of Municipal Lighting," *Electrical World* 9 (12 February 1887): 78.

42. NT, Motor Testimony, 193. In *My Inventions* (p. 72), Tesla said he quit when he was not paid $50,000 he thought he had been promised for redesigning the dynamos. See also NT, *Notebook from the Edison Machine Works*, 248.

43. William Edgar Sackett, ed., *New Jersey's First Citizens* (Paterson, NJ: J. J. Scannell, 1917), 1:507.

44. Entry for Tesla Electric Light and Mfg. Co., New Jersey, vol. 53, p. 159, R. G. Dun & Co. Collection, Baker Business Library, Harvard University (hereafter cited as Tesla Co., R. G. Dun & Co. Collection).

45. TCM, "The Electric Light Industry in America in 1887," *Electrical World* 9 (29 January 1887): 50.

46. Tesla's arc-lighting patents include 334,823; 335,786; 335,787; 336,961; 336,962; 350,954; and 359,748. They are conveniently summarized in TCM, *Inventions, Researches, and Writings*, 451–64.

47. Tesla Co., R. G. Dun & Co. Collection.

48. NT, Edison Medal Speech, 12.

49. Seifer, *Wizard*, 40–41; NT, Motor Testimony, 193–95, 209; Tesla Electric Light and Manufacturing Company, advertisement, *Electrical Review*, 4 September 1886, p. 14.

50. Vail and Lane were not especially effective running a utility; by 1890 they were no longer in business. See Tesla Co., R. G. Dun & Co. Collection.

51. NT, *My Inventions*, 72.

52. O'Neill, *Prodigal Genius*, 65; NT to Institute of Immigrant Welfare, 12 May 1938 in John T. Ratzlaff, comp., *Tesla Said* (Millbrae, CA: Tesla Book Co., 1984), 280.

CHAPTER FOUR
MASTERING ALTERNATING CURRENT

1. Israel, *Edison*, 234.

2. NT, "Thermo-Magnetic Motor," U.S. Patent 396,121 (filed 30 March 1886, granted 15 January 1889); TCM, *Inventions, Researches, and Writings*, 424–28.

3. *Century Electric Company and Edwin S. Pillsbury vs. Wagner Electric Manufacturing Co.*, US Circuit Court of Appeals, 8th District, No. 3419, May 1910, transcript, vol. 2, p. 932 [Item 342, NTM]; O'Neill, *Prodigal Genius*, 65–66; "Alfred S. Brown," [obituary], *New York Times*, 26 September 1891.

4. John B. Taltavall, comp., *Telegraphers of To-Day: Descriptive, Historical, Biographical* (New York: John B. Taltavall, 1893), 19–20.

5. Norvin Green to A. S. Hewitt, 1 September 1887, Western Union Collection, Series 4, Box 204A, National Museum of American History Archives, Washington, DC. I am grateful to Joshua Wolff for sharing this item with me. See also Arthur J. Beckhard, *Electrical Genius Nikola Tesla* (New York: Julian Messner, 1959), 120, 125.

6. See the following documents from *The Papers of Thomas A. Edison*, ed. Robert A. Rosenberg et al. (Baltimore: Johns Hopkins University Press, 1991–): "Multiple Telegraph" [notebook entry], Summer 1873, 2:50–52, esp. note 2; "Automatic Telegraphy," December 1873, 2:119–20; "Article in the *Operator*," 15 July 1874, 2:239–41; and "Speaking Tel[e]g[rap]h," 20 March 1877, 3:271–74, esp. note 1.

7. "William Orton," *Chicago Daily Tribune*, 26 April 1878, p. 2; Seifer, *Wizard*, 42.

8. James D. Reid, *The Telegraph in America and Morse Memorial*, 2nd ed. (New York: John Polhemus, 1886), 601–2; Entries for Mutual Union Telegraph Co., 5 November 1880 and 4 March 1881, New York City, vol. 391, p. 2625, R. G. Dun & Co. Collection.

9. Richard R. John, *Network Nation: Inventing American Telecommunications* (Cambridge, MA: Belknap Press of Harvard University Press, 2010), 149–70, on 158.

10. Reid, *Telegraph in America*, 603–4; Israel, *From Machine Shop to Industrial Laboratory*, 127; Maury Klein, *The Life and Legend of Jay Gould* (Baltimore: Johns Hopkins University Press, 1986), 310–11.

11. Reid, *Telegraph in America*, 604; Entry for Mutual Union Telegraph Co., 27 June 1882, New York City, vol. 391, p. 2625, R. G. Dun & Co. Collection.

12. Reid, *Telegraph in America*, 605; Entry for Mutual Union Telegraph Co., 1885, New York City, vol. 391, p. 2625, R. G. Dun & Co. Collection.

13. NT, Motor Testimony, 196. Given how Tesla respected Peck and Brown, it is highly curious that he did not mention them by name in his autobiography or other recollections. See NT, *My Inventions*, 72 and NT to Institute of Immigrant Welfare, 12 May 1938, in Ratzlaff, *Tesla Said*, 280.

14. NT, Motor Testimony, 196, 210–12, 247, 325–26; Szigeti, 1889 deposition, A398.

15. The address for Mutual Union is from New York City, vol. 391, p. 2625, R. G. Dun & Co. Collection. See also William B. Nellis testimony, NT, Motor Testimony, 122–23, 132.

16. Parker W. Page testimony, NT, Motor Testimony, 415–16. Though the motor was never patented, Tesla did file U.S. Patent 382,845, "Commutator for Dynamo-Electric Machines" (filed 30 April 1887, granted 15 May 1888). See also Martin, *Inventions, Researches, and Writings*, 433–37.

17. NT, Motor Testimony, 212.

18. NT, "Our Future Motive Power," *Everyday Science and Mechanics*, December 1931, pp. 26–28ff., in Ratzlaff, *Tesla Said*, 230–36.

19. TAE, "On the Pyromagnetic Dynamo: A Machine for Producing Electricity Directly from Fuel," *Electrical World* 10 (27 August 1887): 111–13.

20. TCM, *Inventions, Researches, and Writings*, 429–31.

21. NT, Motor Testimony, 207.

22. Ibid., 213.

23. Tesla probably used a DC dynamo that Weston had developed for electroplating that had a rotor with six projecting poles. By replacing the DC commutator with

AC slip rings, Tesla could have drawn two or three separate currents off of this dynamo. See NT, Motor Testimony, 200–201, 210, as well as Thompson, *Dynamo-Electric Machinery*, 280.

24. NT 78, p. 21. "Defendant's Brief, Derivation Electric Motor, in *Westinghouse Electric and Manufacturing Company vs. Dayton Fan and Motor Company*, 1900, Item 78, NTM, Belgrade, Serbia.

25. Louis C. Hunter and Lynwood Bryant, *A History of Industrial Power in the United States* (Cambridge, MA: MIT Press, 1991), 3:210.

26. As Tesla reported years later, Peck and Brown made the connection between the ocean scheme and his motors: "They thought that if the energy could be economically transmitted to distant places by my system and the cost of the ocean plant substantially reduced, this inexhaustible source [of steam] might be successfully exploited" (NT, "Our Future Motive Power," 78). Encouraged by Peck and Brown, Tesla apparently studied the thermodynamics involved and later devised his own system using heat from the Earth's crust rather than the temperature differential in the ocean.

27. "Tesla's Egg of Columbus," *Electrical Experimenter* 6 (March 1919): 774–75ff., on 775.

28. Carlson, *Innovation as a Social Process*, 81–82, 87–108.

29. Hughes, *Networks of Power*, 87–91.

30. Ibid., 95–97.

31. Carlson, *Innovation as a Social Process*, 251–52; Passer, *The Electrical Manufacturers*, 172.

32. "Electric Lighting at the Inventions Exhibition," *Engineering* (London) 39 (1 May 1885): 454–60.

33. Passer, *The Electrical Manufacturers*, 131–32.

34. George Wise, "William Stanley's Search for Immortality," *American Heritage of Invention & Technology* 4 (Spring–Summer 1988): 42–49; Bernard A. Drew and Gerard Chapman, "William Stanley Lighted a Town and Powered an Industry," *Berkshire History* 6 (Fall 1985): 1–40; Laurence A. Hawkins, *William Stanley (1858–1916): His Life and Work* (New York: Newcomen Society, 1951).

35. Carlson, *Innovation as a Social Process*, 259.

36. In January 1887, *Electrical World* observed, "The system of distribution by means of secondary generators [i.e., transformers] has at last obtained a foot-hold in this country, where if all the promises which are made for it are substantiated, it will not require long to establish itself in public favor." See Joseph Wetzler, "The Electrical Progress of the Year," *Electrical World* 9 (1 January 1887): 2–3. A year later, Wetzler wrote, "The prominent feature of the year in electric lighting is the number of systems of distribution by means of induction transformers which have been brought out or elaborated. Prominent among these are the Westinghouse system." See "The Electrical Progress of the Year 1887," *Electrical World* 11 (14 January 1888): 18–19.

37. TAE to Villard, 11 December 1888, LB881112, p. 354, Edison National Historic Site (ENHS), West Orange, NJ.

38. TAE, "Reasons against an Alternating Converter System," N860428, pp. 261–65, ENHS.

39. TAE to Villard, 24 February 1891, Box 63, Folder 475, Villard Papers, Baker Library, Harvard Business School, Boston.

40. Modern studies of Columbus have concluded that although Isabella did need to be convinced to support Columbus, she did not have to pawn her jewels; see John Noble Wilford, *The Mysterious History of Columbus: An Exploration of the Man, the Myth, and*

the Legacy (New York: Alfred A. Knopf, 1992), 93–95 and Miles H. Davidson, *Columbus Then and Now: A Life Reexamined* (Norman: University of Oklahoma Press, 1997), 168–70. In *Admiral of the Ocean Sea: A Life of Christopher Columbus* (Boston: Little, Brown, 1942), Samuel Eliot Morison reported (p. 361) that the story of the egg was first associated with Columbus in an Italian source from 1565 and that the episode took place in 1493 after Columbus returned from his first voyage to the New World and not prior to the journey.

41. "Tesla's Egg of Columbus," 775.

42. This motor is illustrated in two Tesla U.S. patents for an "Electro Magnetic Motor": 381,968 (filed 12 October 1887, granted 1 May 1888) and 382,279 (filed 30 November 1887, granted 1 May 1888). See also NT, Motor Testimony, 215.

43. NT, Motor Testimony, 155, 159–60, 215, 218; NT, U.S. Patent 511,560, "Electrical Transmission of Power" (filed 8 December 1888, granted 2 January 1894).

44. NT, *My Inventions*, 72.

45. Robert C. Post, *Physics, Patents and Politics: A Biography of Charles Grafton Page* (New York: Science History Publications, 1976; Israel, *From Machine Shop to Industrial Laboratory*, 135–36; Parker Page's obituary, *New York Times*, 23 January 1937, 18:4; NT, Motor Testimony, 213.

46. NT, Motor Testimony, 205–6, 289, 307–8, 314–15, 317.

47. NT, "Electric Magnetic Motor," U.S. Patent 381,968, (filed 12 October 1887, granted 1 May 1888).

48. NT, "Electro Magnetic Motor," U.S. Patent 381,969 (filed 30 November 1887, granted 1 May 1888); "Electro Magnetic Motor," U.S. Patent 382,279 (filed 30 November 1887, granted 1 May 1888); "System of Electrical Distribution," U.S. Patent 381,970 (filed 23 December 1887, granted 1 May 1888); "Regulator for Alternate Current Motors," U.S. Patent 390,820 (filed 24 April 1888, granted 9 October 1888); and TCM, *Inventions, Researches, and Writings*, 45–49.

49. The separate patents for power transmission include "Electrical Transmission of Power," U.S. Patent 382,280 (filed 12 October 1887, granted 1 May 1888); "Electrical Transmission of Power," U.S. Patent 382,281 (filed 30 November 1887, granted 1 May 1888); and "Method of Converting and Distributing Electric Currents," U.S. Patent 382,282 (filed 23 December 1887, granted 1 May 1888). With the exception of the claims for the transmission of power, these patents are duplicates of Tesla's polyphase motor patents.

50. NT, Motor Testimony, 160.

51. Carlson, *Innovation as a Social Process*, 87–88.

52. NT, Motor Testimony, 160, 173–74, 210.

53. Ibid., 329.

54. See NT, Motor Testimony, 159, 174–75, 230, 289, 369–72. For a summary of many of these techniques, see NT, Patent 511,560.

55. NT, Motor Testimony, 316–18, 329.

56. The quotes are from NT, Motor Testimony, 308–9, 420. See also NT, Motor Testimony, 164–66, 175, 208, 310, 416–17, 423.

57. These applications included "System of Electric Distribution," U.S. Patent 390,413 (filed 10 April 1888, granted 2 October 1888); "Dynamo Electric Machine," U.S. Patent 390,414 (filed 23 April 1888, granted 2 October 1888); "Regulator for Alternate-Current Motors," U.S. Patent 390,820 (filed 24 April 1888, granted 9 October 1888); and "Dynamo-Electric Machine," U.S. Patent 390,721 (filed 28 April 1888, granted 9 October 1888).

58. NT, Motor Testimony, 416, 418, 426.

59. Carlson, *Innovation as a Social Process*, 251–52. Alexander Graham Bell was also slow to appreciate the commercial potential of the telephone and was guided by his patron and future father-in-law, Gardiner G. Hubbard; see Carlson, "The Telephone as a Political Instrument" and Carlson, "Entrepreneurship in the Early Development of the Telephone: How Did William Orton and Gardiner Hubbard Conceptualize this New Technology?" *Business and Economic History* 23 (Winter 1994): 161–92.

60. Walter Isaacson, *Steve Jobs* (New York: Simon & Schuster, 2011), 110.

CHAPTER FIVE
SELLING THE MOTOR

1. NT, Motor Testimony, 208–9, 314, 322, 426–27; NT, "Electrical Transmission of Power," U.S. Patent 511,915 (filed 15 May 1888, granted 2 January 1894) and "Alternating Motor," U.S. Patent 555,190 (filed 15 May 1888, granted 25 February 1896).

2. Reese V. Jenkins, *Images and Enterprise Technology and the American Photographic Industry, 1839 to 1925* (Baltimore: Johns Hopkins University Press, 1975).

3. William Greenleaf, *Monopoly on Wheels; Henry Ford and the Selden Automobile Patent* (Detroit: Wayne State University Press, 1961).

4. Thomas P. Hughes, *Elmer Sperry: Inventor and Engineer* (Baltimore: Johns Hopkins University Press, 1971), 91–93.

5. Carlson, "The Telephone as Political Instrument, 25–56.

6. *The Scientific American Reference Book* (New York: Munn & Company, 1877), 47–50.

7. Passer, *The Electrical Manufacturers*, 151–64.

8. W. Bernard Carlson, "Nikola Tesla and the Tools of Persuasion: Rethinking the Role of Agency in the History of Technology" (paper presented at Society for the History of Technology, Minneapolis, November 2005).

9. Carlson, *Innovation as a Social Process*, 244, 265.

10. NT, Motor Testimony, 256.

11. "Electrical World Portraits—XI. Prof. W. A. Anthony," *Electrical World* 15 (1 February 1890): 70; NT, Motor Testimony, 214.

12. NT, Motor Testimony, 160, 168–70, 221–22, 247–49, 276–78.

13. W. A. Anthony to D. C. Jackson, 11 March 1888, quoted in Kenneth M. Swezey, "Nikola Tesla," *Science* 127 (16 May 1958): 1147–59, on 1149.

14. William Anthony, "A Study of Alternating Current Generators and Receivers," *Modern Light and Heat*, 24 May 1888, p. 549.

15. NT, Motor Testimony, 252–53; Tesla biography, 1890.

16. NT, "A New System of Alternate Current Motors and Transformers," *AIEE Transactions* 5 (September 1887–October 1888): 307–27 and reprinted in TCM, *Inventions, Researches, and Writings*, 9–25.

17. Discussion of Tesla's paper, *AIEE Transactions* 5 (1887–88): 324–25 in TC 1:23.

18. Elihu Thomson, "Novel Phenomena of Alternating Currents," *Electrical Engineer* 6 (June 1887): 211–15.

19. Discussion of Tesla's paper, *AIEE Transactions* 5 (1887–88): 325–27 in TC 1:24.

20. Remarks of the chairman, *AIEE Transactions* 5 (1887–88): 350 in TC 1:25. A few weeks after the lecture, Tesla was elected an associate member of the AIEE; see "Secretary's Bulletin," June 1888, *AIEE Transactions* 5 (1887–88).

21. NT, Motor Testimony, 328.

22. Ibid., 280.

23. Carlson, *Innovation as a Social Process*, 244; E. Thomson to Charles A. Coffin, 5 May 1888, in LB 4/88–4/89, p. 9, Elihu Thomson Papers, American Philosophical Society, Philadelphia.

24. Henry G. Prout, *A Life of George Westinghouse* (London: Benn Brothers, 1922), 128–29.

25. Thomas B. Kerr testimony, NT, Motor Testimony, 448.

26. "Electro-Dynamic Rotation by Means of Alternating Currents," *Industries*, 18 May 1888, 505–6 and *The Electrician*, 25 May 1888 in TC 1:26–27 and 30–31. Tesla testified that he didn't know anything about Ferraris's work until he read these notices in 1888; see NT, Motor Testimony, 170.

27. Anna Maria Rietto and Sigfrido Leschiutta, "The First Electrical Engineers in Torino," in *Galileo Ferraris and the Conversion of Energy: Developments of Electrical Engineering over a Century* (proceedings of the International Symposium, Turin, October 1997), 407–33.

28. Sigfrido Leschiutta, "The Torino-Lanzo Transmission Experiment," in *Galileo Ferraris and the Conversion of Energy*, 291–305.

29. Adolfo G. B. Hess, "The Monument to Galileo Ferraris in Turin," *Electrical World* 42 (8 August 1903): 215–18.

30. Translation of Galileo Ferraris, "Electro-Dynamic Rotations Produced by Means of Alternate Currents," *Publications of Royal Academy of Sciences of Turin* 23 (1887–88) in "Proofs on Behalf of Ferraris," U.S. Patent Office Interference No. 14,819, Slattery versus Ferraris, Paper No. 53. The copy I consulted was at NTM, catalogued as NT 124. Quote is from p. 22.

31. For example, in his textbook, Anthony J. Pansini credits Ferraris as the inventor of the AC induction motor; see *Basics of Electric Motors: Including Polyphase Induction and Synchronous Motors* (Englewood Cliffs, NJ: Prentice-Hall, 1989), 45. For an extensive discussion in favor of Ferraris, see Giovanni Silva, "Galileo Ferraris, the Rotating Magnetic Field, and the Asynchronous Motor," pamphlet in English based on a longer Italian article in *L'Elettrotecnica*, September 1947.

32. See the articles cited in note 26 from *Industries* and *The Electrician*.

33. Thompson, *Polyphase Electric Currents*, 444.

34. H. M. Byllesby to GW, 21 May 1888, quoted in Passer, *The Electrical Manufacturers*, 277–78.

35. NT, Motor Testimony, 171; Isaacson, *Steve Jobs*, 95–97.

36. NT, Motor Testimony, 330–31; Kerr testimony, NT, Motor Testimony, 449.

37. Wise, "Stanley's Search for Immortality," 46.

38. NT, Motor Testimony, 255.

39. Ibid., 246–51.

40. W. Stanley Jr. to GW, 24 June 1888, in "Complainant's Record on Final Hearing, Volume II: Exhibits," *Westinghouse Electrical and Manufacturing Company versus Mutual Life Insurance Company of New York and H. C. Mandeville*, U.S. Circuit Court, Western District of New York, pp. 592–93. Catalogued in NTM as NT 74.

41. Kerr testimony, NT, Motor Testimony, 449–51.

42. "Agreement of July 7, 1888," in NT 74, 584–87; NT, Motor Testimony, 327.

43. NT, Motor Testimony, 326–27.

44. NT, "Electromagnetic Motor," U.S. Patent 524,426 (filed 20 October 1888, granted 14 August 1894; "Electrical Transmission of Power," U.S. Patent 511,559 (filed 8 December 1888, granted 26 December 1893); "System of Electrical Power Transmission,"

U.S. Patent 511,560 (filed 8 December 1888, granted 26 December 1893); NT, Motor Testimony, 424.

45. [NT] to the *Electrical World*, 1914, Box 18, Folder 4, KSP.

46. NT, Motor Testimony, 237–38, 365–66.

47. NT, Motor Testimony, 333–34; NT, Radio Testimony, 63–65.

48. "The Hercules Mining Machine and Tesla Motor," *Electrical World* 15 (1 February 1890): 77 in TC 2:40.

49. NT, Motor Testimony, 233–34, 283–84, 363–64.

50. Herbert Simon, "A Behavioral Model of Rational Choice," in *Models of Man, Social and Rational: Mathematical Essays on Rational Human Behavior in a Social Setting* (New York: Wiley, 1957); Paul David, "Clio and the Economics of QWERTY," *American Economic Review* 75 (1985): 332–37.

CHAPTER SIX
SEARCHING FOR A NEW IDEAL

1. Ralph W. Pope to NT, 15 August 1889, Notecards, KSP; Jill Jonnes, *Eiffel's Tower: And the World's Fair Where Buffalo Bill Beguiled Paris, the Artists Quarreled, and Thomas Edison Became a Count* (New York: Viking, 2009).

2. O'Neill, *Prodigal Genius*, 99; NT, "On the Dissipation of the Electrical Energy of the Hertz Resonator," *Electrical Engineer* 14 (21 December 1892): 587–88 in Ratzlaff, *Tesla Said*, 22–23; entry for Bjerknes, *Dictionary of Scientific Biography*, 16 vols., ed. C. C. Gillespie (New York: Scribner, 1970–80), 2:167–69.

3. NT, "Some Experiments in Tesla's Laboratory with Currents of High Potential and High Frequency," *Electrical Review* (NY), 29 March 1899, 193–97, 204 in TC 14:74–83, http://www.teslauniverse.com/nikola-tesla-article-some-experiments-in-teslas-laboratory-with-currents-of-high-potential-and-high-frequency (hereafter cited as NT, "1899 Experiments").

4. NT, Motor Testimony, 323–25; *King's Handbook of New York City, 1893*, 2 vols. (Boston: Moses King, 1893; repr., New York: Benjamin Blom, 1972), 1:233.

5. NT, Radio Testimony, 12.

6. NT, Motor Testimony, 320–21.

7. NT, "Alternating-Current Electro-Magnetic Motor," U.S. Patent 433,700 (filed 25 March 1890, granted 5 August 1890); "Alternating-Current Motor," U.S. Patent 433,701 (filed 28 March 1890, granted 5 August 1890); and "Electro-Magnetic Motor," U.S. Patent 433,703 (filed 4 April 1890, granted 5 August 1890).

8. Will of Charles F. Peck, Bergen County Wills 7893B, W 1890, Wills and Inventories, ca. 1670–1900, Department of State, Secretary of State's Office, New Jersey State Archives, Trenton, NJ.

9. NT, "1899 Experiments," 194.

10. NT, Radio Testimony, 1.

11. NT, "Phenomena of Alternating Currents of Very High Frequency," *Electrical World* 17 (21 February 1891): 128–30, on 128 in TC 2:119–22.

12. NT, Radio Testimony, 1–19.

13. NT, "Method of Operating Arc Lamps," U.S. Patent 447,920 (filed 1 October 1890, granted 10 March 1891) and "Alternating Current Generator," U.S. Patent 447,921 (filed 15 November 1890, granted 10 March 1891).

14. NT, "The True Wireless," *Electrical Experimenter*, May 1919, pp. 28–30ff., on 28.

15. von Urbanitzky, *Electricity in the Service of Man*, 195–98.

16. Hugh G. J. Aitken, *Syntony and Spark: The Origins of Radio* (New York: John Wiley, 1976; rep., Princeton: Princeton University Press, 1985), 52–53.

17. Ibid., 53–57. The definitive study of Hertz is Jed Z. Buchwald, *The Creation of Scientific Effects: Heinrich Hertz and Electric Waves* (Chicago: University of Chicago Press, 1994).

18. NT, "The True Wireless," 28.

19. NT, "Alternating Currents of Short Period," 128 and NT, "Experiments with Alternate Currents of Very High Frequency and Their Application to Methods of Artificial Illumination," a lecture delivered before the AIEE at Columbia College, 21 May 1891 in TCM, *Inventions, Researches, and Writings*, 145–97, on 170–71 (hereafter cited as NT, 1891 Columbia lecture).

20. Aitken, *Syntony and Spark*, 54; NT, "Alternating Currents of Short Period," 128; NT, "The True Wireless," 28.

21. In 1868, James Clerk Maxwell published a mathematical analysis of the interaction of capacitors and induction coils in circuits, but there is no indication that Tesla was aware of this prior work; see Julian Blanchard, "The History of Electrical Resonance," *Bell System Technical Journal* 20 (1941): 415–33, on 419–20, http://www.alcatel-lucent.com/bstj/vol20-1941/articles/bstj20-4-415.pdf.

22. Ibid., 417–18.

23. NT, Radio Testimony, 48–50; O'Neill, *Prodigal Genius*, 90. Tesla also filed a patent for the idea of using the vibratory discharge of capacitors; see "Method of and Apparatus for Electrical Conversion and Distribution," U.S. Patent 462,418 (filed 4 February 1891, granted 3 November 1891).

24. Reginald O. Kapp used the metaphor of the electromagnetic pendulum to explain Tesla's oscillating transformer in "Tesla's Lecture at the Royal Institution of Great Britain, 1892," in NTM, *Tribute to Nikola Tesla*, ed. V. Popovich (Beograd: NTM, 1961), A300–A305, on A302.

25. NT, *My Inventions*, 75.

26. NT, "Phenomena of Alternating Currents of Very High Frequency," 128 in TC 2:119.

27. "Alternating Currents of Short Period," *Electrical World* 17 (14 March 1891): 203 in TC 2:138.

28. Elihu Thomson, "Physiological Effects of Alternating Currents of High Frequency," *Electrical World* 17 (14 March 1891): 214.

29. I am careful in using "radio" or "radio waves" since these terms were not introduced until much later. Édouard Branly was the first to use radio as a prefix in 1897 in "radioconductor," his term for the coherer he developed to detect electromagnetic waves. In 1907, Lee de Forest used the term "radio" to describe wireless telegraphy and the word was subsequently adopted by the U.S. Navy so that it became common by the time of the first commercial broadcasts in the United States in the 1920s.

30. J. G. O'Hara and W. Pricha, *Hertz and the Maxwellians* (London: Peter Peregrinus, 1987); Bruce J. Hunt, *The Maxwellians* (Ithaca: Cornell University Press, 1991).

31. Sungook Hong, *Wireless: From Marconi's Black-Box to the Audion* (Cambridge, MA: MIT Press, 2001), 5–9.

32. O'Hara and Pricha, *Hertz and the Maxwellians*, 1.

33. Although these tubes were produced by various European instrument-makers, the best were produced by a glassblower, Heinrich Geissler, in Bonn, Germany. See Joseph F. Mulligan, ed., *Heinrich Rudolf Hertz (1857–1894): A Collection of Articles and*

Addresses (New York: Garland, 1994), 57–58n144, as well as Falk Muller, *Gasentladungs-forschung im 19. Jahrhundert* (Berlin: Diepholz, 2004).

34. William H. Brock, *William Crookes (1832–1919) and the Commercialization of Science* (Burlington, VT: Ashgate, 2008), 236–37.

35. "A Talk by Nikola Tesla," *New York Times*, 24 May 1891, and "Tesla's Experiments with Alternating Currents of High Frequency," *Electrical Engineer*, 27 May 1891, p. 597, both in TC 3:34–35.

36. NT, 1891 Columbia lecture, 173–74.

37. M. K. Wisehart, "Making Your Imagination Work for You," 64.

CHAPTER SEVEN
A VERITABLE MAGICIAN

1. "Statement from the Westinghouse Company," *Electrical World* 17 (24 January 1891): 54.

2. Carlson, *Innovation as a Social Process*, 290–91; Jonnes, *Empires of Light: Edison*, 215–37.

3. O'Neill, *Prodigal Genius*, 78–82. In *Empires of Light* (224), Jonnes suggested that the bankers were also concerned that Westinghouse had paid Tesla significant consulting fees and was spending money defending the Tesla patents in court. I have found no evidence indicating that Tesla was being paid in 1890–91 by Westinghouse. There was no significant patent litigation related to the Tesla motor patents until the late 1890s.

4. Benjamin Garver Lamme, *Electrical Engineer: An Autobiography* (New York: G. P. Putnam's Sons, 1926), 60–61; Henry G. Prout, *A Life of George Westinghouse* (London: Benn Brothers, 1922), 121–25.

5. Francis E. Leupp, *George Westinghouse: His Life and Achievements* (Boston: Little, Brown, 1918), 159.

6. Charles Fairchild to Henry L. Higginson, 6 May 1891, Box XII-3, Folder 1891 CAC, Henry L. Higginson Papers, Baker Library, Harvard Business School, Boston; Carlson, *Innovation as a Social Process*, 291.

7. O'Neill, *Prodigal Genius*, 81–82. The original contract was between Westinghouse and Tesla, Peck, and Brown. Since Peck died in 1890, Tesla would have had to persuade Brown to go along with his decision to tear up the contract.

8. During a deposition taken in 1903 to defend the AC motor patents, Tesla was asked if he had any pecuniary interest in the patents. He replied, "I wish I had." See NT, Motor Testimony, 153–54.

9. Ibid., 235, 163–64.

10. Tesla experimented with other high-resistance, refractory (i.e., heat-resistant) materials including ruby buttons. Since lasers use ruby tubes, several Tesla biographers have claimed that Tesla invented the laser; see Seifer, *Wizard*, 87 and Hunt and Draper, *Lightning in His Hand*, 236–37.

11. "Tesla's System of Electric Lighting with Currents of High Frequency," *Electrical Engineer*, 1 July 1891, p. 9 in TC 2:47; NT, "System of Electric Lighting," U.S. Patent 454,622 (filed 25 April 1891, granted 23 June 1891); NT, "Electric Incandescent Lamp," U.S. Patent 455,069 (filed 14 May 1891, granted 30 June 1891).

12. NT, "Phenomena of Alternating Currents of Very High Frequency," 128 in TC 2:119–22; Elihu Thomson, "Notes on Alternating Currents of Very High Frequency,"

Electrical Engineer, 11 March 1891, 300 in TC 2:137; NT, "Experiments with Alternating Currents of High Frequency," *Electrical Engineer*, 18 March 1891, 336–37 in TC 2:140–41; NT, "Phenomena of Alternating Currents of Very High Frequency," *Electrical World*, 21 March 1891, pp. 223, 225 in TC 2:148–50; Elihu Thomson, "Alternating Currents of High Frequency—A Reply to Mr. Tesla," *Electrical Engineer*, 1 April 1891, pp. 386–87 in TC 2:152–53; and NT, "Phenomena of Currents of High Frequency," *Electrical Engineer*, 8 April 1891, pp. 425–26 in TC 2:157–58.

13. For Thomson's 1890 lecture—also presented at Columbia College—see "Phenomena of Alternating Current Induction, *Electrical Engineer* (London), 25 April and 2 May 1890, 332–35, 345–46 in TC 2:60–65; "Annual and General Meeting of the American Institute of Electrical Engineers," *Electrical World* 17 (9 May 1891): 344 in TC 2:166; "Chairmen of the Institute's Committees," *Electrical Engineering* 53 (May 1934): 828–31, on 830.

14. NT, "System of Electric Lighting," U.S. Patent 454,622 (filed 25 April 1891, granted 23 June 1891); NT, "Electric Incandescent Lamp," U.S. Patent 455,069 (filed 14 May 1891, granted 30 June 1891). For a partial list of Tesla's foreign patents, see *Catalogue of Tesla's Patents* (Belgrade: NTM, 1987). The French patent was filed on 9 May 1891 and the German patent is dated 20 May 1891.

15. In 1891, Columbia College had its campus on 49th Street between Madison and Fourth avenues. From 1857 to 1891, while the Law School was nominally part of Columbia, it was single-handedly run by Dwight as a proprietary school; hence, in all descriptions of Tesla's lecture, the location is given as Professor Dwight's room. See Robert A. McCaughey, *Stand, Columbia: A History of Columbia University in the City of New York, 1754–2004* (New York: Columbia University Press, 2003), 138–40, 182–83. On the history of electrical engineering at Columbia, see Michael Pupin, *From Immigrant to Inventor* (New York: Charles Scribner's Sons, 1922), 276–85 and James Kip Finch, *A History of the School of Engineering, Columbia University* (New York: Columbia University Press, 1954), 68–70.

16. Joseph Wetzler, "Electric Lamps Fed from Space, and Flames That Do Not Consume," *Harper's Weekly* 35 (11 July 1891): 524 in TC 3:104–6; "Tesla's Experiments with Alternating Currents of High Frequency," *Electrical Engineer*, 27 May 1891, p. 597 in TC 3:35.

17. NT, 1891 Columbia lecture, 155–87.

18. "Alternating Currents of High Frequency," *Electrical Review*, 30 May 1891, p. 185.

19. NT, 1891 Columbia lecture, 187–90. Tesla covered the basic idea behind this demonstration in the patent application he filed prior to the lecture; see "System of Electric Lighting," U.S. Patent 454,622 (filed 25 April 1891, granted 23 June 1891).

20. Wetzler, "Electric Lamps Fed from Space," 524; NT, *My Inventions*, 82.

21. By holding the brass balls, Tesla protected his hands from burns. Although not mentioned in the published text of the Columbia lecture, this physiological demonstration is described in newspaper accounts of the lecture; see "A Talk by Nikola Tesla," *New York Times*, 24 May 1891 and "Tesla's Experiments with Alternating Currents of High Frequency," 597, both in TC 3:34–35.

22. "Alternating Currents of High Frequency," *Electrical Review*, 30 May 1891, 184 in TC 3:37.

23. NT, "What Is Electricity?" *Literary Digest* 3 (18 July 1891): 319–20 in TC 3:137–38.

24. "Alternating Current Phenomena," *Engineering*, 12 June 1891, p. 710 in TC 3:46.

25. "Tesla's Experiments," *Telegraphic Journal and Electrical Review* 29 (24 July 1891): 93–94 in TC 3:142–43.

26. "Mr. Tesla's New System of Illumination," *Electrical Engineer*, 1 July 1891, p. 11 in TC 3:48.

27. "Mr. Tesla's High Frequency Experiments," *Industries*, 24 July 1891, p. 86 in TC 3:169.

28. Wetzler, "Electric Lamps Fed from Space," 524.

29. "Naturalization Record of Nikola Tesla," 30 July 1891, Naturalization Index, NYC Courts, http://www.footnote.com/spotlight/1093.

30. NT, Radio Testimony, p. 7.

31. A. E. Dolbear, "Mode of Electric Communication," U.S. Patent 350,299 (filed 24 March 1882, granted 5 October 1886).

32. *Hawkins Electrical Guide* (New York: Theo. Audel, 1915), 9:2264–65.

33. NT, Radio Testimony, 9.

34. Ibid.

35. 1892 Lecture, 226–32, 237–74, 282–85.

36. NT, "Notes on a Unipolar Dynamo," *Electrical Engineer*, 2 September 1891, p. 258 in TC 3:175 and TCM, *Inventions, Researches, and Writings*, 465–74.

37. 1892 Lecture, 233–35.

38. NT, Radio Testimony, 12.

39. NT, Radio Testimony and "The True Wireless."

40. J. A. Fleming, *The Principles of Electric Wave Telegraphy* (New York: Longmans, Green 1906), 426–27.

CHAPTER EIGHT
TAKING THE SHOW TO EUROPE

1. NT, *My Inventions*, 82; "Progress of Mr. Tesla's High Frequency Work," *Electrical World*, 9 January 1892, p. 20 in TC 4:72.

2. Thompson, *Polyphase Electric Currents*, 454–56; "Priority in Alternating Current Motors," *Electrical Engineer* (London), 24 September 1891, p. 292 in TC 4:25.

3. Carl Hering, "Electrical Practice in Europe as Seen by an American—IV," *Electrical World* 18 (19 September 1891): 193–95, on 194 in TC 4:4–8.

4. Dolivo-Dobrowolsky developed both the wye and delta connections commonly used in induction motors but called them the star and mesh connections; see Thompson, *Polyphase Electric Currents*, 52–53, 456–58. For a discussion of modern practice, consult Pansini, *Basics of Electric Motors*, 46–47.

5. Hughes, *Networks of Power*, 129–135; Thompson, *Polyphase Electric Currents*, 456–71; Georg Siemens, *History of the House of Siemens* (Munich: Karl Alber, 1957), 1:120–23.

6. C.E.L. Brown, "Reasons for the Use of the Three-Phase Current in the Lauffen-Frankfort Transmission," *Electrical World* 18 (7 November 1891): 346 in TC 4:34.

7. "Mr. Tesla and Rotary Currents," *Electrical Engineer* (London), 29 January 1892, pp. 111–12 in TC 4:78–79.

8. Hering, "Electrical Practice in Europe," 194. See also "Tesla Motors in Europe," *Electrical Engineer* (NY), 26 September 1892, p. 291 in TC 5:149.

9. "Progress of Mr. Tesla's High Frequency Work," *Electrical World*, 9 January 1892, p. 20 in TC 4:72; "Mr. Tesla's Experiments," *Electrical Engineer* (London), 18 December 1891, p. 578 in TC 4:59; "Adoption of Report," *Journal of the Institution of Electrical Engineers* 20 (10 December 1891): 600.

10. NT, *My Inventions*, 94.

11. Seifer, *Wizard*, 84.

12. See "Mr. Tesla and Rotary Currents," *Electrical Engineer* (London), 29 January 1892, pp. 111–12 in TC 4:78–79. Note the quote is from the reporter and not Tesla.

13. William Crookes, "Some Possibilities of Electricity," *Fortnightly Review* 102, n.s. (1 February 1892): 173–81, on 174 in TC 4:81–85. For an analysis of the context of this paper, see Graeme Gooday, "Liars, Experts, and Authorities," *History of Science* 46 (2008): 431–56, on 441–49.

14. "Notes," *The Electrician*, 5 February 1892 in TC 4:158; "Mr. Tesla's Lectures on Alternate Current of High Potential and Frequency," *Nature* 45 (11 February 1892): 345–47 in TC 4:166–67.

15. NT, *My Inventions*, 82. On Dewar, see his entry in the *Dictionary of Scientific Biography*, 4:78–81.

16. For a photograph of Tesla's apparatus on the stage of the Royal Institution in February 1892, see figure 24, "Hidden in Plain Sight: Nikola Tesla's 'Radiant Energy' Devices," http://amasci.com/tesla/tesray1.html.

17. Reginald O. Kapp, "Tesla's Lecture at the Royal Institution of Great Britain, 1892," in *Tribute to Nikola Tesla*, pp. A300–A305, on A303.

18. "Alternate Currents of High Frequency," *Engineering*, 5 February 1892, pp. 171–72 in TC 4:159–60. Quotes are from 1892 Lecture, 198–99. The audience included J. J. Thomson, Oliver Heaviside, Silvanus P. Thompson, Joseph Swan, John Ambrose Fleming, James Dewar, Sir William Preece, Oliver Lodge, Sir William Crookes, and Lord Kelvin; see "Mr. Tesla before the Royal Institution, London," *Electrical Review* (NY), 19 March 1892, p. 57 in TC 5:27. In *William Crookes*, William H. Brock (p. 264) concluded that Tesla probably read Crookes's paper "On Radiant Matter" [lecture to British Association at Sheffield] in *Chemical News* 40 (29 August and 16 September 1879): 91–93, 104–7, 127–31.

19. "Alternate Currents of High Frequency," 171–72.

20. Ibid.; 1892 Lecture, 212–16.

21. "Mr. Tesla's Lectures on Alternate Currents of High Potential and Frequency," *Nature*, 11 February 1892, p. 345 in TC 4:166.

22. 1892 Lecture, 226–31, on 230. Tesla was probably referring to improving the operation of the Atlantic cable and not to sending messages over the Atlantic wirelessly.

23. Ibid., 232–36, on 232.

24. 1892 Lecture, 236–87; "Alternate Currents of High Frequency," 172.

25. A. P. Trotter, "Reminiscences," SC MSS 66, p. 532, Institution of Electrical Engineers Archives, London.

26. "Alternate Currents of High Frequency," 172; "Electric Light and Electric Force," *The Spectator*, 6 February 1892, p. 193 in TC 4:164.

27. 1892 Lecture, 288–89; NT, Radio Testimony, 95.

28. "Mr. Tesla and Vibratory Currents," *Electrical Engineer* (London), 12 February 1892, p. 157 in TC 4:168.

29. "Mr. Tesla's Lecture," *Electrical Review* (London), 12 February 1892, p. 192 in TC 4:173.

30. NT, *My Inventions*, 82.

31. Trotter, "Reminiscences," 536.

32. J. A. Fleming to NT, [5 February] 1892, *NT Tribute*, LS-13.

33. W. H. Brock, entry for William Crookes, *Dictionary of Scientific Biography*, 3:474–82, on 476; NT, *My Inventions*, 104.

34. E. Raverot, "Tesla's Experiments with Alternating Currents of High Frequency," *Electrical World* 19 (26 March 1892): 210–13 in TC 5:30–34.

35. "Mr. Tesla's Experiments of Alternating Currents of Great Frequency" [translation of Hospitalier's report in *La Nature*], *Scientific American*, 26 March 1892, pp. 195–96 in TC 5:35–38; "Tesla's Experiments," *Electrical Review* (NY) 20 (9 April 1892): 89 in TC 5:48.

36. A. Blondel to Monsieur le President, 20 May 1936, in *Tribute to Nikola Tesla*, LS-69; "Tesla Motors in Europe," *Electrical Engineer* (NY), 26 September 1892, p. 291 in TC 5:149; NT to GW, 12 September 1892, LC; NT to JJA, 6 January 1899, in Seifer, *Wizard*, 210–11.

37. William Crookes to NT, 5 March 1892, in *NT Tribute*, LS-12.

38. NT, *My Inventions*, 94–95.

39. O'Neill, *Prodigal Genius*, 101; Mrkich, *Tesla: The European Years*, 83.

40. NT, *My Inventions*, 104.

41. On Tesla's explanation for this dream, see NT to G. S. Viereck, 17 December 1934, Benson Ford Research Center, Henry Ford Museum, Dearborn, MI. See also NT, *My Inventions*, 105.

42. Mrkich, *Tesla: The European Years*, 85–86.

43. NT to Pajo Mandic, 20 April 1892 in Serbian edition of NTM, *Nikola Tesla: Correspondence with Relatives* (Belgrade: NTM, 1993), 37. For translation into English, see Nicholas Kosanovich, trans., *Nikola Tesla: Correspondence with Relatives* (N.p.: Tesla Memorial Society, 1995), 26. In its personal column, the *Electrical Engineer* reported that Tesla had "sustained a severe bereavement in the death of his mother." See *Electrical Engineer* (NY), 27 April 1892, p. 439 in TC 5:78.

44. Pribic, "Human Side of Tesla," 25.

45. "Tesla Motors in Europe," *Electrical Engineer* (NY), 26 September 1892, p. 291 in TC 5:149; Seifer, *Wizard*, 95; NT to GW, 12 September 1892, Tesla Papers (microfilm), LC, Reel 7.

46. "Honors to Nikola Tesla from King Alexander I," *Electrical Engineer*, 1 February 1893, 125 in TC 6:70; Pribic, "Human Side of Tesla," 25; A.P.M. Fleming, "Nikola Tesla," *Journal of the Institution of Electrical Engineers* 91 (February 1944): 58ff., reprinted in *Tribute to Nikola Tesla*, A215–A230, on A215. Tesla's family did not attend the ceremonies in Belgrade but read about them in the newspaper; see Marica Kosanović to NT, 4 December 1892, in *Correspondence with Relatives*, 40–41 (Serbian ed.) and 29 (English ed.).

47. TCM, "Nikola Tesla," *Century Magazine* 47 (February 1894): 582–85, on 584 in TC 9:1–4.

48. NT, "The True Wireless," 62; Johanna Hertz, arr., *Heinrich Hertz: Memoirs, Letters, Diaries*, 2nd enl. ed., prepared by M. Hertz and C. Susskind, trans. L. Brinner, M. Hertz, and C. Susskind (San Francisco: San Francisco Press, 1977), 323–25. After visiting Hertz in Bonn, Tesla attended the annual meeting of the British Association for the Advancement of Science; see "The Edinburgh Meeting of the British Association," *Electrical World* 20 (1892): 114.

49. NT, *My Inventions*, 82–83.

CHAPTER NINE
PUSHING ALTERNATE CURRENT IN AMERICA

1. "Mr. Nikola Tesla," *Electrical Engineer* (NY), 31 August 1892, p. 202 in TC 5:145.

2. The timing of Tesla's move from the Astor to the Gerlach Hotel is based on letterheads used in 12 and 27 September 1892 letters to GW; NT asked for equipment to be

shipped to the South Fifth Avenue address in the 27 September letter. On the new lab on South Fifth, see Walter T. Stephenson, "Nikola Tesla and the Electric Light of the Future," *The Outlook*, 9 March 1895, pp. 384–86, on 384 in TC 9:116–18. On the Gerlach Hotel, see "Mr. and Mrs. Gerlach Assign. Owners of Hotel Unable to Carry Heavy Debts Any Longer," *New York Times*, 3 June 1894 and Moses King, *King's Handbook of New York City, 1893*, 1:230. The Gerlach is still standing and is today known as the Radio Wave Building; see "The Beautiful New York City Where Tesla Spent 60 Years of His Life," http://www.teslasociety.com/beautifulnyc.htm.

3. Passer, *The Electrical Manufacturers*, 280–81; Thompson, *Polyphase Electric Currents*, 181–82.

4. Lamme, *An Autobiography*, 60–61; Passer, *The Electrical Manufacturers*, 279.

5. Charles F. Scott, "Long Distance Transmission for Lighting and Power," *Electrical Engineer* 13 (15 June 1892): 601–3; Franklin L. Pope, "Electricity," *Engineering Magazine*, August 1892, pp. 710–11 in TC 5:139.

6. Jonnes, *Empires of Light*, 247–61.

7. Trumbull White and William Ingleheart, *The World's Columbian Exposition: Chicago, 1893* (Philadelphia: Monarch Book Company, 1893), 305–6.

8. As White and Ingleheart observed, "The Westinghouse people were slow to indicate that they would exhibit, as they feared that the great incandescent lamp contract would employ all their time and money." See *The World's Columbian Exposition*, 316.

9. NT to GW, 12 September 1892, LC.

10. See NT to Westinghouse Electric Co., 27 September and 2 December 1892, LC.

11. [Oswald] Villard, *Memoirs of Henry Villard*, 2 vols. (Westminster: Archibald Constable, 1904); Alexandra Villard de Borchgrave and John Cullen, *Villard: The Life and Times of an American Titan* (New York: Nan A. Talese/Doubleday, 2001); Carlson, *Innovation as a Social Process*, 291–97.

12. NT to H. Villard, 10 October 1892, Henry Villard Papers, MS Am 1941–1941.3, Houghton Library, Harvard University.

13. Passer, *The Electrical Manufacturers*, 282.

14. deLancey Rankine, *Memorabelia of William Birch Rankine* (Niagara Falls, NY: Power City Press, 1926).

15. Charles F. Scott, "Personality of the Pioneers of Niagara Power," 31 March 1938, Western New York Historical Materials, National Grid USA, Syracuse, NY (hereafter cited as National Grid Collection).

16. "Edward Dean Adams," *Time*, 27 May 1929, http://www.time.com/time/magazine/article/0,9171,927947,00.html; Christopher Kobrak, *Banking on Global Markets: Deutsche Bank and the United States, 1870 to the Present* (New York: Cambridge University Press, 2008), 70–71; Edward Everett Bartlett, *Edward Dean Adams* (New York: Privately printed, 1926), 10–11.

17. Jonnes, *Empires of Light*, 283.

18. EDA, *Niagara Power: History of the Niagara Falls Power Company, 1886–1918* (Niagara Falls: Privately printed for the Niagara Falls Power Company, 1927), 1:146; Steven Lubar, "Transmitting the Power of Niagara: Scientific, Technological, and Cultural Contexts of an Engineering Decision," *IEEE Technology and Society Magazine*, March 1989, pp. 11–18, on 14.

19. Adams, *Niagara Power*, 2:191.

20. NT to GW, 2 December 1892, LC.

21. Lamme, *Autobiography*, 60–64; Hughes, *Networks of Power*, 121–22.

22. Westinghouse Electric and Manufacturing Co., *Transmission of Power: Polyphase System, Tesla Patents* (trade catalog, ca. January 1893).

23. NT, Edison Medal Speech.

24. NT to EDA, 9 January 1893, National Grid Collection. W. C. Unwin was a professor of engineering in London and, like Kelvin, a member of the International Niagara Commission set up by Adams.

25. NT to EDA, 2 February 1893, National Grid Collection.

26. NT to EDA, 6 February 1893, National Grid Collection.

27. F. H. Betts to [EDA], 11 March 1893, quoted in Adams, *Niagara Power*, 2:241.

28. NT to EDA, 12 March 1893, National Grid Collection.

29. NT to EDA, 21 March 1893; NT to E. D. Adams, 12 and 22 March 1893, National Grid Collection; Adams, *Niagara Power*, 2:235–36.

30. Adams, *Niagara Power*, 2:233–35.

31. NT to EDA, 26 March 1893, National Grid Collection.

32. NT to EDA, 11 May 1893, National Grid Collection.

33. TCM, *Inventions, Researches, and Writings*, 477–85.

34. Passer, *The Electrical Manufacturers*, 281–82.

35. Ibid., 290–92; Lamme, *Autobiography*, 64–66; Chas. A. Bragg to NT, 10 November 1893, Folder 7, Box 18, KSP.

36. P. M. Lincoln, "Some Reminiscences of Niagara," *Electrical Engineering*, May 1934, pp. 720–25, on 720; Rankine, *Memorabelia of William Birch Rankine*, 28–30.

37. Norman R. Ball, *The Canadian Niagara Power Company Story* (Erin, Ontario: Boston Mills Press, 2006); William J. Hausman et al., *Global Electrification: Multinational Enterprise and International Finance in the History of Light and Power, 1878–2007* (New York: Cambridge University Press, 2008), 18.

38. David E. Nye, *American Technological Sublime* (Cambridge, MA: MIT Press, 1994), 13–15, 21–23, 135–37.

39. "Tesla's Work at Niagara," *New York Times*, 16 July 1895.

CHAPTER 10
WIRELESS LIGHTING AND THE OSCILLATOR

1. NT, "On Light and Other High Frequency Phenomena," in TCM, *Invention, Researches, and Writings*, 294–373 (hereafter cited as NT, 1893 lecture).

2. "The Tesla Lecture in St. Louis," *Electrical Engineer* 15 (8 March 1893): 248–49 in TC 6:75–76.

3. NT, 1893 lecture, 318–20.

4. NT, Radio Testimony, 87.

5. "The Tesla Lecture in St. Louis," 249; George Heli Guy, "Tesla, Man and Inventor," *New York Times*, 31 March 1895 in TC 9:140–42, on 142.

6. NT, 1893 lecture, 346. Tesla may very well have been referring to Mahlon Loomis as the "enthusiast" for an atmospheric induction telephone. In 1886 Loomis sent signals between two mountains in Virginia that were fourteen miles apart. He had received a patent for his idea in 1872 and unsuccessfully lobbied Congress for funding. See Orrin E. Dunlap Jr., *Radio's 100 Men of Science: Biographical Narratives of Pathfinders in Electronics and Television* (New York: Harper & Brothers, 1944), 58–59.

7. NT, "The True Wireless," 29.

8. NT, 1893 lecture, 346.

9. Ibid., 341, 344–45.

10. Ibid., 347; NT, Radio Testimony, 23–26.

11. NT, Radio Testimony, 23.

12. "Nikola Tesla's Lecture," *Electrical Industries*, 31 October 1893, pp. 5–6 in TC 8:92–3.

13. NT, Radio Testimony, 36–47; TCM, "Tesla's Oscillator and Other Inventions," *The Century Magazine* 46 (April 1895): 916–33, on 920 in TC 9:143–59.

14. NT, "Reciprocating Engine," U.S. Patent 514,169 (filed 13 August 1893, granted 6 February 1894); "Electrical Generator," U.S. Patent 511,916 (filed 19 August 1893, granted 2 January 1894); "Steam Engine," U.S. Patent 517,900 (filed 29 December 1893, granted 10 April 1894); "Stages and Types of the Tesla Oscillator," *Electrical Engineer* 19 (3 April 1895): 301–4 in TC 10:165–68; "Mr. Tesla's Lecture on Mechanical and Electrical Oscillators," *Electrical Engineer* 41 (30 August 1893): 208 in TC 8:59.

15. TCM, "Tesla's Oscillator," fig. 2.

16. See "Nikola Tesla's Work," *New York Sun*, 3 May 1896 in TC 11:64–65, on 64; TCM, "Tesla's Oscillator," 919–20; and Tesla's remarks in "Meeting of the New York Electrical Society," *Electrical World*, 9 December 1893, pp. 444–46 in TC 8:136–39, on 138.

17. W. Garrett Scaife, *From Galaxies to Turbines: Science, Technology, and the Parsons Family* (Philadelphia: Institute of Physics, 2000), 152–426.

18. Tesla compared his oscillator with existing steam turbines in "Meeting of the New York Electrical Society," 138.

19. NT, "Method and Apparatus for Electrical Conversion and Distribution," U.S. Patent 462,418 (filed 4 February 1891, granted 3 November 1891).

20. NT, "Means for Generating Electric Currents," U.S. Patent 514,168 (filed 2 August 1893, granted 6 February 1894).

21. 1892 Lecture, 209–12; Elihu Thomson devised a similar device using a magnet to protect lighting circuits from lightning; see David Woodbury, *Beloved Scientist: Elihu Thomson, Guiding Spirit of the Electrical Age* (New York: Whittlesey House, 1944), 124–25.

22. NT, "Meeting of the New York Electrical Society," 138.

23. NT, "Mechanical Therapy," undated typescript, 184–87, on 185 in "Tesla Papers, Columbia" folder, Box 1, Anderson Collection, and http://www.rexresearch.com/teslamos/tmosc.htm.

24. See Robert Pack Browning et al., eds., *Mark Twain's Notebooks & Journals*, vol. 3, 1883–1891 (Berkeley: University of California Press, 1979), 431. Twain purchased the rights to the Paige typesetter in 1889 and continued to support Paige's efforts, but Paige's machine could not compete successfully with Ottmar Merganthaler's Linotype machine. Along with several other bad investments, the Paige typesetter bankrupted Twain in the 1890s. See Stephen Railton, "MT and the Paige Typesetter," http://etext.virginia.edu/railton/yankee/cymach6.html and John H. Lienhard, "No. 50: The Paige Compositor," Engines of Our Ingenuity, http://www.uh.edu/engines/epi50.htm.

25. NT, "Mechanical Therapy," 186.

26. Earl Sparling, "Nikola Tesla, at 79, Uses Earth to Transmit Signals," *New York World-Telegram*, 11 July 1935, "Nikola Tesla: Mechanical Oscillator," http://www.rexresearch.com/teslamos/tmosc.htm. The Discovery Channel's *MythBusters* program examined Tesla's claim that he had created an "earthquake machine" in the sixtieth episode (aired 30 August 2006). They tested the physical phenomenon known as mechanical resonance on a bridge built in 1927. Even though they felt the bridge vibrating

many yards away, there were no "earth-shattering" effects. See http://dsc.discovery .com/fansites/mythbusters/episode/episode-tab-05.html.

27. NT, Radio Testimony, 48–60; NT, "Means for Generating Electric Currents," U.S. Patent 514,168 (filed 2 August 1893, granted 6 February 1894). In terms of controllers, Tesla's efforts culminated in a series of mercury interrupters "in which a thin centrifugal ribbon of mercury whirls a three-vaned spindle into rapid contact with itself"; for one Tesla expert, these interrupters combined "fine adjustment and high speed with absolute economy of moving parts." See NT, "Electric Circuit Controller," U.S. Patent 609,247 (filed 14 March 1898, granted 16 August 1898) and Jim Glenn, ed., *The Complete Patents of Nikola Tesla* (New York: Barnes & Noble, 1994), 231.

28. NT, Radio Testimony, 52.

29. Ibid., 68, 62.

30. Ibid., 62.

31. TCM, "Tesla's Oscillator," 927.

32. The professional groups included the Society of Architects and the American Electro-Therapeutic Association. See NT, Radio Testimony, 59 and "An Evening in Tesla's Laboratory," *Electrical Engineer* (NY) 18 (3 October 1894): 278–79 in TC 9:82–83.

33. Walter T. Stephenson, "Nikola Tesla and the Electric Light of the Future," *The Outlook*, 9 March 1895, pp. 384–86, on 385 in TC 9:116–18.

34. NT to Pajo Mandic, 30 November 1893, in *Tesla Correspondence with Relatives*, 39.

CHAPTER ELEVEN
EFFORTS AT PROMOTION

1. Wyn Wachhorst, *Thomas Alva Edison: An American Myth* (Cambridge, MA: MIT Press, 1981); Charles Bazerman, *The Languages of Edison's Light* (Cambridge, MA: MIT Press, 1999); Frederick Dalzell, *Engineering Invention: Frank J. Sprague and the U.S. Electrical Industry* (Cambridge, MA: MIT Press, 2010).

2. Stanley M. Guralnick, "The American Scientist in Higher Education, 1820–1910," in *The Sciences in the American Context: New Perspectives*, ed. N. Reingold (Washington, DC: Smithsonian Institution Press, 1979), 99–142; Graeme Gooday, "Liars, Experts and Authorities," *History of Science* 46 (December 2008): 431–56; Olivier Zunz, *Making America Corporate, 1870–1920* (Chicago: University of Chicago Press, 1990).

3. Burton Bledstein, *The Culture of Professionalism: The Middle Class and the Development of Higher Education in America* (New York: W. W. Norton, 1976).

4. "T. Commerford Martin," *AIEE Electrical Engineering* 53 (May 1934): 789.

5. TCM and Joseph Wetzler, *The Electric Motor and Its Applications* (New York: W. J. Johnston Company, 1889).

6. "Scientists Honor Nikola Tesla," *New York Herald,* 23 April 1893 in TC 6:91–93.

7. TCM, *Mr. Martin's Lawsuit: Its Object Cash, No Vindication. Wasted Exertion, How and Why It Failed* (New York: Electrical World, 1891); National Reporter System, *New York Supplement* [of Decisions by New York State Superior, Appeals, and Supreme Courts], (1893–94), vol. 26, pp. 1105–8.

8. NT to Petar Mandic, 8 December 1893 and NT to Simo Majstorovic (cousin), 17 May 1893, both in *Tesla Correspondence with Relatives*, 41, 35.

9. TCM, *Inventions, Researches, and Writings*; "Tesla and His Researches," *New York Times*, 22 January 1894 in TC 8:175.

10. TCM to NT, 6 February 1894, in Seifer, *Wizard*, 129.

11. "Third Edition of 'The Inventions, Researches, and Writing of Nikola Tesla,'" *Electrical Engineer* 19 (6 February 1895): 124 in TC 9:105; TCM, *Nikola Teslas Untersuchungen über Mehrphasenströme und über Wechselströme hoher Spannung und Frequenz* (Halle: A. S. Knapp, 1895); TCM to Elihu Thomson, 16 January 1917, in Harold J. Abrahams and Marion B. Savin, eds., *Selections from the Scientific Correspondence of Elihu Thomson* (Cambridge, MA: MIT Press, 1971), 352.

12. On Bettini, see http://en.wikipedia.org/wiki/Gianni_Bettini. See also TCM to NT, 6 February 1894, in Seifer, *Wizard*, 138.

13. Seifer, *Wizard*, 139; NT to S. S. McClure, 11 March 1893, Clifton Waller Barrett Library, accession #13114, Special Collections Library, University of Virginia.

14. "Robert U. Johnson, Poet, Is Dead at 84," *New York Times*, 15 October 1937.

15. NT to RUJ, 7 December 1893, in Seifer, *Wizard*, 124; NT to RUJ, 8 January 1894, KSP.

16. Seifer, *Wizard*, 123; RUJ, *Remembered Yesterdays* (New York: Little, Brown, 1923), 401.

17. NT, "Zmai Iovan Iovanovich," *The Century Magazine* 48 (May 1894): 130–31; "Luka Filipov: Paraphrased from the Servian of Zmai Iovan Iovanovich, after Literal Translation by Nikola Tesla," *The Century Magazine* 49 (February 1895): 528–30; RUJ, *Songs of Liberty and Other Poems* (New York: The Century Co., 1897).

18. RUJ, *Remembered Yesterdays*, 400.

19. TCM, "Nikola Tesla," *The Century Magazine* 47 (February 1894): 582–86 in TC 9:1–4.

20. TCM to RUJ, 7 February 1894 and TCM to NT, 17 February 94, in Seifer, *Wizard*, 129.

21. NT to RUJ, 15 February 1894, Bakken Museum of Electricity, Minneapolis.

22. NT, *New York Academy of Sciences Lecture*, 31.

23. On Jefferson and Crawford, see http://en.wikipedia.org/wiki/Joseph_Jefferson and http://en.wikipedia.org/wiki/Francis_Marion_Crawford.

24. TCM, "Tesla's Oscillator and Other Inventions," 928.

25. NT to RUJ, 2 May 1894, in Seifer, *Wizard*, 128; NT to KJ, 2 May 1894, in Cheney, *Tesla: Man out of Time*, 95.

26. Martin, "Tesla's Oscillator."

27. Arthur Brisbane, "Our Foremost Electrician," *New York World*, 22 July 1894, p. 1; John Foord, "Nikola Tesla and His Work," *New York Times*, 30 September 1894; and Curtis Brown, "A Man of the Future," *Savannah Morning News*, 21 October 1894, all in TC 9:44–48, 64–67, 84–87; TCM, "The Burning of Tesla's Laboratory," *Engineering Magazine*, April 1895, pp. 101–4, on 101 in TC 9:162–64.

28. Brown, "Man of the Future," 84–85.

29. Franklin Institute, Cresson Medal Citation, 6 December 1893, in NTM, *Tribute to Nikola Tesla Presented in Articles, Letters, Documents* (Beograd: NTM, 1961), D3–D5.

30. H. G. Osborn to Seth Low, 30 January 1894 and TCM to RUJ, 7 February 1894, both in Seifer, *Wizard*, 129–30.

31. RUJ to H. G. Osborn, 7 May 1894, Box 6, Folder 9, KSP; *Tribute to Nikola Tesla*, D6, D7.

32. NT, Radio Testimony, 72; Bartlett, *Edward Dean Adams*, 11.

33. Ernest K. Adams, "Nikola Tesla," *Yale Scientific Monthly*, February 1895, pp. 217–20 in TC 9:102–5; O'Neill, *Prodigal Genius*, 124.

34. "The Nikola Tesla Company," *Electrical Engineer*, 13 February 1895, p. 149 in TC 9:109. While the *Electrical Engineer* reported that the company was capitalized at $5,000, a majority of the directors ran a notice in the *New York Times* on 4 February 1895 (p. 11) calling for a meeting to raise the capitalization to $500,000.

35. See "Tesla Motors in Europe," *Electrical Engineer* (NY), 26 September 1892, p. 291 in TC 5:149 and NT to GW, 12 September 1892, LC.

36. NT to JJA, 6 January 1899 in Seifer, *Wizard*, 210-11.

37. Passer, *The Electrical Manufacturers*, 328; Carlson, *Innovation as a Social Process*, 304.

38. W. Bernard Carlson, "Thomas Edison as a Manager of R&D: The Development of the Alkaline Storage Battery, 1899-1915," *IEEE Technology and Society* 12 (December 1988): 4-12.

39. Guy, "Tesla, Man and Inventor."

40. NT, Radio Testimony, 56.

41. TCM, "Tesla's Oscillator," A28; NT, Radio Testimony, 56.

42. NT, Radio Testimony, 140.

43. Indeed, Lodge generally opposed grounding his apparatus since it interfered with achieving his goal of syntony or tuning, and though Marconi grounded his apparatus after 1896, he was much more concerned with transmitting over longer distances by building larger aerials and increasing the power of his transmitter. See Aitken, *Syntony and Spark*, 193-97.

44. NT, Radio Testimony, 72-73; NT, "Coil for Electro-Magnets," U.S. Patent 512,340 (filed 7 July 1893, granted 9 January 1894).

45. NT, Radio Testimony, 73-74; TCM, "Tesla's Oscillator," A32.

46. TCM, "Tesla's Oscillator," A32.

47. Ibid.

CHAPTER TWELVE
LOOKING FOR ALTERNATIVES

1. Paul R. Baker, *Stanny: The Gilded Life of Stanford White* (New York: Free Press, 1989); Leland M. Roth, *McKim, Mead & White: A Building List* (New York: Garland, 1978).

2. Baker, *Stanny*, 135-37; Stanford White to NT, 25 February 1894, in Seifer, *Wizard*, 159-60; Stanford White to The Players, 25 February 1894, in Baker, *Stanny*, 137.

3. Stanford White to NT, 5 February 1895 and White to NT, 2 March 1895, in Seifer, *Wizard*, 160.

4. "Tesla's Laboratory Burned," *Electrical Review*, 20 March 1895, p. 145 in TC 9:127.

5. Stephenson, "Tesla and the Electric Light of the Future," 384.

6. "Fruits of Genius Were Swept Away," *New York Herald*, 14 March 1895 and TCM, "The Burning of Tesla's Laboratory," *Engineering Magazine*, April 1895, pp. 101-4 in TC 9:119, 162-64.

7. "Fruits of Genius Were Swept Away."

8. KJ to NT, 14 March 1895, in Cheney and Uth, *Tesla*, 53.

9. *New York Sun*, 14 March 1895 in TC 9:121.

10. Guy, "Tesla, Man and Inventor," 142.

11. "Tesla's Laboratory Burned."

12. Rather than use the terms "electroshock" or "electroconvulsive therapy," which refer to specific practices in modern psychiatry, I have deliberately chosen to use the more general term "electrotherapy" since we do not know anything about the precise treatment that Tesla followed. Electroshock therapy was introduced in the 1930s and, though controversial, is still used to treat severe depression. On electroconvulsive therapy, see http://en.wikipedia.org/wiki/Electroshock_therapy.

13. Carlson, *Innovation as a Social Process*, 29.

14. Jennie Melvene Davis, "Great Master Magician Is Nikola Tesla," *Comfort*, May 1896, in Seifer, *Wizard*, 158; undated *New York Herald* article quoted in Cheney, *Tesla: Man out of Time*, 107.

15. "Nikola Tesla's Work," *New York Sun*, 3 May 1896 in TC 11:64–65, on 64.

16. O'Neill, *Prodigal Genius*, 123; "Nikola Tesla's Work."

17. On this redesigned Tesla coil, see NT, *NY Academy of Sciences Lecture*, 41–45. For a photograph of this device, see http://www.electrotherapymuseum.com/2007/Oscillator/images/Tesla%20Oscillator%202.jpg. See also "Tesla's Electric Oscillator," 13 September 1896, *New York Tribune* in TC 11:120. On the portrait, see "Tesla's Important Advances," *Electrical Review*, 20 May 1896, p. 263 in TC 11:68.

18. NT, *NY Academy of Sciences Lecture*, 31.

19. Edward Ringwood Hewitt, *Those Were the Days: Tales of a Long Life* (New York: Duell, Sloan and Pearce, 1943), 199.

20. Eugene W. Caldwell, "A Brief History of the X-Ray," *Electrical Review* 38 (12 January 1901): 78–79; E.R.N. Grigg, *The Trail of the Invisible Light: From X-Strahlen to Radio(bio)logy* (Springfield, IL: Charles C. Thomas, 1965), 3–4, 9–10; David J. DiSantis, "Early American Radiology: The Pioneer Years," *American Journal of Radiology* 147 (October 1986): 850–53, on 850.

21. Hewitt, *Those Were the Days*, 199.

22. NT, *NY Academy of Sciences Lecture*, 32.

23. NT, "On Roentgen Rays—Latest Results," *Electrical Review* 28 (18 March 1896): 147.

24. Edward R. Hewitt to NT, 18 March [1896] and n.d., Box 8, Folder 4, KSP.

25. NT, "Tesla on Roentgen Rays," *Electrical Review* 28 (11 March 1896): 131, 135 in TC 10:151–54; DiSantis, "Early American Radiology," 851.

26. NT, "On the Hurtful Actions of Lenard and Roentgen Tubes," *Electrical Review*, 5 May 1897, reprinted in *NY Academy of Sciences Lecture*, 90; Maja Hrabak et al., "Nikola Tesla and the Discovery of X-rays," *Radiographics* 28 (2008): 1189–92, on 1190–91.

27. See the following articles by NT in *Electrical Review*: "Roentgen Ray Investigations," 22 April 1896; "An Interesting Feature of X-Ray Radiations," 8 July 1896; "Roentgen Rays or Streams," 12 August 1896; NT, "On the Source of Roentgen Rays and the Practical Construction and Safe Operation," 11 August 1897; all reprinted in NT, *X-Ray Vision: Nikola Tesla on Roentgen Rays* (Radford, VA: Wilder Publications, 2007).

28. Carlson, *Innovation as a Social Process*, 322–28.

29. Lisa Nocks, *The Robot: The Life Story of a Technology* (Westport, CT: Greenwood Press, 2007), 3.

30. NT, "Tesla Describes His Efforts in Various Fields of Work," *Electrical Review*, 30 November 1898, pp. 344–45, available at http://www.tesla.hu; NT, *My Inventions*, 102.

31. NT, *My Inventions*, 106; NT, "Tesla Describes His Efforts."

32. NT to Benjamin F. Miessner, 29 September 1915, in Misc. Mss. Collection, Tesla, LC, and reprinted in Leland I. Anderson, ed., *Nikola Tesla: Guided Weapons &*

Computer Technology (Breckenridge, CO: Twenty-First Century Books, 1998), 227–29; Branimir Jovanović, "Nikola Tesla—Hundred Years of Remote Control," in Branimir Jovanović et al., *Nikola Tesla: One Hundred Years of Remote Control* (Beograd: NTM, 1998), 88–101, on 89.

33. NT to Miessner, 29 September 1915.

34. Tesla may have become aware of the naval armaments race from Theodore Roosevelt who was then assistant secretary of the navy. In November 1897, Roosevelt gave a speech at Delmonico's before the Society of Naval Architects calling for a stronger U.S. Navy. Moreover, Tesla was a friend of Roosevelt's sister, Corinne Robinson, and we know Tesla met Roosevelt at least once in 1899; as he told Corinne, "It was a great privilege to meet your brother and to listen to his enlightening conversation." See "Roosevelt on the Navy," *New York Times*, 13 November 1897 and NT to Mrs. Robinson, 6 March 1899, Corinne (Roosevelt) Robinson Papers, MS Am 1785 (1362), Houghton Library, Harvard University, Cambridge, MA.

35. "Pre-dreadnought Battleship," http://en.wikipedia.org/wiki/Pre-dreadnought.

36. "Tesla Declares He Will Abolish War," *New York Herald*, 8 November 1898 in TC 13:138–40, on 139.

37. I am grateful to Antonio Pérez Yuste for calling this distinction to my attention. According to Yuste, the key pioneer in remote control was the Spanish engineer Leonardo Torres y Quevedo; see Yuste, "Early Developments of Wireless Remote Control: The Telekino of Torres Quevedo," *Proceedings of the IEEE*, 96 (January 2008): 186–89.

38. Yuste ("Early Developments," 186) notes that Marconi used Hertzian waves in 1896 to ring a bell on his receiver and that several British patents were filed in 1898 for the remote control of torpedoes and ships.

39. NT, "My Submarine Destroyer," *New York Journal*, 13 November 1898, available at http://www.tesla.hu.

40. NT to Miessner, 29 September 1915. While Leland Anderson thought that Tesla must have done private demonstrations at Madison Square Garden, the sources he cites could be interpreted as listing people who saw demonstrations of the boat at the laboratory. See Anderson, *Guided Weapons and Computer Technology*, 129. See also O'Neill, *Prodigal Genius*, 175.

41. See Thomas H. White, "W. J. Clarke and the United States Electrical Supply Company," in Section 7, "Pioneering U.S. Radio Activities (1897–1917)," United States Early Radio History, available at http://earlyradiohistory.us/sec007.htm; "New Way to Fire Mines," *New York Times*, 7 May 1898; "Tesla's Electrical Control of Moving Vessels or Vehicles from a Distance"; and "High Frequency Oscillators for Electro-therapeutic and Other Purposes," *Electrical Engineer* 26 (17 November 1898): 489–91 in TC 13:176–78. Several biographies assume that Tesla must have displayed his boat at the Madison Square Garden exhibition, but like the experts at the Tesla Museum, I have not been able to find any evidence to support this assumption. Indeed, the description of the boat quoted in the text suggests that although Tesla had a working prototype in his laboratory by November 1898, he did not necessarily have a boat that could be operated in water at the time of the exhibition in May 1898. See Jovanović, "Hundred Years of Remote Control," 90.

42. NT, *My Inventions*, 107; NT to Parker W. Page, 19 October 1898, Box 14, Folder 2, KSP.

43. NT, "Will Abolish War."

44. NT to Parker Page, 1 December 1898, in Jovanović, "Hundred Years of Remote Control," 92–93.

45. Mark Twain to NT, 17 November 1898, LC, and reprinted in Anderson, *Guided Weapons & Computer Technology*, 130–31.

46. "Doubts Value of Tesla Discovery" and "Chary about Tesla's Plans," *New York Herald*, 9 and 10 November 1898, respectively, in TC 13:144–45.

47. As Leland Anderson pointed out, financial problems forced Martin to merge the *Electrical Engineer* with the *Electrical World* in March 1899 and go back to work for his old boss, W. J. Johnston; see Anderson, *NY Academy of Sciences Lecture*, 6.

48. "Mr. Tesla and the Czar"; "Tesla's Electrical Control of Moving Vessels or Vehicles from a Distance"; and "High Frequency Oscillators for Electro-therapeutic and Other Purposes," *Electrical Engineer* 26 (17 November 1898): 486–87, 489–91, 477–81, respectively, in TC 13:174–78.

49. NT, "Mr. Tesla to His Friends," *Electrical Engineer* 26 (24 November 1898): 514 in TC 14:14.

50. "His Friends to Mr. Tesla," *Electrical Engineer* 26 (24 November 1898): 514–15 in TC 14:14–15.

51. NT, "Problem of Increasing Human Energy," 186–87.

52. Fessenden Interference, 18.

53. Ibid., 40.

54. "Tesla Declares He Will Abolish War."

55. Ibid.; NT, *My Inventions*, 109; NT to Samuel Cohen, 19 March 1916, KSP; Jovanović, "Hundred Years of Remote Control," 94–96; "Tesla's Visit to Chicago," *Western Electrician*, 20 May 1899 in TC 14:133–34.

56. Orrin E. Dunlap, "Nikola Tesla at Niagara Falls," *Western Electrician*, 1 August 1896 in TC 11:103.

57. Quotes are from an undated *New York Herald* clipping in Tesla Papers, Butler Library, Columbia University, and reprinted in Cheney, *Man Out of Time*, 105–7. The clipping can be dated to the summer of 1896 because reference is made to it in "Nikola Tesla and Matrimony," *Electrical Review* (London) 39 (14 August 1896): 193 in TC 11:112.

58. Cheney and Uth, *Master of Lightning*, 51; O'Neill, *Prodigal Genius*, 307.

59. "Mr. Tesla Explains Why He Will Never Marry," *Detroit Free Press*, 10 August 1924 and translation of Dragislav Lj. Petkovich, "A Visit to Nikola Tesla," *Beograd Politika* 24, no. 6824 (27 April 1927), both in Contextual 1 Box, Homosexuality Folder, Anderson Collection; TCM to KJ, 8 January 1894, in Seifer, *Wizard*, 126.

60. O'Neill, *Prodigal Genius*, 302; John J. O'Neill to Leland I. Anderson, 2 May 1953, both in Contextual 1 Box, Homosexuality Folder, Anderson Collection.

61. NT to Alfred Schmid, 2 July 1895 and Henry Floy to NT, 11 October 1895, Tesla Microfilm, Reel 6, LC.

62. Leland Anderson, "Notes on conversation with Richard C. Sogge," Fall 1956, Contextual Box 1, Homosexuality Folder, Sogge Notes, Anderson Collection. According to the 1961 AIEE Directory in the IEEE History Center, Sogge became a member in 1935 and was elected a fellow in 1953. Sogge was employed by General Electric in New York as an industry standards consultant.

63. "Nikola Tesla," *AIEE Electrical Engineering* 53 (May 1934): 817.

64. Baker, *Stanny*, 280.

65. George Chauncey, *Gay New York: Gender, Urban Culture, and the Making of the Gay Male World, 1890–1940* (New York: Basic Books, 1994), 36.

66. Richard Neil Sheldon, "Richmond Pearson Hobson: The Military Hero as Reformer during the Progressive Era" (Ph.D. diss., University of Arizona, 1970).

67. Richmond Pearson Hobson, *The Sinking of the Merrimac* (New York: Century Co., 1899; repr., Annapolis, MD: Naval Institute Press, 1987).

68. RUJ to Lieut. Richmond Hobson, 15 August 1898, Box 22, Folder 1, Richmond P. Hobson Papers, LC.

69. Grizelda Hobson, untitled notes, Box 72, Biog.-Anecdotes, Hobson Papers.

70. See two notes, KJ to NT, ca. 1898, one in Notecards, KSP, and the other in Anderson, *Guided Weapons & Computer Technology*, 134. Quote is from NT to RUJ, 6 December 1898, in Seifer, *Wizard*, 212.

71. Richmond [Hobson] to NT, n.d., Box 8, Folder 6, KSP; NT to Hobson, 1 January 1901, 13 April and 14 May 1901, all in Box 22, Folder 1, Hobson Papers; Seifer, *Wizard*, 259.

72. Richmond Hobson to NT, 6 May 1902, in Anderson, *Guided Weapons & Computer Technology*, 134–35.

73. "Nicola Tesla on Far Seeing," *New York Herald*, 30 August 1896 in TC 11:116–118, on 117–18.

74. Ware, *The Orthodox Way*, 114–19.

75. 1904 Essay.

76. Fessenden Interference, 58.

77. "Nicola Tesla on Far Seeing," 117.

78. See the following Tesla patents: "Apparatus for Producing Electric Currents of High Frequency and Potential," U.S. Patent 568,176 (filed 22 April 1896, granted 22 September 1896); "Electrical Condenser," U.S. Patent 567,818 (filed 17 June 1896, granted 15 September 1896); "Apparatus for Producing Ozone," U.S. Patent 568,177 (filed 17 June 1896, granted 22 September 1896); "Method of Regulating Apparatus for Producing Currents of High Frequency," U.S. Patent 568,178 (filed 20 June 1896, granted 22 September 1896); "Method of and Apparatus for Producing Currents of High Frequency," U.S. Patent 568,179 (filed 6 July 1896, granted 22 September 1896); "Apparatus for Producing Electrical Currents of High Frequency," U.S. Patent 568,180 (filed 6 July 1896, granted 22 September 1896); "Apparatus for Producing Electrical Currents of High Frequency," U.S. Patent 577,670 (filed 3 September 1896, granted 23 February 1897); "Apparatus for Producing Currents of High Frequency," U.S. Patent 583,953 (filed 19 October 1896, granted 8 June 1897); "Manufacture of Electrical Condensers, Coils, & c." U.S. Patent 577,671 (filed 5 November 1896, granted 23 February 1897).

79. NT, "1899 Experiments," 76, 79–80.

80. NT, "Electrical Transformer," U.S. Patent 593,138 (filed 20 March 1897, granted 2 November 1897).

81. NT, "System of Transmission of Electrical Energy," U.S. Patent 645,675 (filed 2 September 1897, granted 20 March 1900).

82. "Tesla's System of Electric Power Transmission through Natural Media," *Electrical Review*, 26 October 1898 in TC 13:124–26, on 126.

83. NT, "Problem of Increasing Human Energy," 209–10.

84. Ibid., 210.

85. On transmitter tests at Houston Street, see "A Wonderful Possibility in Electric Power Transmission," *Electrical Review*, 26 October 1898, p. 262 in TC 13:127–28; "Tesla Would Use Air as Conductor," *New York Herald*, 27 October 1897 in TC 13:129. On the West Point distance test, which Tesla recalled occurred in 1897, see NT, Radio Testimony, 27–28, 67, 108.

86. Fessenden Interference, 36–37.

87. Notes on clippings on Prince Albert from *New York Journal* (22 August 1898) and *New York Herald* (Paris ed., 23 August 1898) in Notecards, KSP; O'Neill, *Prodigal Genius*, 175.

88. David Sinclair, *Dynasty: The Astors and Their Times* (New York: Beaufort Books, 1984), 199–208.

89. John Jacob Astor, *A Journey in Other Worlds* (New York: D. Appleton, 1894); "Appraisement of Estate Reveals Astor's Personality," *New York Times*, 22 June 1913, p. SM2.

90. Marc J. Seifer, "Nikola Tesla and John Jacob Astor," in *Proceedings of the Sixth International Symposium on Nikola Tesla*, ed. A. Marincic and M. Stojic (Belgrade, 2006), 31–38, on 32.

91. NT to JJA, 20 December 1895, in Seifer, *Wizard*, 162–63.

92. NT to JJA, 6 January 1899 in Seifer, *Wizard*, 210–11.

93. See NT to JJA, 6 January 1899. Correspondence between NT and Scherff indicates that these men continued to be involved in the Nikola Tesla Company; see NT to Scherff, 30 May 1899; Scherff to NT, 29 June 1899; NT to Scherff, 13 July 1899; and Scherff to NT, 15 July 1899, in John T. Ratzlaff and Fred A. Jost, eds., *Dr. Nikola Tesla . . . Tesla/Scherff Colorado Springs Correspondence, 1899–1900* (Millbrae, CA: Tesla Book Company, 1979), 30, 86, 91–92.

94. Seifer, *Wizard*, 211; "Appraisement of Estate Reveals Astor's Personality"; O'Neill, *Prodigal Genius*, 176.

95. W. M. Dalton, *The Story of Radio, Part I: How Radio Began* (Bristol, UK: Adam Hilger, 1975), 88.

96. NT, "1899 Experiments," 76–77.

97. *Town Topics*, 6 April 1899, p. 10 in TC 14:88; for an example of Tesla's "droolings," consult "More Wonders Worked by Tesla," *New York Herald*, 30 March 1899 in TC 14:85.

98. "Tesla Says: . . . ," *New York Journal*, 30 April 1899 in TC 14:97–104, on 102.

CHAPTER THIRTEEN
STATIONARY WAVES

1. 1904 Essay, 429.

2. See http://en.wikipedia.org/wiki/Colorado_Springs.

3. "Earth Electricity to Kill Monopoly," *New York World* Sunday Magazine, 8 March 1896, p. 17 in TC 10:147–50.

4. NT to Leonard Curtis, n.d., quoted in Aleksandar Marincic, foreword to Nikola Tesla, *The Problem of Increasing Human Energy, with Special Reference to the Harnessing of the Sun's Energy* (Belgrade: NTM, 2006), 6; Hunt and Draper, *Lightning in His Hand*, 105–6.

5. "Tesla as 'The Wizard,'" *Chicago Tribune*, 14 May 1899 and "Tesla's Task of Taming Air," *Chicago Times-Herald*, 15 May 1899, both in TC 14:117–19.

6. "Nikola Tesla Will 'Wire' to France," *Colorado Springs Evening Telegraph*, 17 May 1899 in TC 14:121.

7. Richard L. Hull, *The Tesla Coil Builder's Guide to the Colorado Springs Notes of Nikola Tesla* (Richmond, VA: by the author, 1996), A24–A26; "Tesla's Station Is Ready," *Colorado Springs Evening Telegraph*, 2 June 1899 in TC 14:139; NT, Edison Medal Speech; Hunt and Draper, *Lightning in His Hand*, 13, 110, 114; NT, Radio Testimony, 117–19; Cheney and Uth, *Tesla: Master of Lightning*, 87.

8. On Lowenstein, see his testimony in Anderson, *Guided Weapons & Computer Technology*, 110, 110; Benjamin Franklin Miessner, *On the Early History of Radio Guidance* (San Francisco: San Francisco Press, 1964), 6; and "Inventor of Radio Devices Died with Praises Unsung," *Philadelphia Public Ledger*, 16 November 1922, in Biographical Files, IEEE Archives, Piscataway, NJ. On Willie, see George Scherff to NT, 2 June 1899, in Ratzlaff and Jost, *Tesla/Scherff Colorado Springs Correspondence*, 62. On Gregg, see his letter to Mrs. Nelson V. Hunt, 9 October 1962, in "Tesla's Lab" folder, Colorado Springs series, Anderson Collection.

9. Hull, *Coil Builder's Guide to the Colorado Springs Notes*, A28; Cheney and Uth, *Tesla: Master of Lightning*, 87.

10. NT, Fessenden Testimony, 24.

11. Gregg to Hunt, 9 October 1962; "Tesla's Call from Mars?" *New York Sun*, 3 January 1901 in TC 15:115.

12. 1904 Essay, 429.

13. O'Neill, *Prodigal Genius*, 179.

14. Aitken, *Syntony and Spark*, 103–6; J. A. Fleming, *The Principles of Electric Wave Telegraphy* (London: Longmans, Green, 1906), 357–61; NT, Fessenden Interference, 66, 87–88. Tesla also replaced the iron filings with more coarse nickel chips.

15. 1904 Essay, 429.

16. Ibid. Tesla mentions stationary waves in passing in his entry for 3 July 1899 in CSN, 68.

17. For a simulation, see http://www.walter-fendt.de/ph14e/stwaverefl.htm.

18. Dalton, *The Story of Radio*, 79–80; to see a standing wave created in a string, consult http://www1.union.edu/newmanj/lasers/Light%20as%20a%20Wave/light_as_a_wave.htm.

19. 1904 Essay, 429.

20. NT, CSN, 4 July 1899, 69.

21. Tesla met Popov at the Chicago World's Fair in 1893, and so he may have read Popov's description of this detector in *The Electrician* in 1897. On Popov, see Fleming, *Principles of Electric Wave Telegraphy*, 362–63, 425 and James P. Rybak, "Alexander Popov: Russia's Radio Pioneer," *Popular Electronics*, August 1982, available at http://www.ptti.ru/eng/forum/article2.html. For his 1895 lightning detector, see R. Victor Jones, "The Branly-Lodge 'Coherer' Detector: A Truly Crazy Device That Worked!" available at http://people.seas.harvard.edu/~jones/cscie129/nu_lectures/lecture6/coherers/coherer.html.

22. NT, CSN, 4 July 1899, 69.

23. Ibid., 70.

24. Leland Anderson suggested that Tesla detected these periodic signals as a result of the waves being reflected by the mountains west of Colorado Springs; see Seifer, *Wizard*, 471.

25. "Extremely Low Frequency Transmitter Site Clam Lake, Wisconsin," U.S. Navy Fact File, 28 June 2001, available at http://www.fas.org/nuke/guide/usa/c3i/fs_clam_lake_elf2003.pdf; Lucy Sheriff, "U.S. Navy Cuts ELF Radio Transmissions," *The Register*, 30 September 2004, http://www.theregister.co.uk/2004/09/30/elf_us_navy/.

26. NT, CSN, 4 July 1899, 70, Tesla's British patent No. 8200 of 1905, quoted in James Erskine-Murray, *A Handbook of Wireless Telegraphy*, 2nd ed. (New York: D. Van Nostrand, 1909), 278.

27. 1904 Essay, 430.

28. NT, Fessenden Interference, 75.

29. 1904 Essay, 430.

30. Kenneth L. Corum and James F. Corum estimate that Tesla's receiver was probably a hundred times more sensitive than the receivers used by Marconi and other early wireless experimenters. See "Nikola Tesla and the Planetary Radio Signals" (2003): 3–4, http://www.teslasociety.com. For an example of one of Tesla's more sensitive receivers, see his entry for 12 July 1899, CSN, 89.

31. NT, "Talking with the Planets," *Collier's Weekly* 26 (9 February 1901): 4–5 in TC 15:157–62.

32. Ibid. Tesla noted that he heard the signals on more than one occasion in "A New Century Call-Up from Mars," *Electrical World and Engineer*, 5 January 1901 in TC 15:120. See also "Tesla's Call from Mars?" *New York Sun,* 3 January 1901 in TC 15:115.

33. In later years, however, Tesla regularly stated that he thought the signals came specifically from Mars. See, for example, NT, "Signalling Mars—A Problem in Electrical Engineering," *Harvard Illustrated*, March 1907, pp. 119–21 in TC 18:1–3.

34. Astronomers today believe that the canals on Mars were a result of perceptual psychology and the limited resolution of telescopes available in the late nineteenth century. Unable to get completely clear images of the Martian surface, astronomers allowed their imaginations to convert the blurry images they were seeing into straight lines or canals. For a history of the Martian canals, see William Sheehan, *The Planet Mars* (Tucson: University of Arizona Press, 1996), especially 71–77, as well as W. G. Hoyt, *Lowell and Mars* (Tucson: University of Arizona Press, 1976).

35. Seifer, *Wizard*, 223–24.

36. Corum and Corum, "Tesla and Planetary Radio Signals," 1, 6.

37. Dalton, *Story of Radio*, 92.

38. Corum and Corum, "Tesla and Planetary Radio Signals," 8.

39. "Tesla's Call from Mars?" *New York Sun*, 3 January 1901.

40. Indeed, to get an extra-large amount of power for his more spectacular experiments, Tesla had to wait until after midnight when the power company was no longer supplying current for lighting; only then could he draw on all of the generators at the local station. See Leland Anderson and Inez Hunt, "Lightning over 'Little London,'" *Denver Post*, Empire Magazine, 11 July 1976.

41. Gregg to Hunt, 9 October 1962.

42. Hull, *Coil Builder's Guide to the Colorado Springs Notes*, A28; Gregg to Hunt, 9 October 1962.

43. NT, CSN, 31 July 1899, 119–20.

44. Ibid.; NT, "Problem of Increasing Human Energy," 206.

45. See Marincic commentary for 23 August 1899, CSN, 411. Tesla later claimed in "Problem of Increasing Human Energy" (p. 208) that he had produced hundred-foot-long sparks, but there is no record of sparks of this length in CSN. See also Hull, *Coil Builder's Guide to the Colorado Springs Notes*, 90–91 and NT, CSN, 23 August 1899, 155.

46. NT, Edison Medal Speech; NT to RUJ, 1 October 1899, LC.

47. See NT, CSN, 23 October 1899, 229 and NT, Radio Testimony, 119. Tesla noted the importance of avoiding streamers in his entry for 30 July 1899, CSN, 115. See also Hull, *Coil Builder's Guide to the Colorado Springs Notes*, 90.

48. "Lighthouses: An Administrative History," http://www.nps.gov/maritime/light/admin.htm; Francis J. Higginson to NT, 11 May 1899, Lighthouse Board Correspondence Folder, Box 3, Anderson Collection. This correspondence was found by Anderson in the National Archives.

49. "The 'Herald' to Report Steamships at Sea by Using Marconi's Wireless Telegraph," *New York Herald* (Paris), 9 June 1901, Third Section, page 2, http://earlyradiohistory.us/1901nan.htm.

50. NT to George Scherff, 4 July 1899, in Ratzlaff and Jost, *Tesla/Scherff Correspondence*, 88–89; see also Thomas Perry to NT, 3 August 1899, Anderson Collection.

51. Thomas Perry to NT, 14 September 1899, Anderson Collection.

52. NT to Light-House Board, 27 September 1899, Anderson Collection.

53. L. S. Howeth, *History of Communications-Electronics in the United States Navy* (Washington, DC: Bureau of Ships and Office of Naval History, 1963), chap. 4, http://earlyradiohistory.us/1963hw04.htm.

54. Lowenstein testimony in Anderson, *Guided Weapons & Computer Technology*, 112.

55. NT, Fessenden Testimony, 80–81.

56. Howeth, *History of Communications-Electronics in the United States Navy*, 38–39.

57. 1904 Essay, 430; Herbert Spencer, *Principles of Psychology*, 2nd ed. (London: Williams and Norgate, vol. 1, 1870; vol. 2, 1872), 1:563, cited in C.U.M. Smith, "Evolution and the Problem of Mind: Part I. Herbert Spencer," *Journal of the History of Biology* 15 (Spring 1982): 55–88, on 73; NT to JPM, 5 September 1902, LC.

58. NT, Fessenden Testimony, 6; NT to JPM, 5 September 1902.

59. NT, Fessenden Testimony, 24.

60. Ibid., 30 and NT, CSN, 27 June 1899, 49–50. Leland Anderson has suggested that in developing these techniques for using two frequencies, Tesla should be regarded as the inventor of the AND gate used in the logic circuits of computers; see Anderson, *Guided Weapons & Computer Technology*, 150–51.

61. NT, Fessenden Testimony, 31–33; "Friends of Tesla Said to Fear for His Health," *New York Herald*, 9 October 1899 in TC 14:159.

62. NT, CSN, 23 and 24 July 1899, 103–5.

63. See entries for 22 August, 5 September, and 11 September 1900, CSN, 154, 174–76, 179–80. Tesla mentioned taking the receiver to a nearby lake in NT, Fessenden Testimony, 75–76, 80. Hull also concluded that the longest transmission test at Colorado Springs was one mile; see *Coil Builder's Guide to the Colorado Springs Notes*, 91.

64. NT, CSN, 2 January 1900, 341, 343.

65. NT, Fessenden Interference, 24–25.

66. NT, "Problem of Increasing Human Energy," 210; O'Neill, *Prodigal Genius*, 193.

67. Between July and October 1899, Tesla gave only two brief interviews, and in one he simply refused to talk to the reporter; see "Tesla Talks to the Telegraph," *Colorado Springs Evening Telegraph*, 29 July 1899 and "Tesla's Work in Colorado," *New York Tribune*, 20 September 1899 in TC 14:148, 150.

68. Lowenstein testimony in Anderson, *Guided Weapons & Computer Technology*, 121, 122–23.

69. Hunt and Draper, *Lightning in His Hand*, 107.

70. Scherff to NT, 19 September 1899, NT to Scherff, 13 October 1899, and Scherff to NT, 16 October 1899, in Ratzlaff and Jost, *Tesla/Scherff Correspondence*, 113–14, 124, and 127, respectively; "Inventor of Radio Devices Died with Praises Unsung."

71. NT, "Problem of Increasing Human Energy," 208.

72. NT to Richard Watson Gilder, n.d., Box 100, *Century* Collection, Manuscript and Archives Division, New York Public Library; NT to RUJ, 28 November 1899, MSS 001452 A, Dibner Library, NMAH.

73. Aleksandar Marincic, foreword to *The Problem of Increasing Human Energy*, 17.

74. See the following entries in NT, CSN: 31 December 1899 (p. 323), 329; Entry for Photo XL, 3 January 1900 (p. 357); Plate XIII (p. 324).

75. Plates I–IV and XXXIII–XXXIX in NT, CSN, 298–304, 348.

76. This is the longest streamer that Tesla reports in CSN; see his entry for 31 December 1899, pp. 325, 327. Most sources claim that Tesla created a 135-foot streamer, but I have not been able to find any reference to this length.

77. NT, CSN, 3 January 1900, 357.

78. Plates XI–XIV on 318–28, 322 and XL–XLIII on 350, 357–63, CSN.

79. Plates XV–XXX, CSN, 331–53.

80. NT, CSN, 2 January 1900, 355, 341.

81. Ibid., 351. There is no evidence that Tesla ever lit an entire field of lamps as shown in the movie *The Prestige*.

82. NT, CSN, 1 January 1900, 333; O'Neill, *Prodigal Genius*, 187.

83. "Great Balls of Fire!" *The Economist*, 27 March 2008, http://www.economist .com/science/displaystory.cfm?story_id=10918140; Schiffer, *Draw the Lightning Down*, 165–66.

84. Paul Sagan, *Ball Lightning: Paradox of Physics* (New York: iUniverse, 2004).

85. NT, CSN, 3 January 1900, 359, 361.

86. See NT, CSN, 31 December 1899 and 3 January 1900, 327 and 363. The Corum brothers reported that they created fireballs using a magnifying transmitter; see Kenneth L. Corum and James F. Corum, extract from "Tesla's Production of Electric Fireballs," *TCBA News* 8, no. 3 (1989), http://home.dmv.com/~tbastian/ball.htm.

87. NT, CSN, 2 January 1900, 337.

88. Karl Popper, *The Logic of Scientific Discovery*, 2nd ed. (New York: Harper & Row, 1968).

89. Henry Petroski, *To Engineer Is Human: The Role of Failure in Successful Design* (New York: St. Martin's Press, 1985); Matthew Josephson, *Edison: A Biography* (New York: McGraw-Hill, 1959), 163.

CHAPTER FOURTEEN
WARDENCLYFFE

1. NT: "Art of Transmitting Electrical Energy through the Natural Mediums," U.S. Patent 787,412 (filed 16 May 1900, granted 18 April 1905); "Method of Signaling," U.S. Patent 723,188 (filed 16 July 1900, granted 17 March 1903); "System of Signaling," U.S. Patent 725,605 (filed 16 July 1900, granted 14 April 1905); and "Method of Insulating Electrical Conductors," U.S. Patent 655,838 (filed 15 June 1900, granted 14 August 1900); George Scherff testimony, Fessenden Interference, in Anderson, *Guided Weapons & Missile Technology*, 93.

2. Untitled item, *The Electrician*, 19 January 1900, p. 423 in TC 15:3.

3. Untitled item, *Electricity*, 24 January 1900, p. 35 in TC 15:3.

4. NT, Radio Testimony, 170.

5. NT to GW, 22 January 1900, LC. Also available in Hunt and Draper, *Lightning in His Hand*, 133–34.

6. Seifer, *Wizard*, 238.

7. For instance, see "Decision in Favor of Tesla Rotating Magnetic Field Patents," *Electrical World and Engineer* 36 (8 September 1900): 394–95 in TC 15:87–88.

8. NT to Scherff, 31 May 1899; Scherff to NT, 3 June 1899; NT to Scherff, 10 June 1899; Scherff to NT, 11 September 1899; NT to Scherff, 14 October 1899, all in Ratzlaff and Jost, *Tesla/Scherff Colorado Springs Correspondence*, 70, 71, 73, 110–11, 125, respectively.

9. Seifer, *Wizard*, 241, 243–44.

10. Marincic, foreword to *The Problem of Increasing Human Energy*, 7.

11. NT, Fessenden Interference, 32.

12. O'Neill, *Prodigal Genius*, 195.

13. RUJ to NT and NT to RUJ, both 6 March 1900, in Seifer, *Wizard*, 239–40.

14. NT, "Problem of Increasing Human Energy."

15. Ibid., 175.

16. Ibid., 177, 192–93; John William Draper, *History of the Intellectual Development of Europe* (New York: Harper Brothers, 1891), 2:392. Draper (1811–82) was Professor of Chemistry and Physiology at New York University.

17. NT, "Problem of Increasing Human Energy," 178–80.

18. Ibid., 188.

19. Ibid., 211. In the article, Tesla quoted Goethe's poem "Hope" in German and the English translation was provided in a footnote.

20. For a sampling of this coverage, see the scrapbooks of newspaper clippings, NTM.

21. Physicist, "Science and Fiction," *Popular Science Monthly* 58 (July 1900): 324–26 in TC 15:66–67.

22. TCM, "Newspaper Science," *Science* 12 (2 November 1900): 684–85 in TC 15:110–11. True to his word, as editor of *Electrical World*, Martin asked specialists in electrical engineering at the end of 1900 to vote on who were the greatest inventors and scientists in their field; see "Twenty-Five Great Names in Electrical Science and Invention during the Nineteenth Century," *Electrical World and Engineer* 37 (5 January 1901): 18–19 in TC 15:118–19. Rank-and-file AIEE members placed Tesla at number 7 while professors of electrical engineering put him at 15 and "prominent" AIEE members ranked Tesla at 13.

23. Bledstein, *The Culture of Professions*; Louis Galambos, *The Creative Society—and the Price Americans Paid for It* (New York: Cambridge University Press, 2012).

24. "Electricity a Cure for Tuberculosis," *New York Herald*, 3 August 1900; "Niagara's Power for City Wheels," 17 August 1900; [no title], *The Electrician*, 24 August 1900, all in TC 15:69, 73, 78.

25. Jean Strouse, *Morgan: American Financier* (New York: Random House, 1999); Carlson, *Innovation as a Social Process*, 293–96.

26. Guglielmo Marconi, "Transmitting Electrical Signals," U.S. Patent 586,193 (filed 7 December 1896, granted 13 July 1897) and "Apparatus Employed in Wireless Telegraphy," U.S. Patent 647,008 (filed 13 June 1899, granted 10 April 1900).

27. [Henry Saunders] to Directors of Wireless Telegraph & Signal Co., 15 September 1899, in Reports and correspondence on general activities, 1899–1908, MS. Marconi 178, Marconi Archives, Bodleian Library, University of Oxford.

28. NT to JPM, 13 December 1904, LC.

29. Strouse, *Morgan*, 394–95; George Wheeler, *Pierpont Morgan and Friends: Anatomy of a Myth* (Englewood Cliffs, NJ: Prentice-Hall, 1973), 61–62.

30. NT to JPM, 26 November 1900, LC.

31. NT to JPM, 10 December 1900, LC.

32. NT to JPM, 12 December 190[0], LC.

33. NT to JPM, 12 December 1900, LC.

34. Strouse, *Morgan*, 401–3.

35. NT to the American Red Cross, [7 January 1901], Tesla Collection, Rare Book and Manuscript Library, Columbia University (hereafter cited as Tesla Columbia Collection).

36. "Tesla's Call from Mars?" *New York Sun*, 3 January 1901; "Astronomers Discuss Tesla's Alleged Message from Mars," *New York Journal*, 4 January 1901; "Discredits Tesla's Martian Theory," *New York Herald*, 5 January 1901; "That Message from Mars," *Scientific American*, 19 January 1901; "An Alleged Message from Mars," *Literary Digest*, 26 January 1901, all in TC 15:115–17, 121, 132, 137, respectively.

37. NT to JJA, 11 and 22 January 1901, respectively, in Seifer, *Wizard*, 253–54; "Tesla's Wireless Light," *New York Sun*, 26 January 1901 in TC 15:138; "Tesla's Vacuum Tube Light, *New York Tribune*, 27 January 1901; "Vacuum Tube Lighting," *Electrical World and Engineer*, 2 February 1901, p. 201; "Tesla's Wireless Light," *Scientific American*, 2 February 1901; "Tesla's Vacuum-tube Lighting," *Western Electrician*, 2 February 1901, p. 79; "Nikola Tesla Duplicates the Light of Day," *New York Herald*, 3 February 1901; "Tesla's 'Artificial Sunshine,'" *Public Opinion*, 7 February 1901, 175, all in TC 15:139, 143, 148–56, respectively.

38. "Mr. Tesla's Wireless Telegraphy," *New York Tribune*, 15 February 1901; "Tesla Ready to Try Transatlantic Talk," *New York Journal*, 22 February 1901; "Tesla's New Telegraph," *New York Sun*, 15 February 1901, all in TC 15:167, 171, 166, respectively.

39. Herbert L. Satterlee, *J. Pierpont Morgan: An Intimate Portrait* (New York: Macmillan, 1939; repr., New York: Arno, 1975), 369–70.

40. "Tesla and Wireless Telegraphy," *Literary Digest* 22 (2 March 1901): 257 in TC 16:4.

41. Charles Steele to NT, 15 and 25 February 1901; NT to Steele, 18 February 1908, LC.

42. NT to JPM, 1 March 1901 and Charles Steele to NT, 4 March 1901, both in LC.

43. NT to JPM, 13 October 1904, LC; Strouse, *Morgan*, 412, 418, 426.

44. Strouse, *Morgan*, 495.

45. NT to JPM, 13 December 1904, LC.

46. NT to Charles Steele, 5 March 1901, LC.

47. Details of the laboratory building, which is still standing, were taken from stories from the *Port Jefferson Echo*, 2 August 1901 and February 1902, quoted in Natalie Aurucci Stiefel, *Looking Back at Rocky Point: In the Shadow of The Radio Towers*, vol. 1, http://www.teslasociety.com/warden.htm. See also Leland M. Roth, *McKim, Mead & White: A Building List* (New York: Garland, 1978), entry 818, p. 148; Stanford White to NT, 26 April 1901, in Seifer, *Wizard*, 262.

48. Mervin G. Pallister, "A History of the Incorporated Village of Shoreham," 4 July 1976, and Mary Lou Abata, "History of Shoreham," 1979, both available at http://www.shorehamvillage.org/Shoreham_History/History_home.html; "Mr. Tesla at Wardenclyffe, L.I.," *Electrical World and Engineer* 38 (28 September 1901): 509–10 in TC 16:40; *Port Jefferson Echo*, 2 August 1901, in Stiefel, *Looking Back at Rocky Point*; Leland I. Anderson, "Wardenclyffe—A Forfeited Dream," *Long Island Forum* (August and September 1968), http://www.teslascience.org/pages/dreamhtm; "Tesla Judgment Filed: Inventor Had Paid Lawyer with Promissory Note," *New York Times*, 14 June 1925; O'Neill, *Prodigal Genius*, 205. The name Wardenclyffe proved short-lived and in 1906 the village near Tesla's laboratory was given its present name of Shoreham.

49. "Tesla's Description of Long Island Plant and Inventor of the Installation as Reported in 1922 Foreclosure Appeal Proceedings," appendix 2 in NT, Radio Testimony, 191–98.

50. NT, Radio Testimony, 143. Note that when he was working at Wardenclyffe, Tesla referred to the tower as an "elevated terminal" and that he only used the term "antenna" much later, as in this quote from 1916.

51. See NT, Radio Testimony, 145, and NT, "Apparatus for Transmitting Electrical Energy," U.S. Patent 1,119,732 (filed 18 January 1902, granted 1 December 1914). Robert van de Graaff also realized that a sphere was the best shape for storing large amounts of electrical charge, and so his electrostatic generators were similarly crowned by a metallic sphere. Like Tesla's tower, van de Graaff's largest generators were able to generate charges on the order of 7 million volts. See entry for Robert J. van de Graaff at http://en.wikipedia.org/wiki/Robert_J._Van_de_Graaff.

52. NT, untitled notes, 29 May 1901, original in NTM, copy in NT notes folder, Wardenclyffe Box, Anderson Collection; NT, Radio Testimony, 143.

53. NT to JPM, 13 September 1901, LC.

54. NT to Stanford White, 13 September 1901, Personal Miscellaneous Collection, Manuscript and Archives Division, New York Public Library.

55. "Tesla's Description of Long Island Plant," NT, 200–202. Several drawings of the finished tower show the hemispheric terminal on top studded with smaller hemispheres; see, for example, Smithsonian Neg. 86-604066.

56. NT to JPM, 19 December 1904, LC.

57. "Tesla and Telegraphy," *New York Tribune*, 27 November 1901, and "A New Tesla Laboratory on Long Island," *Electrical World and Engineer* 40 (27 September 1902): 499–500, both in TC 16:54 and 98; "Tesla's Description of Long Island Plant," 200–202; O'Neill, *Prodigal Genius*, 205.

58. "Cloudborn Electric Wavelets to Encircle the Globe," *New York Times*, 27 March 1904 in TC 17:3.

59. "Tesla's Description of Long Island Plant," 203.

60. "Cloudborn Electric Wavelets to Encircle the Globe."

61. "Tesla's Description of Long Island Plant," 203.

62. Ibid.

63. *Port Jefferson Echo*, 22 February 1902, in Stiefel, *Looking Back at Rocky Point*.

64. See "Inventor Tesla's Plant Nearing Completion," *Brooklyn Eagle*, 8 February 1902 in TC 16:61; *Port Jefferson Echo*, February 1902; and *Patchogue Advance*, March 1902, both in Stiefel, *Looking Back at Rocky Point*. Leland Anderson thought that the four tunnels led to an outer circular tunnel that "was perhaps required to establish a large surface area of contact with the groundwater system"; see "Wardenclyffe Design Mystery," in Building and Tunnels Folder, Wardenclyffe Box, Anderson Collection. See also "Dig for Mystery Tunnels Ends with Scientist's Secret Intact," *Newsday*, 13 February 1979, p. 24, and "Famed Inventor, Mystery Tunnels Linked," *Newsday*, 10 March 1979, p. 19, both available at http://www.teslascience.org/pages/twp/tunnels.htm.

65. NT to JPM, 9 January 1902. I am grateful to Vladimir Jelenković of the Tesla Museum for providing a transcript of this letter.

66. See NT, "Apparatus for Transmitting Electrical Energy," U.S. Patent 1,119,732 (filed 18 January 1902, granted 1 December 1914); "System of Transmission of Electrical Energy," U.S. Patent 645,675 (filed 2 September 1897, granted 20 March 1900); and "Apparatus for Transmission of Electrical Energy," U.S. Patent 649,621 (filed 2 September 1897, granted 15 May 1900). Readers seeking a more comprehensive interpretation of how the Wardenclyffe station may have worked should consult Gary Peterson, "Nikola Tesla's Wireless Work," http://www.teslaradio.com/pages/wireless.htm.

67. See NT, Radio Testimony, 152–55. For a diagram of how Tesla connected these components in Colorado Springs, consult Figure 13.5.

68. Entry on Wardenclyffe, http://en.wikipedia.org/wiki/Wardenclyffe.

69. There are no surviving photographs or diagrams of the magnifying transmitter at Wardenclyffe, and this description is based on Tesla's patent for the elevated terminal: "Apparatus for Transmitting Electrical Energy," U.S. Patent 1,119,732 (filed 18 January 1902, granted 1 December 1914). See also NT, Radio Testimony, 145.

70. A. S. Marinic, "Nikola Tesla and the Wireless Transmission of Energy," *IEEE Transactions on Power Apparatus and Systems* PAS-101 (October 1982): 4064–68, on 4066.

71. Alan Bellows, "Tesla's Tower of Power," http://www.damninteresting.com/teslas-tower-of-power/.

72. Gary Peterson, "Wireless Energy Transmission for the Amateur Tesla Coil Builder," http://www.teslaradio.com/pages/wireless_102.htm.

73. NT, Radio Testimony, 155; 1904 Essay, 431.

CHAPTER FIFTEEN
THE DARK TOWER

1. NT to Mrs. Johnson, 13 October 1901, in Seifer, *Wizard*, 272.

2. NT to JPM, 11 November 1901, LC.

3. Gavin Weightman, *Signor Marconi's Magic Box: The Most Remarkable Invention of the 19th Century and the Amateur Inventor Whose Genius Sparked a Revolution* (New York: Da Capo, 2003), 58–65, 75–76.

4. Josephine B. Holman to Marconi, 31 December 1899 and 26 October 1900, Marconi Archives; Hong, *Wireless*, 59–61.

5. Fleming, *Principles of Electric Wave Telegraphy*, 451; Hong, *Wireless*, 58, 72–73.

6. Sungook Hong argues that Marconi was only able to transmit across the Atlantic because Fleming designed such a powerful system (*Wireless*, 53–88). Like Tesla, Fleming used a regular transformer to step up the current and to charge a large capacitor. When this capacitor discharged, the oscillatory current was sent through a second transformer that functioned in a manner similar to Tesla's magnifying transmitter. Unlike Tesla, who added an extra coil between the secondary of his magnifying transmitter and the elevated terminal, Fleming instead added another capacitor and a final transformer that stepped up the current before it went to the aerial. To get a sense of the similarity of the transmitting equipment at Poldhu and that used by Tesla at Colorado Springs, compare the photo of the interior of the Poldhu station (Hong, *Wireless*, p. 75, fig. 3.6) with the interior of Colorado Springs (Figure 13.3). Marconi claimed that there was nothing new in using a Tesla coil in this way and that this circuit was suggested by patents filed by Oliver Lodge and Ferdinand Braun; see G. Marconi, "Syntonic Wireless Telegraphy," lecture delivered at the Society of Arts, 15 May 1901, MS 159, Marconi Papers. See also Weightman, *Marconi's Magic Box*, 91.

7. Weightman, *Marconi's Magic Box*, p. 101.

8. See "Wireless Signals across the Ocean," "Signor Marconi's Career," "Nikola Tesla's Researches," and "T. C. Martin's Views," all in *New York Times*, 15 December 1901.

9. Lee de Forest, *Father of Radio: The Autobiography of Lee de Forest* (Chicago: Wilcox & Follett, 1950), 129.

10. For a modern technical discussion of what Marconi may (or may not) have heard, see John S. Belrose, "Fessenden and Marconi: Their Differing Technologies and

Transatlantic Experiments during the First Decade of this Century" (paper presented at *International Conference on 100 Years of Radio, 5–7 September 1995*), http://www.ieee.ca/millennium/radio/radio_differences.html.

11. TCM to [Elihu Thomson], 17 October 1919, in Abrahams and Savin, *Scientific Correspondence of Elihu Thomson*, 354–55; David O. Woodbury, *Beloved Scientist*, 235–36; Gordon Bussey, *Marconi's Atlantic Leap* (Coventry: Marconi Communications, 2000), 65.

12. Menu from annual dinner of AIEE, 13 January 1902, MS 159, Marconi Archives; "Annual Dinner of the Institute at the Waldorf-Astoria, January 13, 1902, in honor of Guglielmo Marconi," *Transactions of the American Institute of Electrical Engineers*, 1902, pp. 93–121, http://earlyradiohistory.us/1902wt.htm.

13. Three previous quotes are from the 1902 Marconi dinner.

14. Weightman, *Marconi's Magic Box*, 122–26; Bussey, *Marconi's Atlantic Leap*, 70–74.

15. "Tesla's Wireless Telegraph," *New York Sun*, 16 January 1902 in TC 16:59.

16. NT to JPM, 9 January 1902.

17. NT to JPM, 13 October 1904, LC.

18. NT to JPM, 9 January 1902.

19. Noah Wardrip-Fruin and Nick Montfort, eds., *The New Media Reader* (Cambridge, MA: MIT Press, 2003), Section 54, quoted on http://en.wikipedia.org/wiki/World_wide_web#cite_note-3.

20. NT to JPM, 9 January 1902.

21. NT, "Tesla Manifesto," in O'Neill, *Prodigal Genius*, 209.

22. Steven Watts, *The People's Tycoon: Henry Ford and the American Century* (New York: Alfred A. Knopf, 2005), 119.

23. W. Bernard Carlson, "Artifacts and Frames of Meaning: Thomas A. Edison, His Managers, and the Cultural Construction of Motion Pictures," in *Shaping Technology, Building Society: Studies in Sociotechnical Change*, ed. W. E. Bijker and J. Law (Cambridge, MA: MIT Press, 1992), 175–98.

24. NT to JPM, 9 January 1902; NT to JPM, 5 September 1902, LC.

25. NT to JPM, 9 January 1902.

26. Strouse, *Morgan*, 457, 418–69.

27. "Prince Welcomed by Chiefs of Industry," *New York Times*, 27 February 1902.

28. See Fessenden Interference in Anderson, *Guided Weapons & Computer Technology*. On Lowenstein returning to work, see his testimony in Fessenden Interference, 110. Tesla appears to have won this case; see NT to Scherff, 9 August 1902 [or 1903?], in Seifer, *Wizard*, 282.

29. NT to JPM, 5 September 1902, LC.

30. Ibid.

31. J. P. Morgan & Co. to NT, 7 June 1902 and NT to JPM, 5 September 1902, LC; *Port Jefferson Echo*, 21 June 1902, in Stiefel, *Looking Back at Rocky Point*.

32. NT to JPM, 17 September 1902, LC.

33. Charles Steele to NT, 24 September 1902 and 21 October 1902; NT to JPM, 17 September 1902, all in LC.

34. Seifer, *Wizard*, 289; NT to JPM, 1 April 1903, LC.

35. NT to JPM, 3 July 1903, LC; Seifer, *Wizard*, 291; NT to Scherff, 13 October 1905, Tesla Columbia Collection [I used copies of the Tesla-Scherff letters found in the Anderson Collection, but the originals are at Columbia]; NT to Scherff, 11 April 1903, listed in *The Teslian*, September–November 1903, p. C6, in Elmer Gertz Papers, Box 377, Folder 6, LC.

36. Satterlee, *Morgan*, 387–94, 401–2; NT to JPM, 22 April 1903, LC.

37. NT to JPM, 3 July 1903, LC.

38. Satterlee, *Morgan*, 403; J. P. Morgan & Co. to NT, 3 July 1903, and JPM to NT, 17 July 1903, both in LC.

39. "Tesla's Flashes Startling," *New York Sun*, 17 July 1903 in TC 16:140.

40. Tesla estimated that he needed only another $100,000 to complete his work; see NT to William B. Rankine, 19 April 1904, Buildings and Tunnels Folder, Wardenclyffe Box, Anderson Papers.

41. Seifer, *Wizard*, 300.

42. NT, Radio Testimony, 106.

43. Frank Fayant, "Fools and Their Money," *Success Magazine*, January 1907, pp. 9–11, 49–52, http://earlyradiohistory.us/1907fool.htm.

44. Georgette Carneal, *A Conqueror of Space: An Authorized Biography of the Life and Work of Lee DeForest* (New York: Horace Liveright, 1930), 75–83; de Forest, *Father of Radio*, 89–90; James A. Hijiya, *Lee de Forest and the Fatherhood of Radio* (Bethlehem, PA: Lehigh University Press, 1992), 41, 58; Scherff to NT, 26 September 1899, Tesla Columbia Collection.

45. Samuel Lubell, "Magnificent Failure," *Saturday Evening Post*. Lubell's article appeared in three installments in January 1942: 17 January, pp. 9–11ff.; 24 January, pp. 20–21ff.; 31 January p. 27ff. Quote is from 24 January, p. 21.

46. Ibid.

47. Frank Fayant, "The Wireless Telegraph Bubble," *Success Magazine*, June 1907, pp. 387–89ff., http://earlyradiohistory.us/1907fool.htm.

48. De Forest, *Father of Radio*, 130–35; Carneal, *Conqueror of Space*, 146–51.

49. "Wireless Stock Quotations" and "A Perambulating Wireless Telegraph Plant," *Electrical World and Engineer*, 14 and 28 February 1903, pp. 281 and 374, respectively.

50. JPM to NT, 14 December 1905 and 16 February 1906, LC.

51. Hawkins, "Nikola Tesla, His Work, and Unfulfilled Promises," 99, 108 in TC 16:111–20; NT to JPM, 11 December 1903, LC.

52. NT to William B. Rankine, 19 April 1904 in Buildings and Tunnels Folder, Wardenclyffe Box, Anderson Collection; NT to Scherff, 14 June, 3 and 8 August 1905, Tesla Columbia Collection.

53. Petar Mandic to NT, 2 September 1903, in Kosanovich, *Tesla Correspondence with Relatives*, 104.

54. JJA to NT, 6 October 1903, in Seifer, *Wizard*, 295.

55. Entry for Ryan, http://www.vahistorical.org/exhibits/headstales_inventory.htm#ryan; NT to JPM, 13 October 1904, LC; NT to Scherff, 16 November 1903, Tesla Columbia Collection.

56. "Canadian Niagara Power, William Birch Rankine Hydro-Electric Generating Station," http://www.niagarafrontier.com/rankine.html; Frank G. Carpenter, "Wonderful Discoveries in Electricity," *Pittsburgh Dispatch*, 18 December 1904 in TC 16:72–73; NT to JPM, 13 January 1904, LC; Norman R. Ball, *The Canadian Niagara Power Company Story* (Erin, Ontario: Boston Mills Press, 2006).

57. Hobson's love letters to Grizelda can be found in Box 1 of the Hobson Papers, LC; see, in particular, Hobson to Miss Hull, 24 November 1902, 27 May 1903, 25 November [1903?], 26 November 1903, 26 November 1904, and 30 January 1905. See also Grizelda Hull to Hobson, 1 and 14 December 1904.

58. Richmond [Hobson] to [Miss Hull], [22 December 1903], Hobson Papers.

59. "Cloudborn Electric Wavelets to Encircle the Globe," *New York Times*, 27 March 1904, Alfred Cowles, "Harnessing the Lightning," *Cleveland Leader*, 27 March 1904 in Naval History Collection, New-York Historical Society, New York City; NT to JPM, 13 January 1904, LC; "A Striking Tesla Manifesto," *Electrical World* 43 (6 February 1904): 256 in TC 16:159; NT to Scherff, 28 January 1904, Tesla Columbia Collection; NT to RUJ, 24 Jane 1904, in Seifer, *Wizard*, 289.

60. 1904 Essay, 431.

61. "John Sanford Barnes Dead," *New York Times*, 23 November 1911; John S. Barnes, *Submarine Warfare, Offensive and Defensive: Including a Discussion of the Offensive Torpedo System, Its Effects upon Iron-Clad Ship Systems, and Influence upon Future Naval Wars* (New York: D. Van Nostrand, 1869).

62. NT to Rankine, 19 April 1904; Kerr, Page, and Cooper to NT, 8 April 1904 and NT to J. S. Barnes, 14 and 20 April 1904, all in Naval History Collection; NT to Scherff, 21 March 1904, Tesla Columbia Collection.

63. On Schiff, see NT to Scherff, 25 July 1905, in Seifer, *Wizard*, 320. In addition, Tesla courted Messrs. Andrews and Selon in the summer of 1905, but I have been unable to determine who these investors were; see NT to Scherff, 31 July, 1 and 14 August 1905, Tesla Columbia Collection.

64. NT to JPM, 15 and 16 February 1906, JPM to NT, 16 February 1906, NT to JPM, 17 October 1904, all in LC.

65. NT to JPM, 15 December 1905, LC.

66. NT to JPM, 24 September 1903, and NT to JPM, 13 October 1903, both in LC.

67. NT to JPM, 13 October 1904, LC.

68. NT to JPM, 17 October 1904, LC.

69. NT to JPM, 14 December 1904, LC.

70. NT to JPM, 17 February 1905, LC.

71. NT to John Hays Hammond Jr., 18 February 1911, KSP.

72. Seifer, *Wizard*, 318–19; B. A. Behrend, "Tesla and the Polyphase Patents," *Electrical World* 45 (6 May 1905): 828 in TC 18:97; NT to Scherff, 23 January 1905, Tesla Columbia Collection.

73. Hobson to NT, 1 May 1905, Box 8, Folder 6, KSP.

74. "Hobson-Hull Wedding, *New York Times*, 26 May 1905; Grizelda H. Hobson, "Biographical Notes on the Life of R. P. Hobson," 1940, Box 72, Folder Biog.-Anecdotes, Hobson Papers.

75. NT to Scherff, 12 and 14 June, 7, 14, and 18 July, and 8 August 1905, Tesla Columbia Collection.

76. "Tesla on the Peary North Pole Expedition," *Electrical World* 46 (22 July 1905): 130 in TC 17:121.

77. For a sampling of the discussion both favoring and criticizing scalar waves, see Hank Mills, "Tesla's Scalar Fields Still Beaming On!" http://pesn.com/2011/03/26/9501797_Teslas_Scalar_Waves_Replicated_by_Steve_Jackson and "Scalar Weapons: Tesla's Doomsday Machine?" http://skeptoid.com/episodes/4121.

78. NT to G. S. Viereck, 17 December 1934, Benson Ford Research Center, Henry Ford Museum, Dearborn, MI; Frank G. Carpenter, "Wonderful Discoveries in Electricity," *Pittsburgh Dispatch*, 18 December 1904.

79. Sylvia Nasar, *A Beautiful Mind: A Biography of John Forbes Nash, Jr., Winner of the Nobel Prize in Economics, 1994* (New York: Simon & Schuster, 1998), 11.

80. NT, *My Inventions*, 93.

81. NT to Scherff, 11 October 1905, Tesla Columbia Collection.

82. On the possible deal with Frick, see JPM to NT, 14 December 1905; NT to JPM, 24 January and 6 February 1906, LC. Tesla had high hopes that Frick would support him; as he wrote to Scherff after a brief encounter with Frick, "He was most friendly and said that he was sorry he had to go out, but he will talk with me some other day. *I have my man* as sure as the law of gravitation. I know it." See NT to Scherff, 11 November 1905, in Seifer, *Wizard*, 320.

83. NT to Edward P. Mitchell, 11 December 1905, Mitchell Papers, New-York Historical Society; TCM to NT, 24 December 1905, in Seifer, *Wizard*, 321.

84. Scherff to NT, 10 April 1906, in Seifer, *Wizard*, 322.

85. Tom Reiss, "The First Conservative" [on Peter Viereck, G. S. Viereck's son], *New Yorker*, 24 October 2005, pp. 38–47, on 40.

86. NT to Viereck, 17 December 1934. In the original letter, these three paragraphs constituted a portion of one very long paragraph, but I have broken it up here to make it easier to read.

87. Ibid.

CHAPTER SIXTEEN
VISIONARY TO THE END

1. John G. Trump to Walter Gorsuch, 30 January 1943, Freedom of Information Act file (hereater cited FOIA file) for Nikola Tesla, Federal Bureau of Investigation, pp. 174–81, on 175, http://www.scribd.com/.

2. Frank Parker Stockbridge, "Will Tesla's New Monarch of Machines Revolutionize the World?" *New York Herald*, 15 October 1911, in Jeffrey A. Hayes, ed., *Tesla's Engine: A New Dimension for Power* (N.p.: Tesla Engine Builders Association, 1994), 22–36, on 35. In the late 1980s, Canadian researchers flew a model plane that was powered by an electric beam; see William J. Broad, "New Kind of Aircraft Is on Horizon as Designers Try Microwave Power," *New York Times*, 21 July 1987.

3. Tom Crouch, *The Bishop's Boys: A Life of Wilbur and Orville Wright* (New York: W. W. Norton, 1989), 244–45.

4. This explanation of the Tesla turbine is based on discussions with my engineering colleague Robert Ribando. See also William Harris, "How the Tesla Turbine Works," http://auto.howstuffworks.com/tesla-turbine.htm/printable.

5. Stockbridge, "Tesla's New Monarch of Machines," 27.

6. O'Neill, *Prodigal Genius*, 218–21.

7. NT to JJA, 22 March 1909, in Seifer, *Wizard*, 336; "Tesla Says He Has New Power Secret," *New York Herald*, 20 May 1909 in TC 18:146; "Southern Iron Merger Plan," *New York Times*, 2 April 1911; NT, "Fluid Propulsion," U.S. Patent No. 1,061,142 (filed 21 October 1909, granted 6 May 1913) and "Turbine," U.S. Patent No. 1,061,206 (filed 21 October 1909, granted 6 May 1913).

8. O'Neill, *Prodigal Genius*, 222–24.

9. Seifer, *Wizard*, 362–66.

10. See http://www.discflo.com/; http://www.phoenixnavigation.com/ and http://www.teslaengine.org/main.html.

11. "Tesla Has Only Credit," *New York Times*, 18 March 1916.

12. "Tesla's Discovery: Nobel Prize Winner," *New York Times*, 7 November 1915; NT to RUJ, 16 November 1915, in Seifer, *Wizard*, 380.

13. O'Neill, *Prodigal Genius*, 229–37, with quote from 231. For the delay between announcement and award of the medal, see "Tesla-Honors-1" Notecard, KSP. See also NT, Edison Medal Speech.

14. "Developments in Wireless Telegraphy," *Electrical World* 39 (29 March 1902): 540.

15. Friedrich Heilbronner, "Marconi and the Germans" (paper presented at Marconi09, Museo della Tecnica Elettrica, Pavia, Italy, October 2009); Linwood S. Howeth, *History of Communications-Electronics in the United States Navy*, chap. 19, "Operations and Organization of the United States Naval Radio Service during Neutrality Period," sec. 3, "Operation of the Tuckerton and Sayville Stations," http://earlyradiohistory.us/1963hw19.htm.

16. "Tesla Sues Marconi on Wireless Patent," *New York Times*, 4 August 1915; NT, Radio Testimony; Peterson quote is from http://www.tfcbooks.com/teslafaq/q&a_022 .htm; Leland I. Anderson, *Priority in the Invention of Radio—Tesla vs. Marconi* (Brecken-ridge, CO: Twenty-First Century Books, n.d.); A. David Wunsch, "Misreading the Su-preme Court: A Puzzling Chapter in the History of Radio," *Antenna* 11, no. 1 (November 1998), http://www.mercurians.org/1998_Fall/Misreading.htm.

17. Branimir Jovanović, "Nikola Tesla-Research Methodology in the Light of Facts Discovered during Reconstruction of His Work on Bladeless Pumps from 1908–1911" (paper presented at ICOHTEC meeting, Belfort, France, July 1998), 8.

18. NT, "Speed Indicator," U.S. Patent No. 1,209,359 (filed 29 May 1914, granted 19 December 1916); "Frequency Meter," U.S. Patent No. 1,402,025 (filed 18 December 1916, granted 3 January 1922); "Speed-Indicator," U.S. Patent No. 1,274,816 (filed 18 Decem-ber 1916, granted 6 August 1918); "Ship's Log," U.S. Patent No. 1,314,718 (filed 18 De-cember 1916, granted 2 September 1919); and "Flow-Meter," U.S. Patent No. 1,365,547 (field 18 December 1916, granted 11 January 1921).

19. For the third quarter of 1918, Waltham Watch paid Tesla $165.80 in royalties on 829 speedometers sold; see F. C. Graves to NT, 15 October 1918, KSP.

20. Advertisement for Waltham Speedometer, *New York Times*, 8 June 1921, http:// blog.hemmings.com/index.php/2010/06/17/nikola-teslas-pound-per-horsepower -engine/#more-2549.

21. Exhibit C, Trump to Gorsuch, 30 January 1943; NT to U.S. Steel, 26 July 1931 and Agreement between NT and American Smelting and Refining Co., n.d., KSP.

22. "Tesla Judgment Filed: Inventor Had Paid Lawyer with Promissory Note," *New York Times*, 14 June 1925; F. A. Merrick to NT, 2 January 1934, LC; Hugo Gernsback, "Westinghouse Recollections," Westinghouse Broadcasting Company, *Engineering Con-tours* 5, no. 1 (January 1960), in Box 18, Folder 4, KSP. In addition, Tesla was sued by the New York Telephone Company in 1922 for failing to pay a debt amounting to $107.32 and by Brentano's Book Stores in 1941 for nonpayment of $149; see "Tesla-Money-Debts" Notecard, KSP.

23. H. Winfield Secor, "Tesla's View on Electricity and the War," *Electrical Experi-menter* 5 (August 1917): 229ff.; quote is from entry for Girardeau, http://en.wikipedia .org/wiki/%C3%89mile_Girardeau.

24. Jonathan Coopersmith, *The Electrification of Russia, 1880–1926* (Ithaca: Cornell University Press, 1992).

25. NT, *My Inventions*, 99–100. Tesla also mentions negotiations with Lenin in his letter to J. P. Morgan Jr., 29 November 1934, http://www.tfcbooks.com/tesla/1935-00 -00.htm.

26. NT, Radio Testimony, 185.

27. On Tesla and pigeons, see O'Neill, *Prodigal Genius*, 307–17. Tesla left the Waldorf-Astoria in 1922 and moved subsequently to the Hotel St. Regis, the Pennsylvania Hotel,

the Governor Clinton Hotel, and finally the Hotel New Yorker. He left the St. Regis in 1924 after racking up an unpaid bill of $993.41, and in 1941 he owed the New Yorker $172.85. See "Tesla-Money-Debts" Notecard, KSP.

28. For a sampling of these letters, see Vojin Popvic, *Tribute to Nikola Tesla Presented in Articles, Letters, Documents* (Beograd: NTM, 1961), LS 25–63. See also "Foundation of the Nikola Tesla Institution in Belgrade, Yugoslavia (March 2, 1936)," http://www .teslasociety.com/ntinn.htm; "Science: Tesla at 75," *Time*, 20 July 1931, http://www.time .com/time/magazine/article/0,9171,742063,00.html; and O'Neill, *Prodigal Genius*, 275.

29. David Dietz, "Tesla Wiggles Toes," newspaper clipping, 15 July 1936, Box 6, Folder 17, KSP.

30. King Peter of Yugoslavia presented Tesla with the Golden Cross of the White Eagle while the minister of Czechoslovakia awarded him the Gold Cross of the Order of the White Lion.

31. Quoted in Cheney and Uth, *Master of Lightning*, 151–52, but see also 142.

32. "Tesla at 78 Bares New Death-Beam," *New York Times*, 11 July 1934.

33. NT, as told to George Sylvester Viereck, "A Machine to End War," *Liberty*, February 1937, http://www.tfcbooks.com/tesla/1935-02-00.htm.

34. Grindell Matthews greatly admired Tesla and told his biographer that "Whenever I had some little success which might have turned my head, I always thought of Tesla, and realized that I was a mere student sitting at the feet of a great master." E.H.G. Barwell, *The Death Ray Man: The Biography of Grindell Matthews, Inventor and Pioneer* (London: Hutchinson, 1943), 109; Jonathan Foster, "The Death Ray: The Secret Life of Harry Grindell Matthews," http://www.harrygrindellmatthews.com/theDeathRay.asp.

35. NT, "The New Art of Projecting Concentrated Non-dispersive Energy through Natural Media," n.d., http://www.tfcbooks.com/tesla/1935-00-00.htm.

36. All three quotes are from Exhibit F, Trump to Gorsuch, 30 January 1943. On Trump's background, see Louis Smullin, "John George Trump, 1907–1985," in National Academy of Engineering, *Memorial Tributes* (Washington, DC: National Academy Press, 1989), 3:332–37, http://www.nap.edu/openbook.php?record_id=1384&page=332.

37. Paul J. Nahin, *The Science of Radio*, 2nd ed. (New York: Springer-Verlag, 2001), 11–12.

38. Seifer, *Wizard*, 431–34; Leland Anderson, *Nikola Tesla's Residences, Laboratories, and Offices* (Denver: Boyle & Anderson, 1990).

39. NT to J. P. Morgan Jr., 29 November 1934, http://www.tfcbooks.com/tesla/1935 -00-00.htm.

40. Foxworth to Director, 9 January 1943, FOIA file, 8–9; Cheney, *Man Out of Time*, 276.

41. Breckinridge Long to Secretary of State, 12 July 1934, in Cheney and Uth, *Master of Lightning*, 145. On Long, see http://www.breckinridge.com/breckbio.htm.

42. Exhibits Q and D, Trump to Gorsuch, 30 January 1943. On Amtorg's spying activities, see Frank J. Rafalko, *A Counterintelligence Reader*, vol. 3, *Post–World War II to Closing the 20th Century*, chap. 1, p. 22, http://www.fas.org/irp/ops/ci/docs/ci3/index.html.

43. John J. O'Neill, "Tesla Tries to Prevent World War II" (unpublished chap. 34 of *Prodigal Genius*), http://www.pbs.org/tesla/res/res_art12.html; Exhibit H, Trump to Gorsuch, 30 January 1943.

44. O'Neill, "Tesla Tries to Prevent World War II."

45. "Aerial Defense 'Death Beam' Offered to U.S. by Tesla," *Baltimore Sun*, 12 July 1940; William L. Laurence, "'Death Ray' for Planes," *New York Times*, 22 September 1940,

both available at http://www.tfcbooks.com/tesla/1935-00-00.htm#1940-09-22; [Name deleted] to J. Edgar Hoover, 24 September 1940, FOIA file, 3.

46. "Nikola Tesla Dies: Prolific Inventor," *New York Times*, 8 January 1943; O'Neill, *Prodigal Genius*, 276.

47. "Tesla-Mark Twain-Sends $100" Notecard, KSP.

48. E. E. Conroy to Director, FBI, 17 October 1945, FOIA file, 170–73; "Purple Plates—Legacy of Nikola Tesla," http://www.essentia.ca/PurplePlate/purpTesla.htm.

49. Quoted in Aleksandar S. Marincic, "Excerpt: The Tesla Museum," http://www.teslasociety.com/tmuseum.htm.

50. See "2000 Are Present at Tesla Funeral," *New York Times*, 13 January 1943. On Kosanović's decision to cremate Tesla's remains, see "Commemoration for Nikola Tesla's Death Will Be Held by the Serbian Orthodox Church in Belgrade, in Saborna Crkva, on January 23, 2006," http://www.teslasociety.com/ntcom.htm.

51. The starting point of this speculation is that O'Neill claimed in *Prodigal Genius* (p. 277) that FBI agents had entered Tesla's room the day after he died and removed papers related to a secret invention from the safe.

52. Charlotte Muzar, "The Tesla Papers," *The Tesla Journal*, nos. 2&3 (1981–82): 39–42, on 39–40.

53. Foxworth was killed in a plane crash a few days later when he was flying on a secret mission to Dutch Guiana; see Athan G. Theoharis, *The FBI: A Comprehensive Reference Guide* (Westport, CT: Greenwood, 1999), 326–27. On Spanel, see "A. N. Spanel, 83; Inventor, Manufacturer, Activist" [obituary], *Los Angeles Times*, 5 April 1985. See also Foxworth to Director, 9 January 1943 and E. E. Conroy to Director, FBI, 17 October 1945, FOIA file, 8–9 and 170–73.

54. E. E. Conroy to Director, FBI, 17 October 1945.

55. Foxworth to Director, 9 January 1943; D. M. Ladd to [name deleted], 11 January 1943; and Edw. A. Tamm to Ladd, 12 January 1943, in FOIA file, 8–12. On Ladd and Tamm's careers at the FBI, see Theoharis, *The FBI*, 338, 356.

56. Smullin, "John George Trump."

57. George specialized in counterespionage, breaking into the offices, desks, and safes of suspected Nazi spies at night and gathering incriminating evidence for prosecution. He later became the chief instructor of surreptitious-entry techniques for the Office of Strategic Services, leading his own lockpicking and safecracking team in wartime Europe. See his book, *Surreptitious Entry* (New York: Editions for the Armed Services, 1946), as well as http://www.textfiles.com/anarchy/WEAPONS/crimecat.004.

58. Cheney, *Man Out of Time*, 276.

59. Trump to Gorsuch, 30 January 1943, p. 174.

60. Bogdan Raditsa, "Red Ambassadors: Sava Kosanovich of Yugoslavia," *Plain Talk*, March 1948, pp. 6–10, in FOIA file, 211–13.

61. E. E. Conroy to Director, FBI, 17 October 1945; Muzar, "The Tesla Papers," 40.

62. Muzar, "The Tesla Papers," 41–42.

63. E. E. Conroy to Director, FBI, 17 October 1945; D. M. Ladd to the Director, 3 April 1950; L. B. Nichols to Tolson, 30 January 1951; [Paul Snigier] to Clarence Kelly, 20 April 1976; Lt. Col. Allan J. MacLaren, USAF, to FBI Director, 9 February 1981, all in FOIA file, 171–73, 195–96, 253–55, 107–110, 124–25, respectively.

64. Clarence A. Robinson Jr., "Soviets Push for Beam Weapon," *Aviation Week & Space Technology* 106 (2 May 1977): 16–23.

65. SAC, Cincinnati, to Director, FBI, 18 August 1983, FOIA file, 133–35.

66. John Pike, "The Death-Beam Gap: Putting Keegan's Follies in Perspective," October 1992, E-Print, Space Policy Project, Federation of American Scientists, http://www.fas.org/spp/eprint/keegan.htm.

67. David E. Hoffman provides a brief discussion of the problems with laser beam weapons in *The Dead Hand: The Untold Story of the Cold War Arms Race and Its Dangerous Legacy* (New York: Doubleday, 2009), 276.

EPILOGUE

1. Brisbane, "Our Foremost Electrician."

2. Wachhorst, *Thomas Alva Edison*; Watt, *The People's Tycoon*.

3. Illustrative of Tesla's outsider status in the 1950s is that biographers Kenneth Swezey and Leland Anderson found it extremely difficult to get major organizations—the National Academy of Science, the Post Office, the Smithsonian Institution—to take any significant interest in marking the centennial of Tesla's birth in 1956.

4. Arthur H. Matthews, *The Wall of Light: Nikola Tesla and the Venusian Space Ship, the X-12* (Pomeroy, WA: Health Research Books, 1971); Margaret Storm, *Return of the Dove* (Baltimore: Margaret Storm Publication, n.d.).

5. 1892 Lecture, 236.

6. Paul Sagan, *Ball Lightning: Paradox of Physics* (New York: iUniverse, 2004), 307–24.

7. Michael Riversong, "International Tesla Society in Review: People, Politics, and Technology," 2002, http://home.earthlink.net/~rivedu/14tesla.html; "Tesla Engine Builders Association," http://www.teslaengine.org/main.html; "Tesla Universe," http://www.teslauniverse.com/; "Tesla Memorial Society of New York," http://www.teslasociety.com/.

8. Nevill Drury, *The New Age: Searching for the Spiritual Self* (London: Thames and Hudson, 2004), 12.

9. F. David Peat, *In Search of Nikola Tesla*, rev. ed. (London: Ashgrove, 2003).

10. Seifer, *Wizard*, 460–61.

11. *Nikola Tesla: Discovering the Future* (Swiss Tesla Institute, 2008), p. 28, http://swisstesla.com/.

12. The critique that technologists have no soul is very much a 1960s counterculture attack based on a skewed reading of C. P. Snow's *The Two Cultures and the Scientific Revolution* (New York: Cambridge University Press, 1959). For a challenge to this critique, see Samuel P. Florman, *The Existential Pleasures of Engineering* (New York: St. Martin's Press, 1976).

13. Posting by Mike, 12 January 2009, on "Feel the Heat: Tesla Roadshow Hits Miami during Art Basel" Blog, Tesla Motors, http://www.teslamotors.com/blog3/?p=88.

14. Samantha Hunt, *The Invention of Everything Else* (Boston: Houghton Mifflin, 2008); Jeffrey Stanley, *Tesla's Letters: A Play in Two Acts* (New York: Samuel French, 1999; Daniel Michaels, "Long-Dead Inventor Nikola Tesla Is Electrifying Hip Techies," *Wall Street Journal*, 14 January 2010, p. 1. For a list of Tesla-inspired movies, books, and video games, consult "Nikola Tesla in Popular Culture," http://en.wikipedia.org/wiki/Nikola_Tesla_in_popular_culture#Music.

15. McCraw, *Prophet of Innovation*; Christensen, *The Innovator's Dilemma*.

16. On the history of radio, see Hugh G. J. Aitken, *The Continuous Wave: Technology and American Radio, 1900–1932* (Princeton: Princeton University Press, 1985); Susan Douglas, *Inventing American Broadcasting, 1899–1922* (Baltimore: Johns Hopkins University Press, 1987); Tom Lewis, *Empire of the Air: The Men Who Made Radio* (New York: Edward Burlingame Books, 1991); and Hong, *Wireless*. On technological change as an evolutionary process, consult George Basalla, *The Evolution of Technology* (New York: Cambridge University Press, 1988) and John Ziman, ed., *Technological Innovation as an Evolutionary Process* (New York: Cambridge University Press, 2000).

17. "Nikola Tesla: Dr. A.P.M. Fleming's Address," *Electric Times*, 2 December 1943, pp. 656–59, on 659.

18. Anderson, "Stone on Tesla's Priority in Radio," 40.

19. Wisehart, "Making Your Imagination Work for You," 62.

20. Curtis Brown, "A Man of the Future," *Savannah Morning News*, 21 October 1894 in TC 9:84–87, on 85.

21. Eugene S. Ferguson, *Engineering and the Mind's Eye* (Cambridge, MA: MIT Press, 1992).

22. Carlson, "Entrepreneurship in the Early Development of the Telephone"; Isaacson, *Steve Jobs*.

23. For a discussion of how Alexander Graham Bell exploited the ambiguity of thinking about the telephone as equivalent to the human ear, see Michael E. Gorman et al., "Alexander Graham Bell, Elisha Gray, and the Speaking Telegraph: A Cognitive Comparison," *History of Technology* 15 (1993): 1–56.

24. David E. Nye, *Technology Matters: Questions to Live With* (Cambridge, MA: MIT Press, 2006), 3–6.

25. NT to Scherff, 6 September 1899, in Ratzlaff and Jost, *Tesla/Scherff Correspondence,* 109.

26. Here I am reminded of a lesson that I learned from a senior engineer at Corning Incorporated: customers always want you to invent something they can use next year, but the company only makes money if you invent something that the customer will need in five years.

27. B. A. Behrend, "Dynamo-electric Machinery and Its Evolution during the Last Twenty Years," *Western Electrician*, 28 September 1907, pp. 238–40 in TC 18:46–50, on 238.

28. George H. Douglas, *The Golden Age of the Newspaper* (Westport, CT: Greenwood, 1999), 95–116.

29. TCM, "Nikola Tesla," *The Century Magazine* 47 (February 1894): 582–85, on 583.

30. [Kenneth Swezey] to NT, 7 July 1924, Box 17, Folder 9, KSP.

31. NT, "Problem of Increasing Human Energy," 182.

32. NT to JPM, 11 December 1903.

ACKNOWLEDGMENTS

As Hillary Clinton said, quoting an African proverb, "It takes a village to raise a child." Well, if it takes a village to raise a child, then it takes a whole town to write a book like this—and that town better be home to a variety of experts and have a decent technical library!

This book has taken me fifteen years to write, and I have enjoyed support from several sources. First and foremost, I am grateful to the Alfred P. Sloan Foundation for a research grant that covered two years of full-time research and writing. At the foundation, Doron Weber early on recognized the need for a reliable biography of Tesla and has waited patiently for me to produce this volume. At the University of Virginia, my work on Tesla has been supported by a grant from the Bankard Fund for Political Economy, by a Sesquicentennial Leave underwritten by my colleagues in the Department of Engineering and Society, as well as by funds provided by the School of Engineering and Applied Science. The funds from the Engineering School allowed me to maintain the momentum I had gained while working under the Sloan grant, and I am grateful to Professor Ingrid Townsend and Dean Richard Miksad for arranging for this extra support. In addition, a last-minute grant from James Aylor, the dean of the School of Engineering and Applied Science, University of Virginia, has underwritten the inclusion of the book's illustrations in future paperback and digital editions.

Over the years, I have had the opportunity to spend time at several institutions where I could concentrate exclusively on this project. In the winter of 1999, I was a visiting professor in the Science, Technology, and Society Program at Stanford University, and my thanks to Tim Lenoir and Robert McGinn for arranging that visit. In the fall of 2005, the Lemelson Center for Invention and Innovation at the Smithsonian Instituion invited me to be a research fellow, and during that stay I learned a great deal from Arthur Molella, Joyce Bedi, and Maggie Dennis. Also during the fall of 2005, I was a visitor at the Centre for the History of Science, Technology, and Medicine at the University of Manchester; my thanks to John Harwood for arranging this visit as well as to my English family in Lymm who took me in and kept me well fed. In the summer of 2010, I had the good fortune to be a scholar in residence at the Deutsches Museum in Munich where I finished a first draft of this book; my thanks to Helmuth Trischler and Andrea Walther for making that appointment possible.

This book is based on research using documents and artifacts found in a large number of archives and museums. I have enumerated the full list of research repositories used in the Note on Sources, and here I want to acknowledge the institutions and individuals who have made a significant difference in my effort to find Tesla-related materials.

At the head of this list must come the Nikola Tesla Museum in Belgrade, Serbia, which holds all of the papers and material in Tesla's possession when he died in 1943. I have been the beneficiary of assistance and guidance from the staff of the Tesla Museum, especially several of its directors: Alexander Marincic, Branimir Jovanović, Marija Sesic, and Vladimir Jelenković have all encouraged me in my efforts to understand Tesla and provided me with a great deal of useful information. Ivana Zoric, also on the staff of the Tesla Museum, has been a great help in securing the photographs illustrating this book.

Beyond the Tesla Museum, I am grateful for the assistance of several archivists and curators, especially Sheldon Hochheiser at the IEEE Archives, Leonard de Graaf at the Edison National Historic Site, and Marc Greuther at the Henry Ford Museum and Greenfield Village. I also wish to thank Jill Jonnes for introducing me to Robert Dischner at National Grid USA, who in turn provided me with copies of Tesla's correspondence with Edward Dean Adams. In the early stages of the project, I was lucky to have several capable research assistants, particularly John Bozeman and Amie Loyer. For assistance with translating several German sources, I am grateful to Ingrid Townsend and Arthur Byrne. And thanks to Bill Kelsh, Pam Lutz, and Adarsh Ramkrishnan for their help in preparing the illustrations.

I also wish to acknowledge all of the important work that Leland Anderson has done in relation to finding and preserving materials relating to Tesla. For well over fifty years, Leland has devoted much of his energy to ensuring that Tesla is not forgotten, and I detected his quiet efforts working in the background as I used various archival collections and publications.

As a historian, I have turned to several technical experts for help in understanding how Tesla's inventions worked, and I am grateful for the advice I have received from Sean Grimes, Gary Peterson, Paul Nahin, Robert Ribando, David Wunsch, and Antonio Pérez Yuste. I know that there are multiple opinions concerning how Tesla's later inventions relating to broadcasting power worked (or failed to work), and I take full responsibility for the interpretation presented here.

At Princeton University Press, it has been my pleasure to have Ingrid Gnerlich as my editor; Ingrid was incredibly patient in waiting for this book to come to fruition and was always a source of good cheer and sound advice. Ingrid's assistant, Eric Henney, was marvelous in helping me secure permission for all of the illustrations. Jennifer Backer did a superb job in copyediting the manuscript and Tobiah Waldron prepared the wonderful index. Debbie Tegarden has provided the expert skills needed to shepherd my manuscript through the production process, ensuring that all of the details have come together to make this into a book that is a pleasure to read and to hold.

Many friends and professional colleagues have listened to my presentations on Tesla and offered valuable advice. These individuals include Margy Avery, Wiebe Bijker, Oskar Blumtritt, Paolo Brenni, Jack Brown, Lynn Burlingame, Susan Douglas, Robert Fox, Mike Gorman, Anna Guagnini, Vigen Guroian, Meg Graham, Eric Hintz, Jeff Hughes, Richard John, Ron Kline, John Krige, Gunther Luxbacher, Keith Nier, David Nye, Bryan Pfaffenberger, Trevor Pinch, Klaus Plitzner, Stuart Sammis, Alex Wellerstein, Karin Zachmann, and Olivier Zunz. Bernard S. Finn and Michael B. Schiffer reviewed the manuscript for Princeton University Press and provided very helpful comments. My daughters, Julia and Rachel, grew up listening to far too many dinnertime orations about Tesla, but they nonetheless regularly offered comments that kept me on that narrow path between scholarship and crackpot ideas. And I am very grateful to one of my professors, Robert Kohler, who reminded me at a critical point in the writing process not to pull any punches and to develop a thoughtful and compelling framework for understanding Tesla.

This book is dedicated to two people who are very dear to me. Tom Hughes has been a mentor, professionally and personally, for thirty years and taught me much of what I know about inventors. Early on, he recognized the importance of understanding Tesla's work and pushed me hard to make this book happen. Finally, this book is also dedicated to my wife, Jane, who has played countless roles in this book and in my life. For this book, she has been a sounding board for theories, the organizer of research trips, and editor extraordinaire; in my life, Jane has been—and always will be—that anchor of love and hope that keeps me going. Jane protects me from false illusions and encourages me to pursue dreams and ideals.

INDEX

Page numbers in italics refer to illustrations in the book.

Abafi (Jósika), 23

Abbey, Ned, 215

AC motors, 3; alternators and, 44, 119, 122, 133–35, 138, *139*, 155, 160–61, 181, 211; Brown and, 81, 85, 87, 90, 92–93, 96–97, 99, 100, 103–4, 113, 118–19, 193, 207, 294; demonstrations and, 105, 195; development of, 63–64, 67, 70–71, 74, 77, 81, 85–87, 90–98, 145–46, 403, 405; disruptive innovation and, 402; economic effects of, 402; Ferraris and, 110–11, 146; ideal of, 371; illusion and, 412; impact of, 402; Martin and, 105, 195; mental engineering and, 43–45, 63; out-of-phase circuits and, 84–85; Page and, 97; patents and, 52, 55–56, 64, 93–95, 114, 118, 145, 159, 167, 186, 305, 361, 429n6, 439n8; Peck and, 81, 87, 90, 92–93, 100–105, 107–8, 111–13, 116, 118–19, 193, 207, 294, 318, 405–7; polyphase, 92–99, 104–7, 110–15, 144–45, 158–63, 166–75, 257, 402, 428n42, 434n49; Pöschl lectures and, 41; power transmission and, 140; promotion of, 100, 103–15; rotating magnetic field and, 4, 9–10, 52, 54–64, 67, 84–85, 91–98, 106–12, 124, 169–70, 209, 294, 302, 364, 369, 371, 404, 410, 429n5, 429n6, 436n31; selling idea of, 100–116; single-phase, 88, 96, 98–99, 112, 114, 119, 144, 159–61, 166–68, 170,

173; six-wire scheme and, 64–65, 96, 144, 410; slip rings and, 39, *93*, 105, 433n23; split-phase, 95–100, 104, 110–15, 123, 159; Stanley and, 112–13; stator coils and, 37, *39*, 40, 44, 52–55, 64–68, 85, *86*, 92, *93*, 96–97, 100, 106, 115, 119, 166, 209, 404; Szigeti and, 52, 64; Tesla's legacy and, 396, 405, 410; Thomson and, 106–7; Twain and, 186; two-phase, 159, 161, 169–73; utility industry and, 195; waste heat and, 82, 97, 110; Westinghouse and, 108, 112–18; wireless power and, 209

Adams, Edward Dean, 219, 302; Betts and, 169; Bradley and, 169; Cataract Company and, 164–65; Edison Electric Light Company and, 164; Morgan and, 164, 406; Niagara Falls project and, 164–74, 176, 205, 416; Nikola Tesla Company and, 205–7, 255; polyphase alternating current (AC) and, 167–75; railroads and, 206; Rankine and, 164, 205–7, 259, 353–54; White and, 214–15

Adams, Ernest, 205, 348

Adams, Henry, 342

Aitken, Hugh, 122

Alabama Consolidated Coal and Iron, 372

Albert, Prince of Belgium, 153, 256

Alexander I, King of Serbia, 155

Alley, Dickenson, 220–21, 270, 289, 295–301, 305

Allgemeine Elektricitats-Gesellschaft, 168

Allis-Chalmers, 373

Alta Vista Hotel, 265, 278

alternating current (AC): AC motors and, 3–4, 9 (*see also* AC motors); alternators and, 44, 119, 122, 133–35, 138, *139*, 155, 160–61, 181, 211; arc lighting and, 44–45, 61, 67, 87, 95–96, 108, 120, 171; Bláthy and, 61, *62*, 88; Columbia College lecture and, 133–38; conductors and, 66; copper and, 65, 88, 91, 96, 160; demonstrations and, 78, 85, 89–92, 96, 144–45, 160, 166, 172–73, 215; Déri and, 61, *62*, 88; disruptive innovation and, 402; egg of Columbus and, 90–92, 103, 172, 257, 407; electromagnetism and, 82, *86*; Ganz and Company and, 60–63, 87–88, 155, 158, 258, 429n2, 429n6; Gaulard and, 87–89, 108–9, 112; generators and, 6, 44, 61, 67, 85, 119, 181, 188, 277, 333, 429n48; Gibbs and, 87–89, 108–9, 112; grounding and, 139, 141, 157, 403, 449n43; high-frequency phenomena and, 4, 249 (*see also* high-frequency phenomena); incandescent lighting and, 88; increasing transmission length of, 138–42; magnifying transmitters and, 268; manufacturers and, 88, 99, 160, 164, 167–68, 173, 304; Niagara hydroelectric project and, 4, 162–67, 170, 172–76, 205, 214, 237, 256, 342–44, 353–55, 365, 416; polyphase, 92–99, 104–7, 110–15, 144–45, 158–63, 166–75, 257, 402, 428n42, 434n49; power transmission and, 95–96, 105–6, 159, 163, 165, 170, 175, 313; promotion of, 100–116, 158–75; pyromagnetic generator and, 76, 81–84; rise of, 85–90; single-phase, 88, 96, 98–99, 112, 114, 119, 144, 159–61, 166–68, 170, 173; skin effect and, 125, 136; split-phase, 95–100, 104, 110–15, 123, 159; Tesla coils and, 120–25, 128, 218, 248–49, 282, 338, 353, 403, 423n6, 450n17, 462n6; thermomagnetic motors and, 76–82; three-phase, 143–45, 159–61, 169–70, 402, 410; three-wire system and, 65, 96, 144; transformers and, 6, 88–89, 111, 184; two out-of-phase currents and, 84–85;

two-phase, 159, 161, 169–73; vs. direct current (DC), 44–45, 167–68, 170–71; wireless lighting and, 207–13 (*see also* wireless lighting); Zipernowski and, 61, *62*, 88

alternators: 44, 119, 122, 133–35, 138, *139*, 155, 160–61, 181, 211; electromagnets and, 119–20; frequency and, 119, 122, 133–35, 138, 181, 211; heat and, 122; high-power, 155; larger, 155, 160; multiphase current and, 161; new ideals and, 119, 122; synchronous speeds and, 119; wireless lighting and, 181

American DeForest Wireless Telegraph Company, 349

American Electric Manufacturing Company (AEM), 72–73, 431n40

American Institute of Electrical Engineers (AIEE): AC motor and, 51–52, 103, 105–7, 116; Edison Medal and, 240, 376; European delegation of, 117; founding of, 195; lectures at, 51–52, 105–7, 134, 195; Marconi and, 335–37; Martin and, 195

American Marconi, 377–78

American Marconi v. the United States, 378

American Museum of Natural History, 317

American Red Cross, 275, 315

American Smelting and Refining Company (ASARCO), 378

American Union Telegraph Company, 79

American Wireless Telephone and Telegraph Company, 285, 348

America's Cup, 283, 285–88, 311, 332, 348

Amtorg Trading Corporation, 388

Anderson, Leland, 240, 416, 452n62

Anthony, William, 103–7, 116, 134, 294

Apple Computer, 112, 406

Arago, François, 52–53, 55, 110, 428n43

Arago's wheel, 52–*53*

arc lighting: Alley and, 296; alternating current (AC) and, 44–45, 61, 67, 87, 95–96, 108, 120, 171; Brush and, 72; capital needed for, 72–73; direct current (DC) and, 67; economic issues and, 76; Edison and, 71–74, 431n40; Goff and, 72–73; Gramme dynamos and, 41; high-frequency generators and, 120; installation issues and, 72–73; Jablochkoff and, 44, 61; Lontin and, 44; manufacturing and, 72–75; mechanism of, 427n8; on single circuit, 44; oscillators and, 135, 182; patents and, 72–75, 431n46; Rahway and, 73–75, 118; Shallenberger and, 108; Strasbourg system and, 66–69; street lighting and, 120; Tesla Electric Light and Manufacturing Company and, 73, 75, 431n44; Tesla's streamer picture and, 297; Thomson-Houston and, 72

Armstrong, Edwin Howard, 390, 402

Army Signal Corps, 283

Astor, John Jacob, 256

Astor, John Jacob, IV, 214, 417–18; background of, 256–57; Cataract Construction Company and, 256; Delmonico's and, 256–57; Ferncliff laboratory and, 256; financial support from, 255–61, 264, 303, 305, 313, 315, 353, 361, 371; Nikola Tesla Company and, 257; poor response of, 305, 353; science-fiction novel of, 256; telephone and, 256; as tinkerer, 256; turbines and, 371–72; wireless lighting and, 257–59, 315, 353; wireless power transmission and, 255–59

Astor, Mrs. John Jacob, 239

Astor, William Waldorf, 256

Astor Gallery, 335

Astor House, 118, 158

Atlantic & Pacific Telegraph Company, 79

Atlantic Communication Company, 377

Augusta Victoria (ship), 158

Austro-Hungarian Empire, 3, 13–14, 34, 48, 129, 138

automatons, 25, 48, 225, 234, 236–37, 367

Aviation Week & Space Technology magazine, 394

Babbage, Charles, 52–55

Baily, Walter, 55–56, 61

Baker, George F., 79

Baltimore & Ohio Railroad, 79

Baltimore Sun, 388–89

Bankovic, Colonel, 28

Baring Brothers, 130

Barnes, John Sanford, 357

Baruch, Bernard, 347

Batchelor, Charles, 63, 66, 69–70, 430n31

Bauzin, M., 68, 68–69

Behrend, B. A., 376

Bell, Alexander Graham, 8, 10, 21–22, 49, 129, 138, 406, 435n59, 471n23

Bell, Melville James, 22

Bell, Ted, 22

Belmont, August, 130–31

Bennett, James Gordon, 203

Bergmann, Sigmund, 373

Bergstresser, Ralph, 399

Bessemer, Henry, 8, 115

Bettini, Gianni, 197

Betts, Frederick H., 169

Bjerknes, Vilhelm, 117

Bláthy, Ottó, 61, 62, 88

Blondel, André, 153

Boldt, George C., 379

Booth, Edwin, 215

Boston Public Library, 214

Boulton, Matthew, 406

boundary layer, 369, 373

Brackett, Cyrus F., 232

Bradley, Charles S., 169

Bragg, William H., 375

Bragg, W. L., 375

Branly, Édouard, 228, 438n29

Braun, Ferdinand, 377

breakwheels, 268, 279, 286, 287

Brisbane, Arthur, 1–7, 396

British Post Office, 146, 313

Brown & Sharpe Works, 118

Brown, Alfred S.: AC motors and, 81, 85, 87, 90, 92–93, 96–97, 99, 100, 103–4, 113, 193, 207, 294; Anthony and, 103–4, 294; background of, 77–78; business dealings of, 100–107, 113, 116, 119, 205, 207, 259, 302, 305, 405–6; creative synergy with Tesla, 405–6; Edison and, 78–79; egg of Columbus and, 91, 407; lab space issues and, 118; patents and, 79, 100–101, 205, 305, 407; power transmission and, 433n26; pyromagnetic generator and, 81–84; royalties and, 119, 130; telegraph and, 77–80; Tesla Electric Company and, 73–75, 81, 84, 118, 431n44, 431n50; thermomagnetic motor and, 77–81; tin-can motor and, 85, 407; Western Union and, 78–80; Westinghouse deal and, 111–13

Brown, Charles E. L., 144–45

Brush, Charles, 74, 87, 129, 138

Brush Electrical Light Company, 72, 74

Budapest, 49–57

Budd company, 373

Bush, Vannevar, 380

button lamps, 133–34, 149–50, 439n10

Byllesby, Henry M., 111, 113

Canadian Niagara Power, 354–57
capacitors: breakwheels and, *268*, 279, 286, *287*; coherer and, 271; copper and, 279; dielectric properties of bottles and, 281; discharging of, 123–28, 132, 135, 184, 188, 191, 248, 268, 271–72, 278–79, 289, *296*, 297–99, 382, 438n23, 462n6; electromagnetic relays and, 180; electrostatic thrusts and, 123–28, 132, 135, 191; energy storage and, 123, 184, 188–89; Fleming and, 333; grounding and, 271; increasing power transmission length and, 138–42; induction coils and, 123–24, 139–41, 180, 211, 262, 438n21; magnifying transmitters and, 268–69; mathematical analysis of, 438n21; Maxwell and, 438n21; motors and, 96; oscillators and, 188–89; pendulum effect and, 123–24; receivers and, 180; release mechanism for, 184; resonance and, 124 (*see also* resonance); spark gap and, 184; stationary waves and, 262, 268, 271, 279–81, 286; Tesla coils and, 120–25, 128, 218, 248–49, 282, 338, 353, 403, 423n6, 450n17, 462n6; Wardenclyffe and, 321, *324*, 328; wireless lighting and, 180, 184, 188–89
Čapek, Karel, 225
Capote, Truman, 368
Carnegie, Andrew, 314
Carter, Jimmy, 394
Castleton Hotel, Staten Island, 349
Cataract Construction Company, 164–74, 245, 256
Cavendish, Henry, 35
cell phones, 340–41, 403
Century Magazine, The: Colorado Springs experimental station and, 295, 305–6; Johnson, and, 198–203, 241–42, 295, 305–10; Tesla treatise and, 305–10
Chamberlain, Neville, 388
Cheever, Charles, 230
Cheney, Margaret, 417
Chicago, Burlington, & Quincy Railroad, 343
Chicago Tribune, 264
Chicago World's Fair, 160–62, 172, 182–83, 188, 429n2, 455n21
Christensen, Clayton, 57–58, 402, 404
Church of the Great Martyr George, Gospic, 21
City of Richmond (ship), 69

Clark, F. W., 118
Clark, George, 390
Clarke, W. J., 230, 233
Clemens, Samuel. *See* Twain, Mark
Coaney, Charles F., 205, 259
Coffin, Charles A., 88, 98, 107, 162, 357
coherers, 228, 269–72, 275, 348, 438n29
Cold War, 394–98
Colorado Springs, 5; Alta Vista Hotel and, 265, 278; *The Century Magazine* and, 295, 305–6; demographics of, 263–64; elites of, 263–64; El Paso Power Company and, 264, 278; experimental station at, 267; Golka and, 398; Knob Hill and, 266–68, 278; Lowenstein and, 267, 269–70, 278, 292–95, 343–44, 375, 455n8, 463n28; magnifying transmitter and, 267–69, 274, 278–82, 284, 288–300, 303, 306, 323, 327–30, 333; patents and, 303, 328; photographic documentation and, 295–300; Pike's Peak and, 263, 265–67; relocating to, 263–69; stationary wave experimentation and, 263–301 (*see also* stationary waves)
Columbia University: honorary degree from, 204–5; Tesla lectures and, 133–38, 140, 143
Columbia (yacht), 311
Commercial Club, Chicago, 264
commutators: AC vs. DC, 44–45; brushes in, *38*, 39–43; motors and, 37–45, 48, 54, 68, 81, 104–7, 161, 168, 170–71, 433n23; poor wear of, 40–41; Pöschl and, 43; slip rings vs., 39, *93*, 105, 433n23; sparking issues and, 41–43
compressed air: AC motors and, 43–45, 53–55, 59; power transmission and, 165–67, 181, *182*, 185, 327, 362, 369, *370*
computers, 6, 112, 279, 339, 341, 402, 406
Consolidated Edison, 397
conspiracy theories, 390, 394, 398
Corum, James F., 276–78
Corum, Kenneth L., 276–78
Crawford, Francis Marion, 201, 256, 264
creative destruction, 57–59
creativity, 11, 19; AC motors and, 60; Colorado Springs and, 263; economic issues and, 401–2; illusion and, 368 (*see also* illusion); media and, 137; motivation for, 408–13; nature and, 8; refining stage and, 404–5; self-confidence and, 57; society and, 8;

subjective rationality and, 57–59; tension of, 13; Tesla's style of, 7, 244, 368, 376, 403–13; urge of, 408–13; Yankee ingenuity and, 397. *See also* invention

Creusot Works, 258

Croatia, 28, 145; Gospíc, 21, 25–32, 47, 154–55, 426n27; Lika province of, 13–21, 32–33, 239, 380; Napoleon and, 14–15; Smiljan, 3, 5, 13, 16–17, 21, 69; soil of, 13–14

Crocker, Francis B., 134

Crookes, William, 29, 126, 145–55, 176, 204, 313

Crookes radiometer, 29, 151, 199

Crookes tubes, 221–22, 249

cryophorus, 82

Cunningham, David, 66

Curtis, Leonard E., 264, 267

Czito, Julius C., 371

Czito, Kolman, 371

Dana, Charles A., 217

da Vinci, Leonardo, 6

Dean, Bashford, 317

deBobula, Titus, 386

Defense Advanced Research Projects Agency (DARPA), 394

de Forest, Lee: background of, 348; demonstrations and, 350; disruptive innovation and, 402; Marconi and, 334–35; Morgan and, 250–51; physics and, 348; radio and, 149, 285; telegraph and, 349, 438n29; telephone and, 348–49; Tesla's birthday and, 380; Tesla's legacy and, 403; transmitters and, 350; vacuum tubes and, 149; White and, 349–51; wireless communication and, 285, 303, 334, 348–50, 380, 402–3, 438n29

de Laval, Gustaf, 371

Delmonico's, 2–3, 5, 186, 194, 198, 201–2, 256–57, 260, 410, 451n34

delta connection, 44n4

demonstrations: Adams and, 166, 205, 219; AIEE and, 116; alternating current (AC) and, 1–2, 78, 85, 89–92, 96, 144–45, 160, 166, 172–73, 215; Anthony and, 116; Bjerknes and, 117; broadcasting power and, 341–42; Brown and, 85, 90–91, 102, 116; Byllesby and, 111; Chicago World's Fair and, 160–62, 172–73, 182–83, 188; Clarke and, 233; Columbia College lectures and,

133–38, 140, 143; de Forest and, 350; dramatic, 1–2, 90–91, 125–26, 128, 135–38, 144–45, 148, 150–53, 177, 190, 219, 226, 230, 332–33, 409; duplex and, 78; dynamos and, 41; Edison and, 78; egg of Columbus and, 90–92, 103, 172, 257, 407; electric brush phenomena and, 148–49; European shows and, 143–57; Faraday and, 36–37, 53; Fontaine and, 41, 45; Geissler tubes and, 126–28, 132, 135, 221–22, 249; Helmholtz and, 188; high-frequency phenomena and, 138–39; Jacob's Ladder and, 127; Kerr and, 111; London lectures and, 146–52; Madison Square Garden and, 230, 233, 451n40, 451n41; Marconi and, 259–60, 293–94, 311, 313, 332–37, 347–48; Martin and, 105, 195; Matthews and, 382; models and, 26, 29; motors and, 35–37, 41–42, 45, 53, 90–92, 96, 102, 105, 111–12, 116, 149, 195, 407; oscillators and, 135, 138, 148, 208, 215, 235, 271; particle beam weapons and, 382, 387; Peck and, 91–92, 102, 112, 116; plant for, 303; Pöschl and, 35, 41–42; Price and, 105; Prince Albert and, 256; private, 190, 194–95, 199–200, 215, 230, 259, 294, 451n40; public imagination and, 11; quadruplex and, 78; radio-controlled boat and, 226, 230, 233–37, 264; reflected waves and, 156; Ruhmkorff coil and, 126, 217; Shallenberger and, 112; skin effect and, 125, 136; spinning fan and, 151; Stanley and, 89; stationary waves and, 264, 271, 282, 293–95; Tesla coils and, 120–25, 128, 218, 248–49, 282, 338, 353, 403, 423n6, 450n17, 462n6; Tesla's fondness for, 10; turbines and, 26, 372–73; Twain and, 186–87; vacuum tube lamps and, 219; von Miller and, 144–45; Westinghouse and, 160, 172–73; wireless lighting and, 135–37, 177, 188, 190–92, 215, 235, 294; wireless power distribution and, 341–42

Déri, Miksa, 61, *62*, 88

Dewar, James, 147

Diary newspaper, Novi Sad, 17

direct current (DC): Adams and, 167–68; arc lighting and, 67; Edison and, 4, 165; high-voltage ionization and, 382; motors and, 37–41, 44, 186; Niagara

direct current (DC) (*continued*)
Falls and, 167; particle beam weapons and, 382, *383*; Twain and, 186; vs. alternating current (AC), 44–45, 167–68, 170–71

DiscFlo Corporation, 374

Dodge, Flora, 239

Dodge, Mary Mapes, 199, 344

Dolbear, Amos, 139, 232

Dolivo-Dobrowolsky, Michael von, 144–46, 161, 166, 303, 441n4

Dozier, Joseph, 267

Draper, John William, 307

drehstrom (rotary current), 144, 166

Duncan, Curtis, & Page, 93–94

Duveen, James Henry, 312

Dvořák, Antonín, 198

Dwight, Theodore W., 134, 440n15

dynamos: Anthony and, 103–4; arc lighting and, 41; electricity and, 35, 41–45, 63–71, 81–82, 85–88, 95–96, 103, 110, 115, 120, 168, 304, 320–21, 328, 431n42, 433n23; Mary-Ann, 70; power transmission and, 41–42; Weston and, 81; (*see also* generators)

Earth: as compressible fluid, 362–64; electrical potential of, 269–74; high-frequency phenomena and, 138–42; Kennelly-Heaviside layer and, 209, 335; oscillators and, 138–42; resonance and, 211, 248, 269–74, 329, 363; response to electricity pumped into, 362–64

earthquakes, artificial, 187, 446n26

Eastman, George, 101

Eastman Kodak, 101

economic issues: arc lighting and, 72–73, 76; Astor and, 214, 255–61, 264, 303, 305, 313–15, 353, 361, 371, 417; bankruptcy, 311, 374–76, 399, 446n24; copper costs and, 179, 257–58; creativity and, 401–2; disruptive innovation and, 401–3; funding and, 5–6, 130–31, 206, 255–56, 282, 302–5, 316, 318, 343–45, 352–58, 368–69, 386, 388, 393, 402, 445n6; innovation and, 10, 57–59 (*see also* innovation); investors and, 102, 128, 130–32, 163, 179, 206–7, 218, 263, 294–95, 305, 310, 332, 344–45, 348–61, 374, 407, 465n63; laboratory fire and, 216–17; licensing and, 94, 101–3, 145, 153, 155, 205–6,

218, 255, 347, 378; Light-House Board and, 283–84; manufacturing and, 339; Morgan deal details and, 357–58; navy and, 282–84; patent-promote-sell strategy and, 102–3, 132, 371, 397; power transmission and, 328 (*see also* power transmission); price of coal and, 165; royalties and, 94, 101, 111, 113, 130–32, 145, 305, 349, 361, 378, 467n19; Schumpeter and, 10, 57–59, 401, 404, 411; selling patents and, 107–13 (*see also* patents); Simpson and Crawford and, 256, 264; speculative bubbles and, 346–51; undersea cables and, 336; Vanderbilt and, 214, 230; Wall Street and, 2, 78–79, 130, 164, 206, 304, 311, 345, 349–50, 353, 357, 359; Wardenclyffe tower and, 323, 326; wireless communication and, 350

eddy currents, 53–*54*

Edison, Thomas Alva, 2; arc lighting and, 71–73, 431n40; Brown and, 78–79; Brush Electric Light Company and, 72; corporate conspiracy and, 398; direct current (DC) and, 4, 165; French lighting organization of, 63–66; General Electric and, 101–2, 129, 160, 162, 166, 224, 311, 317, 352, 357, 373, 397, 452n62; incandescent lighting and, 4, 63–67, 72–73, 102, 140, 206, 219; Johnson and, 197–98; Martin and, 195; Mary-Ann dynamos and, 70; mathematics and, 64; Menlo Park and, 49, 194–95, 198; Nobel Prize and, 375; Orton and, 406; patents and, 49, 63, 72, 161; Peck and, 79; phonograph and, 49, 107, 195, 301; physics and, 375; Puskás and, 49–50; pyromagnetic generator and, 76; reporter coverage of, 194–95; *S. S. Oregon* dynamo and, 70; telegraph and, 64, 79, 197; telephone and, 49, 78, 195; Tesla's legacy and, 397–98; Thomson-Houston Electric Company and, 72, 74–75, 88–90, 96, 102–3, 107, 129, 160, 162, 166, 311, 317; three-wire system and, 65, 96, 144; United States Electric Lighting Company and, 72; X-rays and, 224

Edison Electric Light Company, 102, 164

Edison Lamp Works, 66, 428n40

Edison Machine Works, 69–71, 73, 76, 120, 416

Edison Medal, 9–10, 240, 376, 391

egg of Columbus, 90–92, 103, 172, 257, 407
Einstein, Albert, 380
Electrical Engineer journal, 5, 133, 137, 146, 151, 195–96, 224, 232–34
Electrical Experimenter journal, 379
Electrical Review journal, 75, 105, 135, 137, 153, 217, 224, 232, 260, 295
Electrical World journal, 89, 105, 143, 195–96, 334, 355, 433n36, 459n22
electric brush effect, 148–49
Electrician, The (journal), 150
electromagnetic waves: Bjerknes and, 117; Branly and, 438n29; capacitor charge/discharge cycle and, 184; coherers and, 270, 438n29; Crookes and, 146, 176; dynamos and, 81; electrostatic thrusts and, 123–28, 132, 135, 191; ether and, 134, 191, 208, 251; Faraday and, 10, 36–37, 39, 53, 82, 147; Geissler tubes and, 127, 135; grounding and, 141, 211, 248; Hertz and, 5, 117, 120–22, 126, 132–35, 138, 146, 156, 209, 222; Hertzian waves and, 125–27, 146, 251, 294, 335, 348, 382, 451n38; Io and, 277; Kennelly-Heaviside layer and, 209, 335; light and, 156, 209; lightning and, 271; linear propagation of, 209, 247; Lodge and, 146, 209; Marconi and, 141, 209, 210, 336; Maxwell and, 117, 122, 125–28, 132–33, 156, 209–10, 219, 364, 438n21; power transmission and, 207–13 (*see also* power transmission); radio waves and, 335, 438n29; receivers and, 130, 139, 140–41, 159, 180–81, 208–11, 225, 236, 247–53, 262, 269–77, 285–95, 301, 329–34, 339–44, 347, 375, 403, 451n38, 456n30, 457n63; remote control and, 5, 33, 48, 218, 225–34, 237, 243, 264, 283, 285, 308–9, 357, 361, 382, 408, 451n37, 451n38; resonance and, 9–10, 117, 124, 178–80, 187, 208, 211, 329, 399, 438n21, 446n26; right-hand rule and, 36, 39; Ruhmkorff coil and, 127; sound waves and, 156; stationary waves and, 270–73, 275; Tesla's legacy and, 397, 403; transverse propagation of, 156; tuning and, 208, 235, 262, 284–88, 303, 313, 332, 343, 403, 449n43 (*see also* frequency); as wide–open field, 118; wireless power and, 6, 30 (*see also* wireless power); X-rays and, 218–25, 237, 248, 330, 353, 394

electrostatic thrusts, 123–28, 132, 135, 191
Electrotechnical Exhibition, 144
electrotherapy, 217, 450n12
Elliott Cresson Gold Medal, 204
El Paso Power Company, 264, 278
Empire State Building, 380
Engineering journal, 88
ether, 134, 191, 208, 251
Eureka moments, 51–52, 404, 428n35, 428n42
Evans, John O., 78–79
execution by electrocution, 1
extremely low frequency waves (ELF), 273

Faraday, Michael, 10, 36–37, 39, 53, 82–83, 147
Federal Bureau of Investigation (FBI), 389–95, 416, 469n51
Federation of American Scientists, 395
Ferncliff laboratory, 256
Ferraris, Galileo, 108–11, 146, 436n31
Fessenden, Reginald, 303, 343, 402, 416
fireballs, 299–300, 394, 398, 458n86
Fish, Frederick P., 377
FitzGerald, Bloyce, 390–94
FitzGerald, George Francis, 125
Fizeau, Armand Hippolyte, 123
Fleming, A.P.M., 403
Fleming, John Ambrose, 149, 152, 333, 337–38, 442n18, 462n6
flying machines, 19, 27–28, 226, 369–74, 386, 410
Fontaine, Hippolyte, 41, 45
Ford, Henry, 341, 397
Fortnightly Review magazine, 146
Four Hundred, 2
Foxworth, Percy E., 391
Francis Joseph the First, 49
Franco-Prussian War, 66
Franklin, Benjamin, 35, 299
Franklin Institute, 176, 181, 204, 271
Freedom of Information Act, 390, 394, 416
free-energy movement, 398–99
Frick, Henry Clay, 365
Frost, Robert, 142
Furlan, Boris, 390

Galvani, Luigi, 35
gambling, 46–48, 307
Ganz, Abraham, 61
Ganz and Company, 60–63, 87–88, 155, 158, 258, 429n2, 429n6

Gaulard, Lucien, 87–89, 108–9, 112
Gehring, G. P., 348
Geissler tubes, 249; Crookes and, 126; electrostatic thrusts and, 135; ionization and, 126; heatless light and, 132; oscillators and, 126–28; X-rays and, 221–22
General Electric, 101, 258; Cataract Construction Company and, 166; Coffin and, 357; Edison and, 102, 129, 160, 162, 311, 317, 397; Hawkins and, 352; Morgan and, 317; Niagara Falls project and, 173; rotating machinery and, 373; Sogge and, 452n62; Thomson-Houston and, 102, 129, 160, 311, 317; turbines and, 373; Westinghouse and, 129; X-ray products and, 224
generators: 5–6, 29, 37–45, 61, 64–67, 74, 76, 81–93, 96, 98, 104–6, 112, 114, 119–24, 139, 144, 159, 161, 166, 168–73, 180–85, 188, 205, 215, 223, 249, 277–78, 298–99, 333, 353, 382–84, 392, 429n48, 430n14, 433n36, 456n40, 461n51; alternating current (AC), 6, 44, 61, 67, 85, 119, 122, 133–35, 138, 139, 155, 160–61, 181, 188, 277, 333, 429n48; electrostatic, 29, 223, 382, 384, 461n51; Gramme, 41; high-frequency phenomena and, 120–24; pyromagnetic, 76, 81–84; rotary motion and, 37; Strasbourg system and, 66–69; Tesla coils and, 120–25, 128, 218, 248–49, 282, 338, 353, 403, 423n6, 450n17, 462n6; Van de Graaff and, 382, 384, 461n51, 474; Y-connection and, 144, 161
George, Willis De Vere, 392
Gerlach Hotel, 2, 158, 180–81, 199, 216, 443n2
Gernsback, Hugo, 379
Gesellschaft für Drahtlose Telegraphie System Telefunken, 377
Gibbs, John, 87–89, 108–9, 112
Gilder, Richard Watson, 295
Globe Stationery & Printing Company, 81
GOELRO, 379
Goethe, 51–52, 54–55, 204, 308–9, 459n19
Goff, Edward H., 72–73
Golka, Robert, 398
Gordon, Robert, 311
Gordon Press Works, 74
Gorsuch, Walter C., 392
Gould, Jay, 79–80, 111
Governor Clinton Hotel, 386, 392

Gramercy Park, 214–15
Gramme, Zenobe T., 41
Gramme dynamo, 41, 44
Grand Street laboratory, 118–19, 124, 138–39, 158
Grant, Ulysses S., 198
Gray, Elisha, 79, 424n21, 471n23
Great Eastern (ship), 195
Great Northern Railroad, 343
Gregg, Richard B., 267, 269, 278
Grenfell, Edward C., 311
grounding: capacitors and, 271; high-frequency experiments and, 143; power transmission and, 141, 208–9, 449n43; return circuit and, 209, 247–55, 262, 294, 329–30; telegraph and, 139; transformers and, 144, 157; transmitters and, 403, 449n43
Gymnasia (high school), 15, 26–29, 425n11

Hamlet (Shakespeare), 400
Hammer, Armand, 388
Hammond, John Hays, Sr., 230
Hamptons, Long Island, 201
Harper's Weekly, 129, 136
Haselwander, F. A., 143–46
Hawkins, Laurence A., 6, 352
Hawkins, Ralph J., 379
Hearst, William Randolph, 1, 224
Heaviside, Oliver, 125–26, 209, 335
Helios Company, 153, 168, 258
Helmholtz, Hermann von, 188
Henry, Prince of Germany, 343
Hering, Carl, 145
Herschel, Charles, 53–55
Hertz, Heinrich, 204; Columbia College lecture and, 132–35, 138; Crookes and, 146; ground current and, 209; high-frequency phenomena and, 127; Lenard and, 221; Marconi and, 132, 259; Maxwellians and, 125–27; power transmission and, 179, 209; spark gaps and, 184; Tesla coils and, 120–22
Hertzian waves, 146, 451n38; electrostatic thrusts and, 125–28, 135; ether and, 251; particle-beam weapons and, 329, 381–90, 394–95, 416; stationary waves and, 294; wireless communication and, 335, 348
Hewitt, Abraham, 221
Hewitt, Edward Ringwood, 221–23
Hewitt, Peter Cooper, 221

Hiergesell, David, 118
Higgins Boat Company, 391
Higginson, Henry Lee, 311
high-frequency phenomena, 158, 310; alternators and, 133–34, 138; arc lighting and, 120; artificial earthquakes and, 187, 446n26; Columbia College lecture and, 133–38; demonstrations and, 138–39, 143–50; Earth and, 138–42; European shows and, 143–50; generators and, 120, 122–24; Hertz and, 127; incandescent lighting and, 132–35; increasing transmission lengths and, 138–42; lectures on, 127, 133–38; lighting and, 4, 132–33; Lodge and, 127; manufacturing and, 206; marketing of, 196, 205–11; oscillators and, 176, 179–84 (*see also* oscillators); particle-beam weapons and, 329, 381–95, 416; patents and, 205; physiological effects of, 124–25, 217; power distribution and, 120, 139; promotion of, 138–39, 143–50, 205–11; resonance and, 124 (*see also* resonance); return circuit and, 248–51; safety and, 136; stationary waves and, 262; Tesla coils and, 5–6, 120–25, 128, 218, 248–49, 282, 338, 353, 403, 423n6, 450n17, 462n6; Tesla's search for new ideal and, 119–28; Thomson and, 134; transformers and, 5, 125, 133, 176, 179–84, 218–19; Wardenclyffe tower and, 321–31, 346, 362, 379, 382; wireless lighting and, 176, 179–93, 197, 207–13 (*see also* wireless lighting)
History of the Intellectual Development of Europe (Draper), 307
Hitler, Adolf, 388
Hoadley, Joseph, 372
Hobson, Richmond Pearson, 241–43, 354, 361–62
HOMAG, 377
Homestead Strike, 1
homosexuality, 240–41
Hong, Sungook, 126
Hoover, J. Edgar, 389, 391
Hospitalier, Édouard, 153
Hotel New Yorker, 380, 389–90
Houston Street laboratory, 187, 218–19, 226, 235, 248–52, 267, 278, 285, 303, 453n85
Hubbard, Gardiner, 406
Hull, Grizelda, 354–55, 361–62

Hull, Robert, 282
Hume, David, 48

illusion: invention and, 8–11, 18, 92, 115–16, 193, 208, 246, 263, 295, 299, 301, 315, 330, 368, 373, 381, 389–95, 403, 406–8, 412; power transmission and, 208; technological change and, 115–16
Illyrian provinces, 14, 425n11
Imperial German Postal Authority, 144
incandescent lighting, 226, 427n8, 444n8; alternating current (AC) and, 88–89, 159–60, 170, 173; button lamps and, 133–34, 149–50, 439n10; Columbia College lecture and, 133–38; direct current (DC) and, 87; dynamos and, 64, 206; Edison and, 4, 63–67, 72–73, 102, 140, 206, 219; flickering of, 115, 159; Ganz and, 61; Gaulard and, 87; General Electric and, 224; Gibbs and, 87; high-frequency phenomena and, 132–34, 135; manufacturing and, 224; patents and, 88, 134; platinum and, 133; railroads and, 66; Ruhmkorff coil and, 126; SE Edison and, 63–66; Stanley and, 88–89; stationary waves and, 290, 292, 297; transformers and, 249 vacuum tube lamps and, 219; Westinghouse and, 160–61; wireless lighting, 140, 182–84, 191–92, 257–58, 292, 306
Industrial Revolution, 43
Industries journal, 137
innovation: adaptive, 58, 402, 404; creative destruction and, 57–59; creative urge and, 408–13; disruptive, 10, 58, 401–4, 409; economic role of, 396; marketplace forces and, 400; motivation for, 408–13; process of invention and, 403–5; Schumpeter on, 10, 57–59; sensitive trigger and, 157, 211; subjective rationality and, 10, 57–59, 411; Westinghouse marketing and, 88; Yankee ingenuity and, 397
Institution of Civil Engineers, 147
Institution of Electrical Engineers, 145, 147
International Electrical Exhibition, 66, 108, 181
International Exhibition in Vienna, 41
International Harvester, 342–43
International Latex Corporation, 391
International Merchant Marine Trust, 346

International Tesla Society, 399
interplanetary communication, 265, 274–78, 300, 315, 380, 398
invention: AC motors and, 52; Bell and, 8; creative urge and, 408–13; cultural pulse and, 407–8; disruptive innovation and, 10, 58, 401–4, 409; Eureka moments and, 51–52, 404, 428n35, 428n42; human progress and, 305–10; illusion and, 8–11, 18, 92, 115–16, 193, 208, 246, 263, 295, 299, 301, 315, 330, 368, 373, 381, 389–95, 403, 406–8, 412; investors and, 102, 128, 130–32, 163, 179, 206–7, 218, 263, 294–95, 305, 310, 332, 344–45, 348–61, 374, 407, 465n63; licensing and, 94, 101–3, 145, 153, 155, 205–6, 218, 255, 347, 378; manufacturing and, 63; mental engineering and, 27, 43–45, 53–55, 59–60, 63, 67–68; motivation for, 408–13; natural/social worlds and, 8; Nobel Prize and, 364, 375, 377, 390; practical work experience and, 60–75; process of, 403–8; protection of, 95 (*see also* patents); refining stage and, 404–5; Tesla on intensity of, 237–38, 244–47; Tesla's legacy and, 396–413; Tesla's style of, 7–11, 403–8; tinkering and, 19, 27–28, 120, 123, 180, 256, 315; underlying ideal of, 31–32; Yankee ingenuity and, 397. *See also* specific item
Invention of Everything Else, The (Hunt), 401
Inventions, Researches, and Writings of Nikola Tesla, The (Tesla and Martin), 196–97, 417
Io, 277–78
Isabella, Queen of Spain, 91, 257, 433n40

Jablochkoff, Paul, 44, 61, 67
Jackson, Dugald C., 104–5
Jack the Ripper, 1
Jefferson, Joseph, 201, 214–15
Jevtic, Bogoljub, 390
Joanneum Polytechnic School, 32, 34, 44–45, 427n54
Jobs, Steve, 99, 112, 406
Joenneum Polytechnic School, 32
Johnson, Katharine: background of, 198; concern over Tesla's health, 216, 239; Delmonico's and, 201; flowers sent by, 198–99; Hobson and, 331; as Madame Filipov, 199, 242; private demonstrations to, 199–200; promotion of Tesla's reputation by, 197–203, 355, 417; Tesla's letters to, 331; University of Nebraska and, 204
Johnson, Robert Underwood: background of, 197–98; *The Century Magazine* and, 198–203, 241–42, 295, 305–6, 309–10; Columbia University and, 204–5; cultural pulse and, 407–8; Edison and, 197–98; Gilder and, 295; Hobson and, 242–43; Library of Congress and, 416; as Luka Filipov, 199, 242, 282; Martin and, 197–204, 408; Nobel Prize letter and, 375; photograph idea of, 199–203; private demonstrations to, 199–200; promotion of Tesla's reputation by, 197–205, 214, 355, 417; as telegrapher, 197; Tesla letters to, 282, 416; Twain and, 198, 201; world views of, 407–8
Jósika, Miklós, 23
Journey in Other Worlds, A (Astor), 256–57
J.P. Morgan & Company, 344
Judah, Tim, 13

Karl-Ferdinand University, 48
Karlovac, Croatia, 28–29
Kazakhstan, 394–95
Keegan, George J., 394–95
Kemp, George, 333–35
Kennelly-Heaviside layer, 209, 335
Kerr, Thomas B., 111, 113
Kevlar, 374
Kipling, Rudyard, 199
Knight, Walter H., 372
Knob Hill, Colorado Springs, 266–68, 278
Kosanović, Sava, 390–93, 425n16
Kosovo, 13
Krajina, 13, 30, 410, 426n27
Kulišić, Kosta, 46–47

Ladd, D. Milton, 391
LaGuardia, Fiorello, 390
Lane, Robert, 73–75, 84, 431n50
lasers, 330, 394, 439n10, 470n67
League of Nations, 386
Lenard, Philipp, 221
Lenin, V. I., 379
Leonhardt, Charles, 118
Liberty Street lab, 107, 111, 113, 195
licensing, 94, 101–3, 145, 153, 155, 205–6, 218, 255, 347, 378

Light-House Board, 283–84
lighting: (*see also* arc lighting); button lamp and, 133–34, 149–50, 439n10; Columbia College lecture and, 133–38; high-frequency phenomena and, 4, 132–33, 135; incandescent, 4, 61, 63–64 (*see also* incandescent lighting); installation issues and, 72–73; ionization and, 76, 121, 126, 184, 382; Lane and, 73–75, 84, 431n50; mercury vapor lamp and, 221; phosphoresence and, 150, 178, 191, *192*, 200–201, 208, 221, 245; promotion of, 193–213; SE Edison and, 63–66; single-wire, 149–50; Statue of Liberty and, 283; vacuum tube lamps and, 219, 330; Vail and, 73–75, 84, 431n50; wireless, 128 (*see also* wireless lighting)
lightning, 51, 417, 446n21, 455n21; arresters for, *268*; inspiration from, 157; sparks and, 18; stationary waves and, 270–74, 278, 281–82, 289, 299, 301; streamers and, 213; Wardenclyffe and, 346, 398
Lincoln automobile, 378
Linotype machine, 446n24
Literary Digest, 137, 316
Lloyd's of London, 332
Lodge, Oliver, 125, 127, 135, 138, 146, 209, 228, 271, 380, 449n43, 462n9
Logos, 31
London lectures, 146–52
Long, Breckinridge, 386
Long Island Railroad, 319
Longstreet, John, 79
Lontin, Dieudonné François, 44
Loomis, Mahlon, 445n6
Lowell, Percival, 276
Lowenstein, Fritz, 267, 269–70, 278, 292–95, 343–44, 375, 455n8, 463n28
"Luka Filipov" (Zmaj), 199
Lynn Electric Light Company, 89

Macak (cat), 18–19
Madison Square Garden, 214, 230, 233, 451n40, 451n41
magneto, Pixii, *38*
magnifying transmitters: alternating current (AC) and, 268; bank of capacitors for, 268–69; breakwheel and, 268–69, 279, 286; Colorado Springs and, 267–69, 274, 278–82, 284, 288–300, 303, 306, 323, 327–30,

333; conductor length and, 323; effect on horses, 289; fireballs and, 458n86; Fleming and, 333–34; Geissler tubes and, 128; operation of, 278–82, 284; patents and, 303; primary coil and, 267, 279; secondary coil and, 268; Wardenclyffe and, 321–31, 346, 362, 365, 379, 382, 462n69; wasted energy and, 282; wireless communication and, 267–69, 274, 278–82, 462n6; witnessing tests of, 288–95
malaria, 28
Mandic, Nikola (grandfather), 14
Mandic, Nikolai (uncle), 15
Mandic, Pajo (uncle), 15, 48, 155, 443n43
Mandic, Petar (uncle), 48, 193, 353
Manhattan Storage Company, 392–94
Marconi, Guglielmo: AIEE dinner and, 335–37; America's Cup and, 283, 285–88, 311, 332, 348; Army Signal Corps and, 283; Atlantic Communication and, 377; British Post Office and, 313; de Forest and, 334–35; demonstrations and, 259–60, 293–94, 311, 313, 332–37, 347–48; disruptive innovation and, 402; electromagnetism and, 336; Fleming and, 149, 152, 333, 337–38, 403, 442n18, 462n6; Germany and, 377; ground current and, 209; Hertz and, 132, 259; Kemp and, 333–35; litigation issues and, 377–78; loss of capital and, 332–33; Martin and, 331–37; mathematics and, 336; message privacy and, 285; messages across English Channel by, 261, 273, 332; patents and, 311–13, 315, 332, 338, 377–78, 462n6; physics and, 336, 375; Pickard's jamming and, 285; promotion of, 334; radio and, 5–6, 141–42, 285, 403; Radio Corporation of America and, 397; results of, 260–61; St. John's, New Foundland, and, 333–34; stationary waves and, 276–77; Telefunken and, 377; telephone and, 334; Thomson and, 335; transatlantic messages and, 332–37; transmission distance and, 332–37; transmitters and, 141, 152, 285, 333; U.S. Navy and, 332; Waldorf-Astoria Hotel and, 335; wireless communication and, 5–6, 132, 141–42, 152, 209, 259–61, 265, 273, 276–77, 283–88, 293–94, 303–4, 311–15, 330–38, 342, 344, 347–48,

Marconi, Guglielmo (*continued*)
375–78, 397, 402–3, 416, 449n43,
451n38, 456n30, 462n10, 462n6; Wireless Telegraph and Signal Company
and, 311, 332, 378, 459n27
Marconi Cable and Wireless, 397
Markula, Mike, 406
Mars (Lowell), 276
Martians, 213, 264–65, 276, 278, 315, 334,
398, 456n33, 456n34
Martin, Thomas Commerford: AC motors
and, 105, 195; AIEE and, 134; attacks
Tesla, 232–34, 309–10; background
of, 195; financial problems of, 452n47;
inventor ranking and, 459n22;
double-cross by, 331–37; Johnsons and,
197–203, 408; Marconi and, 331–37;
Mars and, 276; newspaper science and,
309–10; polyphase motor and, 105–6;
as publicity agent, 195–97, 200–204,
208, 211–14, 224, 408, 417; radio-
controlled boat feud and, 309; Tesla
biography and, 196–97; Tesla's nervous
breakdown and, 365; on women, 239
Mary-Ann dynamos, 70
mass-production, 340–41, 397, 402
mathematics, 104, 124, 132; Edison and,
64; Ferraris and, 108–10; Hertz's
results and, 125–26; Josif (uncle) and,
14, 26; Marconi and, 336; Maxwell's
theory and, 125–26, 128, 438n21;
Milutin (father) and, 17, 26; Nash and,
364; predictive power of, 128; Pupin
and, 336; Schumpeter and, 58; Tesla
and, 26, 34, 48, 64, 106, 128, 307, 405;
Thomson and, 123
Mather Electric Company, 104
Matthews, Arthur H., 398
Matthews, Harry Grindell, 382
Maxwell, James Clerk: electromagnetism
and, 117, 122, 125–28, 132–33, 156,
209–10, 219, 364, 438n21; mathemati-
cal analysis of induction and, 125–26,
128, 438n21; Tesla coils and, 122
Maxwellians, 125–28, 133, 210
McAllister, Ward, 2
McClure, S. S., 197
McCormick Harvesting Machine
Company, 343
McKay, Mrs. Clarence, 239
McKim, Charles Follen, 214, 245
Mead, William Rutherford, 214, 245
Menlo Park, 49, 194–95, 198

mental engineering, 27, 43–45, 53–55,
59–60, 63, 67–68
mercury interrupters, *321*, 328, 447n27
mercury vapor lamp, 221
Merganthaler, Ottmar, 446n24
Merington, Marguerite, 239–40
Merrimac mission, 241–42
messenger boys, 229, 389
Metropolitan Life Tower, 372, 372–73
Metropolitan Museum of Art, 317
Military Frontier Administration Author-
ity, Austria–Hungary, 32
Millikan, Robert A., 380
MIT Society of Arts, 105
Montfort, Nick, 339
Moore, D. McFarlane, 219
Morgan, Anne, 239
Morgan, Jack, 373–74, 386–87
Morgan, J. P., 239, 416–17; Adams and,
164, 406; background of, 311; capi-
talistic dominance of, 311; corporate
conspiracy and, 398; death of, 373; de
Forest and, 250–51; deformed nose
of, 312; Edison General Electric and,
311; as the Great Man, 313–18, 332,
338, 343–47, 353–55, 358, 361, 373;
International Harvester and, 342–43;
International Merchant Marine Trust
and, 346; Library of Congress and,
416; loss of support from, 346–51;
manufacturing and, 339; Marconi patents and, 311–12; Niagara Falls project
and, 164; Northern Pacific Railroad
and, 343; patents and, 313, 315–18,
344, 346–47, 352, 355; presence of,
312; radio-controlled boats and, 230;
railroads and, 311, 341–43; receiver
profits and, 339; Schiff and, 357; shipping and, 342; speculative bubbles
and, 346–51; steel industry trust and,
314–17, 341; telegraph and, 317–18; telephone and, 312, 317, 339; Tesla's legacy
and, 398; Tesla's letters to, 312–18,
331–32, 337–42, 346, 358–61, 412–13;
Thomson-Houston Electric Company
and, 311; United States Steel and, 314,
317, 343, 348, 378; Wardenclyffe tower
and, 323, 326; wireless communication
and, 311–18, 337–47, 350–61, 365, 402,
406, 412; wireless lighting and, 316–17;
World Telegraphy System and, 339
motors, 3; AIEE lecture and, 105–7;
alternating current (AC) and, 9 (*see*

also AC motors); alternators and, 44, 119, 122, 133–35, 138, *139*, 155, 160–61, 181, 211; Arago's wheel and, 52–54; Baily and, 55–56, 61; basic mechanism of, 37–41; brushes and, *38*, 39–43, 148, 161, 168, 170–71; capacitors and, 96; commutators and, 37–45, 48, 54, 68, 81, 104–7, 161, 168, 170–71, 433n23; copper and, 109–10, 115, 122, 140, 428n43; demonstrations and, 35–37, 41–42, 45, 53, 90–92, 96, 102, 105, 111–12, 116, 144, 195, 407; direct current (DC) and, 37–41, 44, 186; electromagnetism and, 36–40, 44, 53–55, 108; flywheels and, 43, 76–77, 183; Haselwander and, 143–46; ideal of, 371; manufacturing and, 87, 103, 107–8, 144–45, 153, 155, 258; patents and, 52, 55–56, 64, 76–77, 93–94, 105–6, 114, 118, 145, 159, 167, 186, 305, 361, 432n16, 434n49, 439n8; phase current and, 64; polyphase, 92–99, 104–7, 110–15, 144–45, 158–63, 166–75, 402, 428n42, 434n49; rotary motion and, 37, 43; Shallenberger and, 108, 111–14; shoe-polish tin, 85, *86*, 92, 96, 144, 294, 302, 407; single-phase, 88, 96, 98–99, 112, 114, 119, 144, 159–61, 166–68, 170, 173; slip rings and, 39, *93*, 105, 433n23; spark-free, 43–45; split-phase, 95–100, 104, 110–15, 123, 159; stator coils and, 37, *39*, 40, 44, 52–55, 64–68, 85, *86*, 92, *93*, 96–97, 100, 106, 115, 119, 166, 209, 404; Strasbourg system and, 57, 66–69, 85, 118, 140, 294, 416, 430n20; synchronous speeds and, 119; thermomagnetic, 76–82; three-phase, 143–45, 159–61, 169–70, 402, 410; turbines and, 9, 26, 61, 159, 166–68, 183, 187, 353, 368–74, 378, 386, 408; two-phase, 159, 161, 169–73; waste heat and, 82, 97, 110; wireless power transmission and, 342
Mrkich, Dan, 417–18
Muir, John, 199
Mutual Union, 78–81, 111
Muzar, Charlotte, 390, 393
My Inventions (Tesla), 12

Nahin, Paul, 386
Nantucket Lightship, 283
Napoleon, Bonaparte, 14–15
National Electric Light Association (NELA), 176, 195

Nature journal, 148
Netter, Raphael, 226
New Age philosophy, 399–401
"New Art of Projecting Concentrated Non-dispersive Energy through Natural Media, The" (Tesla), 382
Newton, Isaac, 9
New World Symphony (Dvořák), 198
New York Edison Company, 372
New York Herald, 2, 196, 203, 216, 231, 311
New York Journal, 1, 224, 261
New York State Department of Taxation, 393
New York Sun, 217–18, 222, 337, 346
New York Times, 174, 178, 196, 203, 217, 222, 230, 326, 334, 375, 381
New York World, 1, 203, 380
New York Yacht Club, 311
Niagara Falls: Adams and, 164–74, 176, 205, 416; Cataract Construction Company and, 164–74, 245, 256; direct current (DC) and, 167; General Electric and, 166, 173; hydroelectric project of, 4, 162–67, 170, 172–76, 205, 214, 237, 256, 342–44, 353–55, 365, 416; Morgan and, 164; polyphase alternating current (AC) and, 167–75; power transmission and, 162–76, 205, 214, 237, 256, 342–44, 353–55, 365, 416; Stillwell and, 165; Thomson-Houston and, 166; two-phase alternating current (AC) and, 169–73; Westinghouse and, 162–63, 165–74
Nikola Tesla Company: Adams and, 205–7, 255; Astor and, 257; Coaney and, 205; manufacturing plans of, 205–7, 255, 257, 259, 343–44, 365, 375, 454n93; patent selling and, 205–7; Rankine and, 205, 207; stock issuance of, 205–6
Nikola Tesla Corner, 380
Nikola Tesla Institute, 389
Nikola Tesla Museum, 382, 393, 415–16
Nobel Prize, 364, 375, 377, 390
Northern Pacific Railroad, 164, 343
Northern Securities, 343
Novi Sad, Serbia, 17
nuclear physics, 273, 384, 386, 394–95
Oerlikon, 144, 168
Oersted, Hans Christian, 35–37, 53
Office of the Alien Property Custodian (OAPC), 392–93, 415
O'Neill, John, 27, 130–31, 239–40, 292, 347, 388, 417–18, 439n3

Operator, The (journal), 195
Oregon (ship), 70
Orthodox Church, 10, 14–15, 31–33, 155, 198, 247, 364, 390, 409, 411, 425n18, 469n50
Orton, William, 79, 406
oscillators: arc lighting and, 182; artificial earthquakes and, 187, 446n26; capacitors and, 188–89; Chicago's World Fair and, 182; compact, 218–19; copper and, 181; demonstrations and, 215, 235, 271; development of, 181–83; Earth and, 138–42, 269–74; ground currents and, 248; high-frequency phenomena and, 176, 179–84, 218–19; Lodge and, 271; mobile, 380; patents and, 181–84, 189, 248, 257, 305; power transmission and, 208; promotion of, 193–213; remote-controlled boats and, 236; resonance and, 9–10, 117, 124, 178–80, 187, 189–91, 208, 211, 399, 438n21, 446n26; steam power and, 181–85, 193, 215, 232, 446n18; Tesla coils and, 5–6, 120–25, 218, 282, 338, 353, 403, 423n6, 450n17, 462n6; transformers and, 124–26, 132–40, 148, 156–57, 172, 178–80, 184, 188–89, 208, 218–19, 223–24, 235, 438n24; tuberculosis and, 310; wireless lighting and, 176–92; wireless power transmission and, 178–92; witnessing tests of, 288–95; X-rays and, 224
Ottoman Empire, 13, 15
ozonizers, 353

Padrewski, Ignace, 199
Page, Charles Grafton, 93, 93–94
Page, Parker W., 240, 264, 343, 357, 432n16: AC motors and, 97; background of, 93–94; patents and, 93–95, 98; split-phase motors and, 97–98
Paget, P. W., 333
Paige, James W., 186
Paige typesetter, 446n24
Palo Alto Research Center (PARC), 112
Pantaleoni, Guido, 89, 111
Parsons, Charles, 371
particle beam weapons: abolishing war and, 398; Chamberlain and, 388; "Chinese Wall" claims of, 382, 394–95; as death ray, 382, 393, 398; deBobula and, 386; FBI and, 389, 391; Fitzgerald and, 390–94; Hertzian waves

and, 329, 381–95, 416; Hitler and, 388; Hoover and, 389; impracticality of, 386, 395; League of Nations and, 387; methodology of, 381–82; pilots and, 382, 386; possibility of, 395; Semipalatinsk Test Site and, 394–95; Soviet Union and, 388, 395; Strategic Defense Initiative (SDI) and, 394–95; teleforce principle and, 388–89
Patent Office, U.S., 95, 98, 104, 107, 230, 436n30
patent-promote-sell strategy, 102–3, 132, 371, 397
patents, 10, 360, 453n78; AC motors and, 52, 55–56, 64, 93–95, 114, 118, 145, 159, 167, 186, 305, 361, 429n6, 439n8; Adams and, 169–70; arc lighting and, 72–75, 120, 431n46; Atlantic Communication and, 377; before lectures, 134; Betts and, 169; Bradley and, 169; broad principles and, 94–95; Brown and, 79, 81, 100–101, 116, 193, 205, 302, 305, 407; carbon button lamp and, 133; circuit controller and, 188; Colorado Springs and, 303, 328; complete systems and, 95; comprehensive, 95; creative process and, 416; Dolbear and, 139, 348; Dolivo-Dobrowolsky and, 144–45, 161; Duncan, Curtis, & Page and, 93–94; dynamo commutator and, 81; Edison and, 49, 63, 72, 161; electrical transmission cable and, 310; expiration of, 375; Ferraris and, 143; Fessenden and, 416; foreign, 134, 144–45, 153, 332, 440n14, 451n38; French law and, 63; GE and, 169; Haselwander and, 144; 20; high-frequency phenomena and, 120, 134, 205; incandescent lighting and, 88, 134; individual component design and, 95; infringement of, 80, 102, 145, 166–68, 377; licensing and, 94, 101–3, 145, 153, 155, 205–6, 218, 255, 347, 378; life of, 361; litigation over, 55, 99, 130, 169, 353, 376–78, 403, 416, 428n40, 429n6, 439n3; Loomis and, 445n6; magnifying transmitters and, 303; manufacturing and, 101–2, 218; Marconi and, 311–13, 315, 332, 338, 377–78, 462n6; mercury interrupters and, 447n27; Morgan and, 313, 315–18, 344, 346–47, 352, 355; oscillators and, 181–84, 189, 248, 257, 305; Page and, 93–95, 98, 264, 343, 357, 432n16; Peck

and, 79, 81, 93, 100–101, 116, 132, 193, 205, 207, 302, 305, 406–7; performance data and, 295; polyphase motor and, 95, 105–6, 145, 167–68, 434n49; power transmission and, 169, 249, 313, 332, 434n49; pumps and, 372; Puskás and, 49; pyromagnetic generator and, 81; radio-controlled boat and, 230–32, 243; royalties and, 94, 101, 111, 113, 130–32, 145, 305, 349, 361, 378, 467n19; self-regulating dynamo and, 88; selling, 107–13, 116, 132, 205–7, 371, 397, 405–6; Serrell and, 74; sources for, 416–17; Stanley and, 88; stationary waves and, 263, 303; telegraph relay and, 80; telephone and, 101, 186, 348; Tesla's first in America, 21; thermomagnetic motor and, 76–77; Thomson and, 169; three-phase current and, 144–45, 161; tuning technique and, 286; turbines and, 372; U.S. Patent Office and, 95, 98, 104, 107, 230, 436n30; vibratory discharge of capacitors and, 438n23; Wardenclyffe and, 328, 330, 355, 462n69; Western Union and, 80; Westinghouse and, 108, 110–15, 130–32, 155, 166–67, 169, 439n3; wireless communication and, 313, 377, 403; wireless lighting and, 218, 248, 255, 258, 316, 357, 377; wireless power transmission and, 406; Y-connection and, 144, 161; ZBD system and, 88

Pearce company, 353

Pearson's Magazine, 252

Peck, Charles: AC motors and, 81, 87, 90, 92–93, 100–108, 111–13, 116, 193, 207, 294, 318, 405–7; Anthony and, 103–4, 294; background of, 78; Brown and, 100; business dealings of, 100–105, 107–8, 111–13, 116, 119, 193, 205, 207, 302, 305, 313, 405–7; creative synergy with Tesla, 405–6; death of, 119, 207, 302, 406; Edison and, 79; egg of Columbus and, 90–92, 257, 407; Gould and, 80; illness of, 118–19; Mutual Union Telegraph Company and, 78; patents and, 79, 100–101, 132, 205, 207, 305, 407; polyphase motors and, 92–93; power transmission and, 433n26; pyromagnetic generator and, 81–84; royalties and, 119, 130; telegraph and, 78–80; Tesla Electric Company and, 73–75, 81, 84,

118, 431n44, 431n50; thermomagnetic motor and, 77–81; tin-can motor and, 407; using two out-of-phase currents and, 84; Western Union and, 78–79; Westinghouse deal and, 111–13

Pegler, Westbrook, 391

Perry, John, 125

Peter II, King of Yugoslavia, 390, 393

Peterson, Gary, 378

Philadelphia lecture, 176–77

Phoenix Navigation and Guidance Inc. (PNGinc), 473

phonograph, 49, 107, 195, 301

phosphorescence, 150, 178, 191, *192*, 200–201, 208, 221, 245

photography: Alley and, 220–21, 270, 289, 295–301, 305; *The Century Magazine* and, 199–203, 306; as documentation, 295–301; fluorescent light and, 200; induction coil sparks and, 152; magnifying transmitter and, 295–300; oscillators and, *255*, 289–90, 305; phosphorescent bulbs and, *192*, *200*; shadowgraphs and, 223–24; Tonnelle & Company and, 201, 219–20, 295; Wardenclyffe and, 320–21, 326, 355; X-rays and, 218–25

Physical-Medical Society of Wurzburg, 222

Picasso, Pablo, 262

Pickard, Greenleaf Whittier, 285

Picou, R. W., 64

Pierce–Arrow automobile, 378

Pike, John, 395

Pike, Lancelot E., 348, 350–51

Pike's Peak, 263, 265–67

Pixii, Hippolyte, 37, *38*

Plato, 8–9

Players, The (club), 186, 214–15, 217

polyphase motors, 428n42; alternating current (AC) and, 92–97, 167–75, 145; patents and, 95, 105–6, 145, 167–68, 434n49; promotion of, 104–7, 111, 114; selling Adams on, 167–75; Westinghouse and, 158–63, 166

Popov, Alexander, 271–72

Popular Science Monthly, 309

Port Jefferson Bank, 345, 361

Pöschl, Jacob, 34–35, 41–43, 45, 57, 427n7

Poulsen arc, 185

power transmission: alternating current (AC) and, 95–96, 105–6, 159, 163, 165, 170, 175, 313; Astor and, 255–59;

power transmission (*continued*)
Canadian Niagara Power and, 354–57; disruptive innovation and, 402–3; dynamos and, 41–42; Earth and, 138–42, 179, 211, 345; experimentation with, 178–81; Fontaine and, 41; ground current and, 208–9; Hertz and, 179; high-frequency phenomena and, 139; human development and, 305–10, 341–42; ice to reduce power loss and, 310; illusion and, 208; industrial significance of, 328, 334; Kennelly-Heaviside layer and, 209, 335; Maxwellians and, 210; Niagara Falls project and, 162–76, 205, 214, 237, 256, 342–44, 353–55, 365, 416; ocean steam and, 81–82, 87, 433n26; oscillators and, 208; particle–beam weapons and, 329, 381–95, 416; patents and, 169, 313, 434n49; photographing, 297; Pöschl and, 41–42; resonance and, 9–10, 201, 207–13; return circuit and, 209, 247–55, 262, 294, 329–30; stationary waves and, 262, 294, 302 (*see also* stationary waves); streamers and, 211–12; teleforce principle and, 388–89; Tesla treatise and, 305–10; Wardenclyffe and, 321–31, 341–42, 346, 362, 379–80, 382; wireless, 4, 209, 218, 247–55, 262, 288 (*see also* wireless power transmission); X-rays and, 224–25; Y-connection and, 144, 161
"Pozdrav Nikoli Tesli" (Zmaj), 155–56
Preece, William, 145–46
Prestige, The (film), 401, 424n22
Price, Charles, 105
Prince, Cameron B., 399
Principles of Psychology (Spencer), 285–86
"Problem of Increasing Human Energy, The" (Tesla), 306–10
Proceedings of the Royal Academy of Sciences of Turin, 108
"Process of De-Gassifying, Refining, and Purifying Metals" (Tesla), 378
promotion: AC motors and, 100–116; alternating current (AC) and, 100–116, 158–75; demonstrations and, 105 (*see also* demonstrations); European tour and, 143–57; investors and, 102, 128, 130–32, 163, 179, 206–7, 218, 263, 294–95, 305, 310, 332, 344–45, 348–61, 374, 407, 465n63; Marconi and, 334; photography and, 199–203, 306;

reputation building and, 193–213; scientific press and, 305; weakness for publicity and, 352
Pulitzer, Joseph, 203
Pupin, Michael, 134, 336
Purple Plates, 399
Puskás brothers, 49–50, 63, 69, 428n32, 428n34, 428n35, 430n31
Pyle National, 373
pyromagnetic generator, 76, 81–84

quantum physics, 399
Queen Bee newspaper, Serbia, 46–47

radio, 379, 438n29; Aitken and, 122; Armstrong and, 390; Branly and, 438n29; capacitors and, 403; Clark and, 390; de Forest and, 149, 285; extraterrestrial signals and, 265, 274–78, 277, 300, 315, 380, 398; Fleming and, 149, 403; FM, 209–10; frequency range and, 124; induction and, 403; as joint effort, 403, 406; Kennelly-Heviside layer and, 209, 335; Lowenstein and, 293; Marconi and, 5–6, 141–42, 285, 403; navigation by, 303; Tesla coils and, 403; wireless communication and, 125 (*see also* wireless communication)
radio-controlled boats: Chicago Commercial Club and, 236–37; demonstrations of, 230, 236–37; description of model, 226–30; development of, 225–30; end to war and, 231–37; military and, 5, 33, 48, 225–34, 237, 243, 264, 283, 285, 308–9, 357, 361, 382, 408; multiple vessel control and, 235–36; patents and, 230–32; telegraphy and, 229–30; tuning and, 235; Twain and, 231–32
Radio Corporation of America, 397
radio waves, 335, 438n29
railroads, 4; Adams and, 164, 206; air brakes and, 88; Barnes and, 357; elevated, 245; incandescent lighting and, 66; Morgan and, 311, 341–43; signal systems and, 88; steel and, 8; telegraph and, 79; Villard and, 162, 164; Warden and, 319; Wardenclyffe and, 320; Westinghouse and, 88
Rankine, William Birch, 164, 173, 259; Adams and, 174, 205–7, 353–54; Canadian Niagara Power and, 354; death

of, 365; Nikola Tesla Company and, 205, 207; wireless communication and, 353–54

Ratzlaff, John T., 416–17

Rau, Louis, 66

Rayleigh, Lord, 151, 156

receivers: capacitors and, 180; coherers and, 228, 269–72, 275, 348, 438n29; Earth's resonance and, 269–74; experimentation in, 130, *139*, 140–41, 159, 225, 236, 247–53, 375, 403, 451n38, 456n30, 457n63; lady's parasol and, 340; manufacturing and, 339, 343–44, 347; mobile, 340–41; Morgan and, 339–40; oscillators and, 180–81; power transmission and, 207–13 (*see also* power transmission); resonance and, 140; stationary waves and, 262, 269–77, 285–95, 301; tuning and, 208, 235, 262, 284–88, 303, 313, 332, 343, 403, 449n43; Wardenclyffe and, 329–34, 339–44, 347

"Reflection of Short Hertzian Waves from the Ends of Parallel Wires, The" (de Forest), 348

religion: Christianity, 10, 13, 31–32, 35, 198, 233, 247, 359; Logos and, 31; Milutin and, 14–17; natural ideal and, 31–32; Orthodox Church and, 10, 14–15, 31–33, 155, 198, 247, 364, 390, 409, 411, 425n18, 469n50; Tesla's background and, 8, 10, 14, 17, 20–21, 31–33

remote control, 218, 451n37, 451n38; automatons and, 48, 225, 234, 236–37, 367; borrowed mind and, 308; radio-controlled boat and, 5, 33, 48, 225–34, 237, 243, 264, 283, 285, 308–9, 357, 361, 382, 408; telautomatics and, 225, 308

Republic, The (Plato), 8–9

resonance: artificial earthquakes and, 187, 446n26; Bjerknes and, 117; capacitors and, 124; cell phones and, 403; Earth and, 211, 248, 269–74, 329, 363; oscillators and, 9–10, 117, 124, 178–80, 187, 189–91, 208, 211, 399, 438n21, 446n26; power transmission and, 9–10, 201, 207–13; principle of, 124; radio and, 403; receivers and, 140; return circuit and, 209, 247–55, 262, 294, 329–30; Schumann field and, 399; stationary waves and, 269–74, 281; television and, 403; Tesla coils and, 124;

transmitters and, 140, 338; tuning circuits and, 403; wireless lighting and, 201, 207–13; wireless power transmission and, 178–80, 187, 189–91

return circuit: balloons and, 252–54; earth's resonance and, 248; ground current and, 209, 247–55, 262, 294, 329–30; high-frequency phenomena and, 248–51; solving puzzle of, 247–55

Richardson, Henry H., 214

Richmann, Georg, 299

right-hand rule, 36, 39

Robinson, Corinne Roosevelt, 239

Roentgen, Wilhelm Conrad, 221–22

Rolls–Royce, 378

Roosevelt, Eleanor, 390

Roosevelt, Franklin, 390

Roosevelt, Teddy, 239, 256, 343, 350, 451n34

rotating magnetic field and, 4, 9–10, 52, 54–64, 67, 84–85, 91–98, 106–12, 124, 169–70, 294, 302, 364, 369, 371, 404, 410, 429n5, 429n6, 436n31

Royal Institution, 147, 151, 189

Ryan, Thomas Fortune, 353

Saint-Gaudens, August, 199

Sampson, William T., 241

Scherff, George, 276, 303, 305, 344, 357, 361, 365, 408, 416, 454n93

Schiaparelli, Giovanni Virginio, 276

Schieffelin, Annette Markoe, 312

Schiff, Jacob, 357

Schmid, Albert, 115, 161, 168

Schneider & Co., 153

Schumann cavity, 273, 278, 399

Schumpeter, Joseph, 10, 57–59, 401, 404, 411

Schwarz, Caroline Clausen, 344

Schwarz, F.A.O., 345

Science journal, 309

Scientific American journal, 101

Scott, Charles, 159–60

Scribner's Monthly, 198

secondary coil, 121–24, 268, 286–90, *298*, *324*, 327–28

Seifer, Marc J., 276, 417–18

Semipalatinsk Test Site, 394–95

Serbia: Gospić, 21, 25–32, 47, 154–55, 426n27; honors to Tesla and, 155–56; migration and, 13; Novi Sad, 17; Orthodox Church and, 10, 14–15, 31–33, 155, 198, 247, 364, 390, 409, 411,

Serbia (*continued*)
    425n18, 469n50; Tesla's background
    and, 3, 5, 10, 13–20, 23, 30–33, 46–47,
    145, 155–56, 186, 196, 199, 209, 239,
    319, 353, 390, 409–10, 415
Serrell, Lemuel W., 74
sexuality, 239–41, 376, 409, 412
shadowgraphs, 223–24
Shallenberger, Oliver B., 108, 111–14
shock treatment, 217–18
Siemens & Halske, 67
Simpson and Crawford, 256, 264
single-phase alternating current (AC):
    Brown and, 96; motors and, 88, 96, 98–
    99, 112, 114, 119, 144, 159–61, 166–68, 170,
    173; Thomson and, 98; Westinghouse
    and, 159–60; ZBD systems and, 88
single-wire lamps, 149–50
Sing Sing, 1
six-wire scheme, 64–65, 96, 144, 410
skin effect, 125, 136
Slaby, Adolph, 313, 377
slip rings, 39, *93*, 105, 433n23
Slovenia, 47
Sluzhebnik (Serbian liturgy book), 17
Smith, Emile, 240
Smith, L.M.C., 392
Société de Physique, 152
Société Electrique Edison, 63–66, 69, 71
Société International des Electriciens, 152
Sogge, Richard C., 240
Songs of Liberty (anthology), 199
South Fifth Avenue laboratory, 158, 180,
    182, *190*, *198*, 215, 389, 405
Soviet Union, 379, 388; Cold War and,
    394–98; particle-beam weapons and,
    395; Tesla papers and, 395
Spanel, Abraham N., 391
Spanish-American War, 1, 230, 241, 256
spark gaps: Earth's resonance and,
    271–72; Hertz and, 120, 184; ionization
    and, 184–85; lightning detectors and,
    271–72; quenched, 184; Tesla coil and,
    120–23; wireless power transmission
    and, 184–85
Spencer, Herbert, 285–86
split-phase motors, 95–100, 104, 110–15,
    123, 159
Sprague, Frank, 65
Srbobran newspaper, 17
Stanley, William, Jr., 78, 88–89, 111–13
stationary waves: capacitors and, 262,
    268, 271, 279–81, 286; Colorado

Springs laboratory and, 263–69;
    conductors and, 271, 273, 289; demon-
    strations and, 264, 271, 282, 293–95;
    Earth's resonance and, 269–74;
    electromagnetism and, 270–73, 275;
    high-frequency phenomena and, 262;
    interplanetary messages and, 274–78;
    Marconi and, 276–77; ocean and, 273;
    patents and, 263, 303; photographic
    documentation and, 295–300; reso-
    nance and, 270, 281; Schumann cavity
    and, 273, 278; telegraph and, 269, 271,
    273, 283, 285–89, 295; transmitters
    and, 262, 267–69, 274, 276, 278–82;
    tuning and, 284–88; wireless commu-
    nication and, 262, 265, 273–79, 295,
    300–303; wireless power transmission
    and, 262, 294, 302; witnessing tests
    of, 288–95
Statue of Liberty, 283
steam power, 5; alternators and, 181;
    bladeless turbines and, 353, 369–73;
    Buffalo and, 165; flywheels and, 43;
    Fontaine and, 41; Globe Stationery
    and Power Company, 81; Marconi,
    and, 277; Newcomen and, 426n43;
    ocean scheme for, 81, 87, 433n26;
    oscillators and 181–85, 193, 215, 232,
    446n18; pyromagnetic generator
    and, 81–84; rotary motion and, 37,
    43; ships and, 195, 283, 327, 338, 342;
    triple-expansion engines and, 226;
    Watt and, 19, 406; Westinghouse and,
    165, 304, 320, 328
Steele, Charles, 316–18
Stillwell, Lewis, 165
St. John's, Newfoundland, 333–34
St. Louis lecture, 177–78
stock tickers, 79, 354
Stone, John Stone, 396, 402–3
Storm, Margaret, 398
St. Paul & Northern Pacific Railway
    Company, 164, 357
Strasbourg motor, 57, 66–69, 85, 118, 140,
    294, 416, 430n20
Strategic Defense Initiative (SDI),
    394–95
streamers: copper and, 297; fire hazard
    of, 264, 281; high-voltage electricity
    and, 135, 177, 211–12, *255*, 281–82, 296–
    97, 299, 306, 323, 456n47, 458n76;
    photographing, 295–300; power
    transmission and, 211–12; Tesla coils

and, 120–25, 128, 218, 248–49, 282,
338, 353, 403, 423n6, 450n17, 462n6
Stumpf, Carl, 48
subjective rationality, 10, 57–59, 411
Swezey, Kenneth, 380, 390, 411, 416
Swiss Tesla Institute, 399
Szigeti, Anthony, 114, 241; arc lighting
and, 74; compass and, 131; Edison
and, 63–64; leaves Tesla, 131–32, 240,
362; motors and, 50–52, 57, 62–64,
92, 428n40, 428n42, 430n14; Puskás
and, 63; Strasbourg project and,
66–67, 430n20; talents of, 118, 293;
Tesla Electric Company and, 81

tabloid newspapers, 1, 239, 316, 352, 408
Tamm, Edward A., 392
Tate, Alfred O., 70
T connection, 159
telautomatics, 225, 308
teleforce principle, 388–89
telegraph: American Union and, 79;
Atlantic & Pacific and, 79; Brown and,
77–80; brush response and, 148; call
boxes and, 229; Central Telegraph
Office of Hungary, 50; direct current
and, 37; Edison and, 64, 79, 197;
Evans and, 78–79; Gould and, 79–80;
Gray and, 79; grounding and, 139;
Johnson and, 197; Marconi and, 132,
335–37 (see also Marconi, Guglielmo);
Maxwellians and, 210; messenger boys
and, 229, 389; Mutual Union and,
78–81, 111; obsolescence of, 5; Peck
and, 78–80; railroads and, 79; relays
and, 80; stock tickers and, 79, 354;
transatlantic cable for, 195, 312; Western Union and, 77–80, 94, 101, 406
Telegraphic Journal and Electrical Review,
137
telephone, 5, 261, 345; American Bell
Telephone Company and, 101; atmospheric induction, 445n6; Bell and, 8,
10, 21–22, 49, 138, 219, 406, 435n59,
471n23; coherer circuit and, 275; Dolbear and, 139; Edison and, 49, 78, 195;
marketing of, 8; Morgan and, 312, 317,
339; patents and, 101, 186, 348; Puskás
and, 49–50, 63, 428n34, 428n35; Western Union and, 101
television, 244, 384, 403
Tesla, Ana (grandmother), 14
Tesla, Angelina (sister), 15

Tesla, Dane (brother), 15, 21–23, 366, 410
Tesla, Djuka (mother), 13–15, 17, 19, 21,
47, 145, 154–55, 156
Tesla, Janja (aunt), 14
Tesla, Josif (uncle), 14, 26, 34
Tesla, Marcia (sister), 18
Tesla, Milka (sister), 15
Tesla, Milutin (father): death of, 447;
education promise of, 29–32; Tesla's
background and, 14–18, 21–22, 25,
28–30, 34, 46–47
Tesla, Nikola: AC motors and, 3 (see also
AC motors); active imagination of,
19–20, 24–26, 28, 30–31, 33, 51–52,
55–59, 237–38, 244–47, 403–13; AIEE
lecture and, 105–7; arrest of, 47; ashes
of, 393; background of, 3, 12–33;
bankruptcy of, 374–76; billiards, 46,
51, 70; biographical concepts for, 6–11;
birthday parties for, 380–81; birth of,
13, 17–18; brother's death and, 21–22;
Budapest and, 49–57; celebrity of,
175, 194, 197, 303, 352; childhood of,
12–33; Columbia College lecture and,
133–38; confiscated papers of, 391–95;
cultivated image of, 194–95; death of,
389–95; death of father, 447; demonstrations of, 1–2 (see also demonstrations); disproving theory of relativity
and, 380; disruptive innovation and,
401–3; as draftsman, 47; dreams of,
26–28, 225, 410; early education of,
26–29, 32–33; early interest in electricity, 18–19, 28–29, 34–41; Edison
Machine Works and, 69–73, 76, 120,
416; Edison Medal and, 9–10, 240,
376, 391; education at Gymnasia
and, 26–29; electrotherapy and, 217,
450n12; emotional challenges of,
20–26, 50, 216–18, 221, 237, 364–67,
380, 389–90, 399, 412; establishing
reputation of, 194–213; European
tour of, 143–57; father's promise of
engineering education, 29–32; flattering exam certificates of, 45–46; flying
machines and, 19, 27, 226, 369, 386,
410; gambling of, 46–48, 307; Ganz
and Company and, 60–63, 87–88, 155,
158, 258, 429n2, 429n6; germ phobia
of, 389; gold medals of, 380; as Grand
Officer of the Order of St. Sava, 155;
as hero of free-energy movement,
398–99; high-frequency

Tesla, Nikola (*continued*)
experiments and, 5, 119–28, 132–39, 143–50, 158, 176, 178–84, 193, 196, 205–11, 217–18, 248–51, 262, 310, 328; honorary degrees and, 4, 204–5; human development equation of, 307; on human progress, 305–10; idealism of, 9–11; illusion and, 8–11, 18, 92, 115–16, 193, 208, 246, 263, 295, 299, 301, 315, 330, 368, 373, 381, 389–95, 403, 406–8, 412; impact of *Abafi* on, 23–24; on intensity of invention, 237–38, 244–47; introspective nature of, 23; Joanneum Polytechnic School and, 32, 34, 44–45, 427n54; Johnsons and, 197–203; laboratory fire and, 216–17; lack of artistic abilities, 26–27; left-handedness of, 26–27; legacy of, 396–413; Library of Congress and, 416; as local hero, 26; London lectures and, 146–52; loss of military scholarship, 46; love of pigeons, 380; love of reading, 22–23, 28, 30; as magician, 129–42; on marriage, 5, 238, 244, 247, 452n59; mathematics and, 26, 34, 48, 64, 106, 128, 307, 405; meaning of surname, 14; mental engineering and, 27, 43–45, 53–55, 59–60, 63, 67–68; modern impact of, 5–6; nervous breakdowns of, 6, 50, 154–57, 365–68, 412; New Age philosophers and, 399–401; newspaper notoriety of, 203–5; Niagara hydroelectric project and, 4, 162–67, 170, 172–76, 205, 214, 237, 256, 342–44, 353–55, 365, 416; Nobel Prize and, 375; obsessions of, 22–23; oscillators and, 176 (*see also* oscillators); outsider status of, 397–98, 409–10, 470n3; overdue Waldorf–Astoria bill of, 375, 379; parental approval and, 21–22, 46; particle-beam weapons and, 329, 381–95, 416; passion for cards, 46, 47; patents and, 21 (*see also* patents); Philadelphia lecture and, 176–77; physical appearance of, 2–3; physics and, 28, 34, 48, 64, 375, 384, 398–99; in popular culture, 397–401; practical work experience and, 60–75; Puskás brothers and, 49–50, 63, 69, 428n32, 428n34, 428n35, 430n31; radio-controlled boat and, 5, 33, 48, 225–34, 237, 243, 264, 283, 285, 308–9, 357, 361, 382, 408; religion and, 8,

10, 14, 17, 20–21, 31–33, 247, 364, 409, 411; search of for new ideal, 117–28; SE Edison and, 63–66; self-doubt and, 364–67; senility of, 389–90; sensationalism and, 136–37, 177, 195, 309–10; sexual issues and, 239–41, 376, 409, 412; shock treatment and, 217–18; six-wire system and, 64–65, 96, 144, 410; Smiljan and, 3, 5, 13, 16–17, 21, 69; smoking and, 46, 307; sojourn at Westinghouse of, 114–15; St. Louis lecture and, 177–78; Strasbourg system and, 57, 66–69, 85, 118, 140, 294, 416, 430n20; student carousing and, 46; style of invention of, 7–11, 403–8 (*see also* invention); Szigeti and, 50–52, 57, 62–64, 66–67, 74, 81, 92, 114, 118, 131–32, 240–41, 293, 362, 428n40, 428n42, 430n14, 430n20; tears up Westinghouse contract, 131–32, 439n7; teenage illnesses of, 28–30; teleforce principle and, 388–89; Telefunken and, 377; telephone repeater and, 63; three–phase alternating current (AC) and, 145; as tinkerer, 19, 27–28, 120, 123, 180, 315; treatise of, 305–10; views on the future, 4–5; will of, 390; willpower of, 23–26; wireless communication and, 4–5, 146, 176 (*see also* wireless communication); wireless power transmission and, 6, 30 (*see also* wireless power transmission); as the Wizard, 1–5, 151, 175, 197, 198, 203, 215, 238–39, 306, 314–15, 318, 343, 346, 352, 375, 380, 389, 398, 416; women and, 203, 237–44; workings of human mind and, 5; work load of, 215–16; world telegraphy and, 337–46; X-rays and, 218–25, 237, 248, 330, 353, 394
Tesla, Nikola (grandfather), 14
Tesla, Stanka (aunt), 14, 28
Tesla coils: current interrupter and, 120–21; Hertz and, 120–22; high-frequency phenomena and, 5–6, 120–25, 128, 218, 248–49, 282, 338, 353, 403, 423n6, 450n17, 462n6; invention of, 120–25; Maxwell and, 122; pendulum effect and, 123–24; resonance and, 124; Ruhmkorff coil and, 120
Tesla Electric Light and Manufacturing Company, 352–53; Brown and, 73–75, 81, 84, 118, 431n44, 431n50; Lane and, 73–75, 84, 431n50; Peck and, 73–75,

81, 84, 118, 431n44, 431n50; Vail and, 73–75, 84, 431n50
Tesla Engine Builders Association, 374, 399
Tesla Memorial Society of New York, 399, 423n6
Tesla Motors, 397, 400–401
Tesla Project, The, 417
Tesla Propulsion Company, 372
Tesla Symposium, 398–99
Tesla Turbine Builders Association, 374
Tesla Universe website, 399
thermomagnetic motor, 76–82
Thomas A. Edison Construction Department, 72
Thomson, Elihu, 217, 446n21; AIEE dinner and, 335; arc lighting and, 74, 87; high-frequency phenomena and, 125, 133–34; Marconi and, 335; motors and, 93, 98, 106–7, 169; Wetzler on, 129, 138
Thomson, William, Lord Kelvin, 123, 131, 148, 167, 170, 204, 313, 442n18
Thomson-Houston Electric Company: alternating current (AC) and, 88–90, 96; arc–lighting and, 72, 74–75; Coffin and, 88, 98, 107, 162, 357; General Electric and, 102, 129, 160, 162, 166, 224, 311, 317, 352, 357, 373, 397, 452n62; motors and, 102–3; Niagara Falls project and, 166; ZBD system and, 88
three-wire system, 65, 96
tin can motor, 85, 86, 92, 96, 144, 294, 302, 407
tinkering: Astor and, 256; Tesla and, 19, 27–28, 120, 123, 180, 315
Tito, Marshall, 393
Tonnele & Company, 201, 220–21
transatlantic cable, 195, 312
transformers, 6, 208, 249, 338, 433n36; alternating current (AC), 87–90, 97–98, 105, 108–14, 159, 161, 172–73; 184; Brown and, 144; capacitors and, 188 (see also capacitors); coherers and, 228, 269–72, 275, 348, 438n29; European tour and, 148–49, 156–57; Fleming and, 333; Gaulard and, 87–89, 108–9, 112; Gibbs and, 87–89, 108–9, 112; high-frequency phenomena and, 5, 125, 133, 176, 179–84, 218–19; induction, 433n36; oscillating, 124–26, 132–40, 148 (see also oscillators); phase difference and, 109–10; primary coil and, 109, 121, 123, 268, 279, 281, 288, 324, 329; ring, 61–62, 429n6; secondary

coil and, 109, 120–24, 268, 286–90, 298, 324, 327–28; stationary waves and, 268, 270, 279, 286; T connection and, 159; Tesla coils and, 5–6, 120–25, 218, 282, 338, 353, 403, 423n6, 423n7, 450n17, 462n6; Wardenclyffe and, 321, 328, 333; Y connection and, 144
transmitters: breakwheel and, 268, 279, 286; coherers and, 228, 269–72, 275, 348, 438n29; current produced by, 209, 247–48, 294, 298, 344, 384; de Forest and, 350; distance and, 294–95; Earth's resonance and, 269–74; fireballs and, 299–300, 394, 398, 458n86; Fleming and, 152; Geissler tubes and, 128; ground current and, 247; grounding of, 141, 208, 403; induction coils and, 140, 180, 333; Lowenstein and, 292; magnifying, 128, 267–69, 274, 278–82, 284 (see also magnifying transmitters); Marconi and, 141, 152, 285, 333; power of, 449n43; power transmission and, 207–13 (see also power transmission); radio-controlled boats and, 5, 33, 48, 225–30, 237, 243, 264, 283, 285, 308–9, 357, 361, 382, 408; receivers and, 130, 139, 140–41, 159, 180–81, 208–11, 225, 236, 247–53, 262, 269–77, 285–95, 301, 329–34, 339–44, 347, 375, 403, 451n38, 456n30, 457n63; remote control and, 228 (see also remote control); resonance and, 140, 338; Scherff and, 362; Soviets and, 379; stationary waves and, 262, 267–69, 274, 276, 278–82; streamers and, 135, 177, 211–12, 255, 281–82, 296, 297, 299, 306, 323, 456n47, 458n76; telegraphs and, 139 (see also telegraph); tuning and, 208, 235–36, 262, 279, 284–88, 303, 313, 332, 343, 403, 449n43; Wardenclyffe and, 329 (see also Wardenclyffe); wireless lighting and, 180, 184–85, 189, 191; wireless power and, 209–10, 247–53
Trotter, A. P., 150
Trump, John G., 368, 384, 392–93, 468n36
tuning: privacy and, 284–88; selectivity and, 284–88; transmitters and, 208, 235–36, 262, 279, 284–88, 303, 313, 332, 343, 403, 449n43; wireless communication and, 262, 284–88
turbines: automobile industry and, 373–74; aviation industry and, 373–74;

turbine (*continued*)
  bladeless, 368–74; boundary layer
  and, 369, 373; carbon-fiber, 374;
  demonstrations and, 372–73; flying
  machines and, 369, 386; ideal of, 371;
  Kevlar, 374; motors and, 9, 26, 61, 159,
  166–68, 183, 187, 353, 368–74, 378, 386,
  408; pumps and, 374; steam, 183, 353,
  446n18; titanium, 374; viscosity and,
  369, 371; water, 61
Twain, Mark, 389, 417–18, 426n47;
  bankruptcy of, 446n24; cultural pulse
  and, 408; Grant and, 198; Johnson
  and, 198, 201, 214; Library of Congress and, 416; novels of, 28; Paige
  typesetter and, 446n24; The Players
  club and, 215; private demonstrations
  to, 186–87, 221; radio-controlled boats
  and, 231–32; X-rays and, 221–22

Umbria (ship), 145
Union County Electric Light and Manu-
  facturing Company, 75
United States Electrical Supply Company,
  230
United States Electric Lighting Company,
  72
United States Steel, 314, 317, 343, 348, 378
University of Nebraska, 204
University of Virginia, 214
Unwin, W. C., 167

vacuum bulbs, 29, 221
vacuum tubes, 127, 149, 219–21, 224, 330
Vail, Benjamin A., 73–75, 84, 431n50
Van de Graaff, Robert, 382, 384, 461n51
Vanderbilt, William K., 214, 230
vegetarianism, 307
very low frequency (VLF) waves, 276–77
Viereck, George Sylvester, 365–67
Villard, Henry, 162–64
viscosity, 369, 371
Volta, Alessandro, 35
von Miller, Oskar, 144–45
von Steinhill, Carl August, 139

Waldorf-Astoria Hotel, 256–57, 260, 303,
  320, 330–31, 335, 375, 379, 412, 467n27
Wallace, Henry A., 390, 391
Wall Street: Adams and, 164, 206; Baker
  and, 79; Baring Brothers panic and,
  130; Barnes and, 357; Belmont and,
  130; capital issues and, 164, 206,
  304, 311, 345, 353, 357, 359; de Forest
  and, 349–50; Morgan and, 164, 311,
  345, 359; nouveau riche of, 2; Ryan
  and, 353; stock tickers and, 79, 354;
  Western Union takeover and, 78–79;
  wireless communication and, 349
Waltham Watch Company, 378
Warden, James S., 319, 361
Wardenclyffe, 5, 368, 373; capacitors and,
  321, *324*, 328; demonstrations and, 318;
  financial woes of, 361; high-frequency
  phenomena and, 328; inspection trips
  to, 319–20; layout of, 320–22; location
  of, 318–19; magnifying transmitters
  and, 362, 365, 462n69; manufacturing
  profits and, 371; mortgaging of, 379;
  new funding issues and, 344–46;
  particle-beam weapons and, 382, 384;
  patents and, 328, 330, 355, 462n69;
  plans of, 318; power transmission and,
  341–44; Tesla's view on operations of,
  328–30; tower of, 321–31, 346, 362,
  379, 382; Warden and, 319; White and,
  318; wireless communication and,
  329–35; wireless power transmission
  and, 379–80
Wardrip-Fruin, Noah, 339
Ware, Kallistos, 31
Washington Square Arch, 214
Waterside Power Station, 372
Watt, James, 19, 406
Western Union, 94, 101; Brown and,
  78–80; Gould and, 79–80; Orton and,
  79, 406; patents and, 80; Peck and,
  78–79
Westinghouse, George, 88, 108, 112, 114,
  130, 159, 304, 397, 416
Westinghouse Electrical Manufacturing Company: AC motors and, 108,
  112–18; Adams and, 173; Baring
  Brothers panic and, 130; Behrend
  and, 376; Brown and, 111–13; Chicago
  World's Fair and, 160–61; Edison
  General Electric and, 129–30; expansion of, 129–30; experimentation costs
  of, 130–31; incandescent lighting and,
  160–61; innovation and, 88; Niagara
  Falls project and, 162–74; patents and,
  108, 110–15, 130–32, 155, 166–67, 169,
  439n3; Peck and, 111–13; polyphase
  system and, 160–61; railroads and, 88;
  reorganization of, 130; sales growth
  in, 129; Schmid and, 168; single-phase

alternating current (AC) and, 159–60; split-phase motors and, 159–60; Stillwell and, 165; Tesla's legacy and, 397; Tesla's sojourn at, 114–15

Weston, Edward, 81

Wetzler, Joseph, 129, 136, 138, 179, 195

White, Abraham S., 349, 349–51

White, Lawrence, 215

White, Stanford, 214–15, 241, 245, 318, 323, 326

Whitney syndicate, 258

Wilhelm I, Kaiser of Germany, 66, 430n19

Wilhelm II, Kaiser of Germany, 343, 377

Willie (assistant), 267

Wilson, Woodrow, 386

Winslow, Lanier & Company, 164

Winslow, Mrs. E. F., 344

wireless communication: America's Cup and, 283, 285–88, 311, 332, 348; Army Signal Corps and, 283; Astor and, 305; balloons and, 261, 267; British Post Office and, 313; coherers and, 228, 269–72, 275, 348, 438n29; de Forest and, 285, 303, 334, 348–50, 380, 402–3, 438n29; disruptive innovation and, 402–3; economic issues and, 350; Fessenden and, 303; Fleming and, 149, 152, 333, 337–38, 403, 442n18, 462n6; Germany and, 377; handheld devices for, 340; Hertzian waves and, 335, 348; importance of, 308–9; interplanetary, 265, 274–78, 300–301, 315, 334–35, 380, 398, 460n36; Kennelly-Heaviside layer and, 209, 335; lack of distance tests and, 294–95; Light-House Board and, 283–84; litigation over, 376–78; Loomis and, 445n6; Lowenstein and, 267, 269–70, 278, 292–95, 343–44, 375, 455n8, 463n28; magnifying transmitters and, 267–69, 274, 278–82, 462n6; Marconi and, 5–6, 132, 141–42, 152, 209, 259–61, 265, 273, 276–77, 283–88, 293–94, 303–4, 311–15, 330–38, 342, 344, 347–48, 375–78, 397, 402–3, 416, 449n43, 451n38, 456n30, 462n10, 462n6; Martians and, 213, 264–65, 276, 278, 315, 334, 398, 456n33, 456n34; messages across English Channel and, 261, 273, 332; Morgan and, 311–18, 337–47, 350–61, 365, 402, 406, 412; patents and, 313, 377, 403; photographic documentation and, 295–300; privacy and, 284–88, 355–56; radio–controlled boat and, 5, 33, 48, 225–34, 237, 243, 264, 283, 285, 308–9, 357, 361, 382, 408; radio frequency (RF) voltage and, 275–76; Rankine and, 353–54; secrecy and, 284–88; selectivity and, 284–88; speculative bubbles and, 346–51; Spencer and, 285–86; stationary waves and, 262, 265, 273–79, 295, 300–303; telegraph and, 126, 132, 140–42, 146, 148, 230, 258–61, 283–87, 304, 306, 311–18, 332, 334, 336–37, 346, 349, 378, 402, 438n29; telephone and, 141, 288, 312; telescoping mast and, 267–68, 282; Tesla and, 4–5, 146, 176, 179, 188, 211, 252, 259–62, 265, 273–78, 284–86, 288, 295; trans-Atlantic, 152, 304–5, 312, 316, 332–37, 350, 442n22; tuning and, 262, 284–88; undersea cables and, 336; Wall Street and, 349; Wardenclyffe and, 329–35, 346, 362, 379, 382; witnessing tests of, 288–95; world telegraphy and, 337–46

wireless lighting, 128; alternators and, 181; Astor and, 257–59, 315, 353; capacitors and, 180, 184, 188–89; Columbia College lecture and, 133–38; conductors and, 179, 181; copper and, 179; demonstrations and, 136, 177, 188, 190–92, 215, 235, 294; Helmholtz and, 188; high-frequency phenomena and, 176, 179–92, 197; London lectures and, 146; magnifying transmitters and, 292; manufacturing and, 218; Martin and, 195; Morgan and, 316–17; oscillators and, 176–92; patents and, 218, 248, 255, 258, 316, 357, 377; perfecting, 187–92; promotion of, 193–213, 218; public reaction to, 195; resonance and, 201, 207–13; transmitters and, 180, 184–85, 189, 191; vs. incandescent lighting, 140

wireless power transmission, 174, 398; Astor and, 255–59; demonstrations of, 303, 342; disruptive innovation and, 402–3; Earth and, 179, 345; experimentation with, 178–81; Hertz and, 179; ideal of, 124; illusion and, 208; increasing transmission length of, 138–42; industrial significance of, 328, 334; lightning and, 18, 51, 157, 213, *268*, 270–74, 278, 281–82, 289, 299, 301, 346, 398, 417, 446n21,

wireless power transmission (*continued*)
455n21; magnifying transmitters and,
128, 267–69, 274, 278–82, 284, *287*,
288–300, 303, 306, 323–24, 327–30,
333, 362, 365, 458n86, 462n6, 462n69;
Martians and, 213; methodology for,
209, 218, 247–55, 262, 288, 294, 297,
329–30, 361; Morgan and, 342, 347,
355, 358, 361; oscillators and, 179, *189*;
patents and, 406; photographing,
297; promotion of, 209, *210*, 218–19;
radio-controlled boat and, 5, 33,
48, 225–34, 237, 243, 264, 283, 285,
308–9, 357, 361, 382, 408; receivers
and, 130, *139*, 140–41, 159, 180–81,
208–11, 225, 236, 247–53, 262, 269–77,
285–95, 301, 329–34, 339–44, 347, 375,
403, 451n38, 456n30, 457n63; remote
control and, 225–30 (*see also* remote
control); resonance and, 178–80, 187,
189–91; as response to Marconi, 342;
return circuit and, 209, 247–55, 262,
294, 329–30; spark gaps and, 184–85;
speculative bubbles and, 346–51;
stationary waves and, 262, 267, 288,
294, 297, 302; streamers and, 135, 177,
211–12, *255*, 281–82, 296, 297, 299,
306, 323, 456n47, 458n76; teleforce
principle and, 388–89; Tesla's dream
of, 4–6, 30–33, 364–69, 373, 379, 388,
406–12; Trump and, 368; Wardenclyffe

tower and, 303, 321–31, 346, 362, 367,
379–82; X-rays and, 224–25
wireless speculative bubble, 346–51
Wireless Telegraph and Signal Company,
311, 332, 459n27
Wollaston, W. H., 82
Woolworth Building, 373
World Wide Web, 339, 417
Wozniak, Steve, 406
Wright, John, 79
Wright brothers, 397
Xerox, 112
X-rays: Edison and, 224; Geissler tubes
and, 221–22; health effects of, 224; os-
cillators and, 224; power transmission
and, 224–25; Roentgen and, 221–22;
Ruhmkorff coils and, 223; Strategic
Defense Initiative (SDI) and, 394–95;
Tesla's experimentation on, 218–25,
237, 248, 330, 353

Yankee ingenuity, 397
Y-connection, 144, 161, 441n4
Yugoslavia, 380, 389–90, 393, 415–16

Zadar, Croatia, 17
Zagreb, Croatia, 17, 155
ZBD system, 88
Zenneck, Jonathan, 377
Zipernowsky, Károly, 61–62, 88
Zmaj, Jovan Jovanović, 155–56, 199